STATISTICAL PHYSICS
OF MATERIALS

STATISTICAL PHYSICS OF MATERIALS

L. A. Girifalco

School of Metallurgy and Materials Science
University of Pennsylvania

A Wiley-Interscience Publication

John Wiley & Sons, New York . London . Sydney . Toronto

Library of Congress Cataloging in Publication Data:

Girifalco, L   A   1928–
Statistical physics of materials.

"A Wiley-Interscience publication."
Bibliography: p.
1. Solids. 2. Materials. 3. Statistical mechanics. I. Title.

QC176.G55   530.1'32   73-2622
ISBN 0-471-30230-9

Printed in the United States of America

10 9 8 7 6 5 4 3 2 1

To

CATHY

and

SANDY, TONY, MARY, JOHN, BOB,

TERRY, STEVE,

and **DORI**

and to my

MOTHER and **FATHER**

PREFACE

Recently, large areas of metallurgy, physics, and chemistry have been gradually coalescing into a broader and more integrated view of solids than was usual in the individual disciplines. The unifying theme of the rather heterogeneous activities that are now called "materials science" is that properties of materials can be understood in terms of their structure. In fact, the rich variety of properties exhibited by materials reflects an entire hierarchy of structures and their relations, including electronic, crystallographic, defect, and morphological structures.

Materials can be studied from the viewpoint of the metallurgist, physicist, or chemist. In any case, it is a truism that an understanding of properties and structure must be based on quantum theory and statistical mechanics. In view of this, it is remarkable that, although many excellent texts on the quantum theory of solids exist, few are devoted to the statistical mechanics of the solid state. The purpose of this book is to provide the statistical-mechanical base for the study of solids. It is written for beginning graduate students of metallurgy, physics, or chemistry.

The importance of statistical mechanics for materials science is obvious; a glance at the table of contents of this book reveals the variety of subjects for which the statistical approach acts as a fundamental and unifying base. This fact is sometimes overlooked; too often statistical mechanics is regarded only as a tool for arriving at useful formulas such as the mass action law or the temperature dependence of the vacancy concentration. Hopefully this book demonstrates that statistical mechanics is also a methodology that contributes to a basic understanding of materials.

Most of the subject matter in this book is part of a two-term course sequence in the School of Metallurgy and Materials Science at the University of Pennsylvania. Students of diverse backgrounds have taken these courses quite successfully, and have contributed more than they know to the contents and style of this book. In fact, I owe a special debt to the students in my Special Topics course in the spring of 1967. The portions of this book and of the present course that are devoted to point defects grew out of that course, and the growth was greatly aided by the students' continual interaction as well as their term papers.

L. A. GIRIFALCO

Philadelphia, Pennsylvania
January 1973

ACKNOWLEDGMENTS

I express my appreciation to my colleagues in the School of Metallurgy and Materials Science, who so ably combine excellence in teaching with creativity and depth in research. Their influence has improved my own efforts; many points in this book were discussed with them, and their advice has been invaluable.

Also I thank Mary Sulena for her remarkable work in typing the manuscript. She was marvelously accurate and infinitely patient, and she greatly simplified the mechanics of creating the manuscript.

L. A. G.

CONTENTS

STATISTICAL PHYSICS
OF MATERIALS

chapter 1

PRINCIPLES OF STATISTICAL MECHANICS

1.1 THE DEFINITION OF STATE

The prime article of faith in scientific theory is that all experimentally observed phenomena may be derived from a small number of general principles. The acceptance of this idea implies that, in spite of the constant changes evident in physical systems, they possess some immutable, absolute attributes from which all of their interesting properties can be calculated. A central concept in such a calculation is that of the state of an isolated system. If all of the experimentally measurable properties of a system are known at all times, then these properties would certainly constitute an adequate description of the state of the system. However, such a definition of state would be useless for theoretical purposes and is contrary to the assumption that all properties can be related to a few general laws. The most fruitful general definition of state is this one: the state of a physical system is the minimum amount of information required for a calculation of the properties of the system.

The types of information adopted as basic in this definition differ among the various branches of physical theory, and an appreciation of the various specific definitions of state is necessary for the intelligent pursuit of scientific studies.

1.2 THERMODYNAMIC STATE

The first and second laws of thermodynamics provide a basis for constructing relations among the macroscopic properties of physical systems. These properties include temperature, pressure, volume, specific heat, compressibility, and so forth. In a one-component, one-phase system, two variables (e.g., temperature and pressure) are sufficient to define the thermodynamic state. All other thermodynamic properties are then determined by the equations of thermodynamics. The number of variables needed to define the thermodynamic state of a multiphase, multicomponent system is given by the phase rule as

$$f = C - P + 2 \tag{2.1}$$

where f is the required number of variables, C is the number of components, and P is the number of phases.

In general, there are three different types of variables associated with a thermodynamic system. An appreciation of the nature of these variables is essential for an understanding of the connection between thermodynamics and statistical mechanics. These three kinds of thermodynamic quantities are called:

1. External parameters.
2. Mechanical variables of the system.
3. Nonmechanical variables of the system.

The external parameters define the conditions under which the system exists and the coupling of the system to its environment. They include such quantities as external electric or magnetic fields and the pressure exerted by a movable piston.

The mechanical variables of the system are all quantities that can be interpreted in mechanical terms. These include the energy and the internal pressure.

The nonmechanical variables are those that are peculiar to thermodynamics and have no analogue with the concepts of ordinary mechanics. These include temperature, entropy, and the quantities derived from them.

Two important points must be noted. First, thermodynamics deals with equilibrium systems, so that time and the concept of temporal causality do not enter into it at all. Equilibrium thermodynamic systems are macroscopically static, and the definition of state given here is incapable of describing systems whose properties are changing in time.

Second, no actual numerical computations of the thermodynamic properties can be made from the specification of the state of a system. For example, if we are given the temperature and pressure of a gas, thermodynamics tells us that the volume is determined, but it cannot tell us the numerical value of this volume. To get the volume, an equation of state is needed, and this can only be obtained from experimental data or from some other source outside the framework of thermodynamic theory. In this sense, the specification of state in thermodynamics is incomplete.

1.3   MICROSCOPIC MECHANICAL STATE

All matter consists of a number of atoms or molecules that are composed of electrons, neutrons, protons, and the like. Macroscopic material systems, therefore, are large collections of particles and may be treated by the methods of many-particle mechanics. In classical mechanics, all relevant mechanical properties are determined by the positions and momenta of the particles, so that if we know the position and momenta of every particle in a macroscopic, isolated system at some particular time, we can compute all properties at all times. In quantum mechanics, all relevant properties can be computed at all times from the wave function, the wave function itself being a function of all particle coordinates.

The definition of state in many-particle mechanics is therefore extremely detailed. Classically, it is necessary to specify all coordinates and momenta at some particular time. In quantum mechanics, the wave function at a particular time is needed, the wave function being obtained from a second-order differential equation involving all particle coordinates.

It is evident that the mechanical definition of state differs from the thermodynamic one in several important ways. The mechanical definition is microscopic and requires knowledge concerning all of the particles of the system. Also, it includes the concept of temporal causality, the state at any time being determined by the state at an earlier time. Finally, it permits, at least in principle, actual numerical computations to be made. Given the definition of state and the values of certain fundamental constants, such as the mass and charge of the electron and Planck's constant, calculations can be made (at least in principle) of the values of all the mechanical properties. The thermodynamic definition of state, on the other hand, is macroscopic, requires only a few variables, does not include temporal causality, and cannot yield numerical values of thermodynamic functions.

1.4   THE PROGRAM OF STATISTICAL MECHANICS

Our belief in the ultimate unity of scientific knowledge leads us to conclude that there must be some connection between the thermodynamic and mechanical definitions of state for material systems. We will be concerned only with systems containing a very large number of particles. In practice, it is, of course, impossible to obtain an actual specification of the microscopic state, and even if such a specification were available, the equations of motion would be so complicated that they could not be solved. Also, we are not particularly interested in such detailed information. We are, however, very much interested in the possibility of interpreting thermodynamic data in microscopic terms, and, in particular, it would be very desirable to obtain a method of computing macroscopic quantities from atomic properties, thereby making up for the inherent deficiency of thermodynamics. In view of these considerations, any bridge connecting the macroscopic and microscopic descriptions of equilibrium systems must be of such a nature that:

1. A great deal of microscopic information is erased so that we are not encumbered with an enormous number of microscopic parameters.
2. Enough microscopic information is retained so that atomic properties can be used to compute bulk properties.
3. Temporal causality, operating on a microscopic level, does not appear at the macroscopic level.
4. The laws of thermodynamics have a microscopic interpretation.

It is the program of statistical mechanics to construct such a bridge.

Now, consider a system at equilibrium containing a large number N of atoms or molecules, such as a crystal, a jug of water, or a gas enclosed in a bottle. If the system were completely isolated so that there were no interactions with its environment, the system would be in one quantum state and remain so. Such a system is not of much interest for several reasons. First, no real system can be completely isolated from its environment. Second, there would be no way of making measurements on an isolated system. The act of measurement automatically couples the system to its surroundings. It is true that in making a measurement, one tries to disturb the system as little as possible, but the coupling, although it may be weak, is always there. It will therefore be assumed that our system interacts with its environment but that this interaction is a weak one and that the environment is sufficiently constant so that the measured properties of the system do not change in time. For example, we could place our system in a constant temperature bath with which it continually exchanges varying but small amounts of energy. In addition, the system might be subject to varying but small perturbations caused by stray electric or magnetic fields, cosmic rays, and the products of natural radioactivity.

A system coupled to its environment in this way cannot be in a single, uniquely defined quantum state. Because of the random perturbations, the system is continually being kicked from one quantum state to another, and in the course of a time, small by macroscopic standards, it goes through an enormous number of different microscopic states a great many times. With this in mind, it is easy to construct a relation between microscopic and macroscopic mechanical parameters by taking the macroscopic variable to be a time average of the microscopic variables. If, for example, the energy of the system is E_i when the system is in the ith quantum state, and if out of some long time t the system spends a time t_i in the ith state, then the thermodynamic energy U is given by

$$U = \frac{1}{t} \sum_i t_i E_i \qquad (4.1)$$

where the summation is taken over all states that the system passes through in time t. This definition requires some comment.

First, t_i/t should be independent of t. Otherwise, U would depend on the time. This requirement could be satisfied by taking the limit of (4.1) as $t \rightarrow \infty$ and proving that the limit exists. However, the time between atomic interactions is extremely short, and we will take the above definition as valid, provided that t is large relative to atomic interaction times. The time it takes to make a thermodynamic measurement is always long enough to satisfy this condition.

Second, there is no way of knowing precisely what quantum states the system passes through during the time t and how much time it spends in each state. We can be certain that in any finite time it does not pass through all of them but that it does pass through a great many. Furthermore, because interactions with the surroundings are weak and approximately constant, we have a right to expect that most of the time the system is in one of a group of states whose energies E_i are not too different from the average energy U. Also, we should expect that all states that have energies in a small but finite range ΔU centered on U contribute to the average defined by (4.1), and that this contribution decreases with increasing deviation of E_i from U. In other words, the ratio t_i/t decreases monotonically with increasing $|U - E_i|$. In fact, with practically no loss in accuracy, the system can be taken to pass through all possible microscopic states, provided only that t_i/t decreases sufficiently rapidly with increasing $|U - E_i|$. Note that the ratio t_i/t can be interpreted as a probability because it gives the probability that, at any instant, the system will be in the state i with energy E_i.

1.5 ENSEMBLES

If the values of microscopic parameters are given, then the macroscopic properties could be computed from equations such as (4.1), provided that the ratios t_i/t were known. Unfortunately, there is no practical method for obtaining these ratios, and the time average approach is of no use in developing a workable statistical mechanics. To get around these difficulties, an alternative formulation of the problem of connecting macroscopic phenomena to microscopic physics has been given. The physical and conceptual content of this alternative can best be seen by considering an enormously large number of systems rather than just one. These systems are all replicas of the actual physical system under consideration and are all subject to the same kind of external conditions. Now, stack all these systems upon one another so that they form one great, continuous mass and place imaginary boundaries around each individual system. Such a collection is called an ensemble.

The nature of the boundaries separating the member systems of the ensemble must now be considered. If every boundary is taken to be completely impervious to all physical influences, the ensemble is said to be microcanonical. If we allow the boundaries to freely transmit energy, then we have a canonical ensemble. If, in addition, the boundaries are permeable to matter so that the systems can exchange atoms and molecules as well as energy, the ensemble is called grand canonical. Obviously, a number of different kinds of ensembles can be defined by altering the nature of the boundaries separating their members. In addition to those mentioned, the pressure ensemble, in which changes in pressure can be transmitted from one system to another, is of some importance.

The interactions among the various ensembles is always very weak because the number of atoms near the bounding surfaces is much smaller than the number in each member system, and intermolecular forces act only over very small distances.

At this point, the ensemble idea might seem rather artificial. But a great deal of comfort can be derived from the fact that a real macroscopic system can be thought of as an ensemble of smaller systems. Simply make a mental subdivision of the system into a large number of cells so that each cell contains very many atoms but is still small compared to the dimensions of the entire system. Such a subdivision is always possible because of the enormous number of atoms contained in any bulk material, and the behavior of any cell will be typical of the behavior of the system as a whole.

Clearly, in all ensembles except the microcanonical, fluctuations will occur in the values of some of the physical variables of the member systems. In a canonical ensemble, for example, the energy at any given time will be different for different member systems. Essentially, what has been done in constructing an ensemble is that fluctuations caused by external influences have been replaced by fluctuations caused by internal interactions among various parts of the system. In doing this, the ensemble is taken to be a supersystem practically isolated from the rest of the world. In constructing a particular kind of ensemble, we agree to consider the fluctuations of certain variables and neglect others. For example, in a pressure grand canonical ensemble, fluctuations in pressure, number of atoms, and energy are all taken into account while in a canonical ensemble only energy fluctuations are considered, and in a microcanonical ensemble, all fluctuations are ignored. For a system at equilibrium, fluctuations are generally small. and they can be safely ignored unless they are essential to our problem. As an illustration of this, consider the problem of constructing an ensemble to represent a crystal containing two elements in solid solution. If we are interested only in the physical thermodynamic quantities such as the specific heat, then the canonical ensemble is perfectly adequate. If, however, we are also interested in the range of composition for which the solid solution is stable, then the grand canonical ensemble is much more convenient. The choice of ensemble is generally dictated by the nature of the problem at hand.

1.6 ENSEMBLE AVERAGES

With the picture that we have developed so far, the basic assumption of the statistical mechanical method can be stated as follows: macroscopic properties are given by ensemble averages of suitable microscopic properties. Of course, this statement must be supplemented by a set of directions as to how to take the ensemble average, and this will be done later. The validity of this assumption requires that an ensemble average is equal to the corresponding time average discussed in Section 4. This equivalence is called the ergodic hypothesis. A great deal of work has been done in trying to prove the ergodic hypothesis, and while this work was not entirely successful, it seems to be very plausible. The description of time averages and ensembles given in the previous pages were meant to give an aura of reasonableness and credibility to the method of ensembles from a description of the physical situations they represent. But, from the point of view of rigor, we are starting from some unproved assumptions, and their ultimate justification will be taken to be their agreement with experimental data, their internal consistency, and the convenience with which they may be applied.

Now let us consider a canonical ensemble representing some specific physical system in thermodynamic equilibrium. If we label the possible quantum states of a typical member system with an index i, then we can define a set of numbers $n_1, n_2, \ldots, n_i, \ldots$, such that n_i is the number of systems in the ensemble in the ith quantum state. In a canonical ensemble, only the total energy of the ensemble is specified and, therefore, a large number of sets $\{n_i\}$ can be imagined that are consistent with our requirements. Furthermore, for each particular set of n_i, a large number of quantum states of the ensemble are possible. A particular set of n_i will be called a state distribution of the ensemble, while a particular quantum state of the ensemble will be called a complexion. The ensemble can have a variety of state distributions, and each state distribution can have many complexions.

The quantum state of the ensemble is a composite of the quantum states of the individual members. Furthermore, the member systems interact with each other very weakly, and they are of macroscopic dimensions so that they are distinguishable from one another. This means that for a given set $\{n_i\}$ defining a state distribution, the total number of complexions is just the number of ways of distributing all the member systems among all possible states of a single system in such a way that n_1 are in state 1, n_2 in state 2, and so on. This is true, because the weak interactions among member systems allow the existence of a spectrum of states that is the same for all members, and the distinguishability of systems has the consequence that a permutation of two systems among two states yields a new ensemble wave function. As a further consequence of distinguishability, there is no limitation on the number of member systems that may be in a particular quantum state.

The number of complexions for a given distribution n_1, n_2, n_3, . . . is then equal to the number of ways of placing marbles in boxes so that n_1 are in the first box, n_2 in the second box, and so on. In our case, the member systems correspond to the marbles, while the quantum states correspond to the boxes. The number of complexions $W\{n_i\}$ for a given distribution $\{n_i\}$ is therefore given by elementary combinatorial analysis by

$$W\{n_i\} = \frac{X!}{\prod_i n_i!} \tag{6.1}$$

where X is the total number of systems in the ensemble.

One way of writing the ensemble average for the energy is to take advantage of the fact that a given distribution $\{n_i\}$ has the same energy for all complexions $W\{n_i\}$ corresponding to that distribution. We will define the ensemble average of the energy to be

$$U = \frac{\sum W\{n_i\}E\{n_i\}}{X \sum W\{n_i\}} \tag{6.2}$$

where $E\{n_i\}$ is the ensemble energy for the distribution $\{n_i\}$, and the summations are taken over all possible distributions. It must be stressed that (6.2) can be correct only if every ensemble quantum state has an equal statistical weight. By this, we mean that if we enumerate all the possible quantum states for a canonical ensemble, consistent with the requirement that the total energy be XU, the probability that the ensemble is in a particular state is the same for all states. In terms of the time average picture, this means that the system passes through all possible states with equal frequencies, that is, the system spends an equal amount of time in each state. This assumption is called the hypothesis of equal a priori probabilities.

Because we require the total energy to be a constant and the fluctuations in energy among the member systems to be small, at least on the average, we should expect that there is some distribution $\{n_i\}$ for which the number of complexions is very large, relative to that for the other distributions. This is indeed so, and it can be proved mathematically that if all distributions except the most probable one are ignored, a negligible error is made in computing the ensemble average. This proof will not be undertaken here but will be accepted as reasonable, since it merely states that the most probable energy is equal to the average energy. Later, it will be shown that fluctuations are quite small, on the average, and this may be taken as partial justification for the present procedure. Our work is considerably simplified by using the most probable distribution, since then the ensemble average can be written as

$$U = \frac{1}{X} \sum_i n_i E_i \qquad (6.3)$$

where E_i is the energy of a member system of the ensemble in state i, and the summation is now taken over all quantum states of a typical member system. We really should have put some sort of bar or superscript on the n_i to indicate that they now represent most probable quantities, but we will spare ourselves this complication in notation since we will have no more use for averages of the type of (6.2).

1.7   THE MOST PROBABLE DISTRIBUTION FOR A CANONICAL ENSEMBLE

To make the definition of the ensemble average useful, it is necessary to specify the kind of ensemble needed to represent the physical system, and to relate the n_i to microscopic quantities. The physical system will be taken to be immersed in a constant temperature bath with which it can exchange only energy. The ensemble is then canonical, and the most probable set of n_i maximizes the number of complexions $W\{n_i\}$ subject to the conditions that the total energy of the ensemble is constant, and the total number of member systems in the ensemble is constant. A procedure for doing this is provided by the method of undetermined multipliers of Lagrange. The n_i are varied in such a way that variations in $W\{n_i\}$, U, and X vanish. This gives the most probable n_i in terms of the energy levels of the system. It is more convenient to take the variation of $\ln W\{n_i\}$ because then the factorials are easier to handle. Our problem then reduces to finding a set of n_i such that

$$\delta \ln W\{n_i\} = \delta \left[\ln X! - \sum_i \ln n_i! \right] = 0 \qquad (7.1)$$

$$\delta(UX) = \delta \sum n_i E_i = 0 \qquad (7.2)$$

$$\delta X = \delta \sum n_i = 0 \qquad (7.3)$$

The reason for taking a variation in the logarithm of the number of complexions rather than in $W\{n_i\}$ itself was so that the factorials could be simplified by Stirling's approximation, which states that for any large integer Y

$$\ln Y! = Y \ln Y - Y \qquad (7.4)$$

This approximation is extremely accurate for large integers, and the ensemble can always be taken large enough so that all n_i are large. Using (7.4), (7.1) becomes

$$\delta\left[X \ln X - X - \sum_i (n_i \ln n_i - n_i) \right] = 0 \qquad (7.5)$$

or

$$\delta X \ln X + X \delta \ln X - \sum_i (\delta n_i \ln n_i + n_i \delta \ln n_i)$$

$$= \delta X \ln X + \delta X - \sum_i (\delta n_i \ln n_i + \delta n_i)$$

$$= \ln X \sum_i \delta n_i - \sum_i \delta n_i \ln n_i$$

$$= \sum_i \delta n_i \ln \frac{X}{n_i} = 0 \qquad (7.6)$$

In arriving at (7.6), use has been made of the fact $X = \sum n_i$ and $\delta X = \sum \delta n_i$. Now, just multiply (7.2) and (7.3) by two constants $-\beta$ and $-\alpha$, respectively, which will be determined later, and add the results to (7.6) to get

$$\sum_i \delta n_i \ln \frac{X}{n_i} - \sum_i \alpha \delta n_i - \sum_i \beta \delta(n_i E_i) = 0 \qquad (7.7)$$

or, since we are varying only n_i and not energy levels

$$\sum_i \delta n_i \left[\ln \frac{X}{n_i} - \alpha - \beta E_i \right] = 0 \qquad (7.8)$$

By introducing the undetermined multipliers $-\alpha$ and $-\beta$, we have made all terms in the sum independent, so each term must equal zero, and therefore

$$\ln \frac{n_i}{X} = -\alpha - \beta E_i \qquad (7.9)$$

or

$$\frac{n_i}{X} = e^{-\alpha} e^{-\beta E_i} \qquad (7.10)$$

Equation (7.10) is called the canonical distribution function and gives the most probable set of n_i in terms of the energy spectrum of a typical system in the ensemble. Usually, the ratio n_i/X is reduced to a single symbol f_i, which can be variously interpreted as the probability that a particular member of the ensemble is in a state i, the probability of picking out of the ensemble a member in state i, or the probability that at any particular time, a physical system in thermal equilibrium is in the state i. Note that f_i is the probability of finding a system in a given energy level E_i, only if there is no degeneracy so that there is only one system eigenfunction per energy level. If this is not so, and we wish to describe a distribution-in-energy, it is only necessary to group together states with the same energy. If there are ω_j eigenfunctions with energy E_j, and if we denote energy distributions by p_j, then

$$p_j = \omega_j e^{-\alpha} e^{-\beta E_j} \tag{7.11}$$

is the probability that the system is in an energy level E_j. When using the state distributions f_i, all summations must be taken over states of the system. When using energy distributions p_j, all summations must be taken over energy levels.

The constants α and β can be formally determined as follows: write (7.10) as

$$e^{\alpha} f_i = e^{-\beta E_i} \tag{7.12}$$

and sum over all states. Since $\sum f_i = 1$, we get

$$e^{\alpha} = \sum_i e^{-\beta E_i} \tag{7.13}$$

This determines α in terms of β, so that

$$f_i = \frac{e^{-\beta E_i}}{\sum_i e^{-\beta E_i}} \tag{7.14}$$

Now, use (7.14) to get the average energy of the ensemble

$$U = \sum_i f_i E_i = \frac{\sum_i e^{-\beta E_i} E_i}{\sum_i e^{-\beta E_i}} \tag{7.15}$$

This determines β from the average energy. Clearly, α and β have an essential role in building the bridge between microscopic physics and thermodynamics. As we shall see, these constants have a straightforward thermodynamic interpretation. The sum in the denominator of (7.14) is particularly significant for thermodynamics. It is given a special symbol Z and is called the partition function

$$Z = \sum_i e^{-\beta E_i} \tag{7.16}$$

so that the canonical distribution function is often written as

$$f_i = \frac{1}{Z} e^{-\beta E_i} \qquad (7.17)$$

The canonical ensemble is the most widely used ensemble of statistical mechanics and is adequate for treating a great variety of problems. The methods given in this section, however, are easily generalized to other types of ensembles. Before discussing these generalizations, the development of the canonical ensemble will be completed by connecting the statistical method with thermodynamics and interpreting α and β in terms of thermodynamic functions.

1.8 THE CANONICAL ENSEMBLE AND THERMODYNAMICS

The parameters α and β refer to the ensemble as a whole and not to the individual member systems and, therefore, they must provide the link between statistical mechanics and thermodynamics. In fact, β and α are intimately related to the temperature and Helmholtz free energy, respectively. To show that β has the property of temperature, consider two physical systems, A and B, in thermal equilibrium with each other. Since each system is in equilibrium, then each can be represented by separate canonical ensembles for which the distribution functions are

$$f_i^A = \frac{1}{Z^A} e^{-\beta^A E_i^A} \qquad (8.1)$$

$$f_j^B = \frac{1}{Z^B} e^{-\beta^B E_j^B} \qquad (8.2)$$

The two systems taken together, however, also form an equilibrium system for which a canonical distribution can be written as

$$f_{ij}^{AB} = \frac{1}{Z^{AB}} e^{-\beta^{AB} E_{ij}^{AB}} \qquad (8.3)$$

where f_{ij}^{AB} is the probability that part A of the combined system is in state i, while part B is in state j, Z^{AB} is the partition function for the combined system, and $E_{ij}^{AB} = E_i^A + E_j^B$. But the probability of a joint event is the product of the probabilities of the separate events, so that

$$f_{ij}^{AB} = f_i^A f_j^B \qquad (8.4)$$

and multiplying (8.2) by (8.1) and then equating the product to (8.3) gives

$$\frac{1}{Z^A Z^B} e^{-(\beta^A E_i^A + \beta^B E_j^B)} = \frac{1}{Z^{AB}} e^{-\beta^{AB}(E_i^A + E_j^B)} \tag{8.5}$$

the partition functions being given by

$$Z^A Z^B = \sum_{ij} e^{-(\beta^A E_i^A + \beta^B E_i^B)}$$

$$Z^{AB} = \sum_{ij} e^{-\beta^{AB}(E_i^A + E_j^B)}$$

Now rearrange (8.5) and write it as

$$(\beta^A - \beta^{AB})E_i^A + (\beta^B - \beta^{AB})E_j^B = \ln \frac{Z^{AB}}{Z^A Z^B} \tag{8.6}$$

This equation is correct for any choice of all the possible energies E_i^A and E_j^B. Since the right-hand side is a constant, the left-hand side must also be constant. Because E_i^A and E_j^B are both positive and have an infinite number of values, the only way that the left-hand side can be independent of the subscripts i and j is to have the coefficients of E_i^A and E_j^B equal zero. Therefore

$$\beta^{AB} = \beta^A = \beta^B \tag{8.7}$$

and

$$Z^{AB} = Z^A Z^B \tag{8.8}$$

Thermal equilibrium among systems, therefore, requires that the ensemble parameter β be the same for both systems. Since this is precisely the fundamental nature of the temperature in thermodynamics, β will be taken to be a monotonic function of the temperature T.

$$\beta = \beta(T) \tag{8.9}$$

To find the actual functional dependence of β on T, the relationship between the entropy and the canonical distribution must be considered. A change in the equilibrium value of the energy can be produced by varying both the f_i and the E_i so that, in general,

$$dU = \sum_i (E_i \, df_i + f_i \, dE_i) \tag{8.10}$$

From thermodynamics, however,

$$dU = T \, dS - dW \tag{8.11}$$

where dS is the entropy change associated with the energy change given by (8.10) and dW is the work done on the system during this change. The energy levels E_i can be altered only by doing work on the system. Therefore, if the statistical equation (8.10) and the thermodynamic equation (8.11) are to be identified

$$dW = - \sum_i f_i \, dE_i \tag{8.12}$$

and

$$T \, dS = \sum_i E_i \, df_i \tag{8.13}$$

From (7.17) for the canonical distribution

$$E_i = -\frac{1}{\beta} \ln Z - \frac{1}{\beta} \ln f_i \tag{8.14}$$

Substituting this into (8.13) and solving for dS gives

$$dS = -\frac{1}{\beta T} \sum_i \ln f_i \, df_i \tag{8.15}$$

where use has been made of the fact that $\sum df_i = 0$.

Since the entropy is a state function, dS is a perfect differential, and if our statistical method is to be valid, the right-hand side must also be a perfect differential. The sum can be written as

$$dx = d\left(\sum_i f_i \ln f_i \right) = \sum_i \ln f_i \, df_i \tag{8.16}$$

where x is defined by the above equation, so that (8.15) can be written as

$$dS = -\frac{1}{\beta T} \, dx \tag{8.17}$$

where dx is a perfect differential. If only we could be sure that (βT) were not a function of x, we could integrate (8.17) immediately to get the entropy. We can show that this is true by writing $-(\beta T)^{-1} = f(x)$ and making use of the additive property of the entropy. Write (8.17) for two systems A and B in the form

$$dS_A = f(x_A) \, dx_A$$

$$dS_B = f(x_B) \, dx_B$$

For the combined system AB, we have

$$dS_{AB} = f(x_{AB}) \, dx_{AB}$$

Using the definition of x and (8.4), it is easy to show that

$$dx_{AB} = dx_A + dx_B$$

also

$$dS_{AB} = dS_A + dS_B$$

and therefore

$$f(x_A)\,dx_A + f(x_B)\,dx_B = f(x_{AB})(dx_A + dx_B) = f(x_A + x_B)(dx_A + dx_B)$$

or

$$[f(x_A) - f(x_A + x_B)]\,dx_A + [f(x_B) - f(x_A + x_B)]\,dx_B = 0 \quad (8.18)$$

But dx_A and dx_B are independent differentials. Therefore this equation can be correct only if

$$f(x_A) = f(x_B) = f(x_{AB}) = f(x_A + x_B) \qquad (8.19)$$

But this equation can be true only if f is not a function of x, because the nature of the two systems is completely arbitrary, except for the requirement that they be in thermal equilibrium. Equation 8.17 may therefore be integrated directly. Using the definition of x, we get

$$S = -\frac{1}{\beta T}\sum_i f_i \ln f_i \qquad (8.20)$$

except for an additive constant, which we take to be zero.*

Now that the entropy is expressed in terms of the probability functions, we can proceed to relate β to the temperature. From (8.14), we can solve for $\ln f_i$, substitute it into (8.20), and write the entropy as

$$S = \frac{1}{T}\sum_i f_i E_i + \frac{1}{\beta T}\sum_i f_i \ln Z \qquad (8.21)$$

or

$$S = \frac{U}{T} + \frac{\ln Z}{\beta T} \qquad (8.22)$$

From the thermodynamic equation $A = U - TS$, where A is the Helmholtz free energy

$$S = \frac{U}{T} - \frac{A}{T} \qquad (8.23)$$

Comparing these last equations gives

$$A = -\frac{\ln Z}{\beta} \qquad (8.24)$$

Thermodynamics tells us that

$$S = -\left(\frac{\partial A}{\partial T}\right)_{N,V} \qquad (8.25)$$

---

* The value of this constant is, of course, immaterial for thermodynamics, since only differences in entropy are measurable. Taking it to be zero simplifies the form of the statistical mechanical ⸌quations without detracting anything of physical significance.

where the derivative is to be evaluated at constant volume and constant composition. The entropy can therefore be found by differentiating (8.24) with respect to temperature. The condition of constant N is automatically satisfied, and the condition of constant V can be satisfied by keeping all E_i constant. Therefore

$$S = \frac{d}{dT}\left(\frac{1}{\beta}\right) \ln Z + \frac{1}{\beta} \frac{d \ln Z}{dT} \tag{8.26}$$

From the definition of the partition function (7.16) this reduces to

$$S = -\frac{1}{\beta} \frac{d\beta}{dT} U + \frac{d}{dT}\left(\frac{1}{\beta}\right) \ln Z \tag{8.27}$$

Now, compare this expression for the entropy to (8.22). Obviously

$$-\frac{1}{\beta} \frac{d\beta}{dT} = \frac{1}{T}$$

$$\frac{d}{dT}\left(\frac{1}{\beta}\right) = \frac{1}{\beta T} \tag{8.28}$$

When integrated, both of these equations show that

$$\beta = \frac{1}{kT} \tag{8.29}$$

where k is some constant whose value is to be determined. The functional dependence of β on temperature given above shows that β^{-1} is a statistical-mechanical temperature. The purpose of the constant k is merely to convert the conventional units of temperature to those for β^{-1}, which, as is evident from its place in the probability distribution function, has the units of energy. The numerical value of k is obtained by working out the statistical thermodynamics of some specific system and comparing it with experiment. For example, a statistical derivation of the ideal gas law shows that k is Boltzmann's constant. Being a universal constant, k is the same for all types of physical systems.

In arriving at the interpretation of β, we have accomplished much more. An expression has been obtained for the entropy (8.20) and for the Helmholtz free energy (8.24), which can now be written as

$$S = -k \sum f_i \ln f_i \tag{8.30}$$

and

$$A = -kT \ln Z \tag{8.31}$$

with

$$Z = \sum_i e^{-E_i/kT} \tag{8.32}$$

Furthermore, from (8.31) and (7.13) the constant α is given by

$$\alpha = -\frac{A}{kT} \tag{8.33}$$

and the canonical distribution becomes

$$f_i = e^{(A-E_i)/kT} \tag{8.34}$$

Equations 8.30 and 8.31 are of special importance. The first of these embodies the intimate relationship between entropy and probability concepts and is often taken as a definition of entropy in terms of statistics from which the distribution function is derived. The second of these is the foundation of statistical thermodynamics. If the energy levels of a system are known, the partition function can be determined. Equation 8.31 then gives the Helmholtz free energy. The other thermodynamic functions can then be obtained from thermodynamic equations. For example,

$$S = -\left(\frac{\partial A}{\partial T}\right)_V \tag{8.35}$$

$$U = A - T\left(\frac{\partial A}{\partial T}\right)_V \tag{8.36}$$

while the heat capacity at constant volume C_V and the pressure P are given by

$$C_V = -T\left(\frac{\partial^2 A}{\partial T^2}\right)_V \tag{8.37}$$

$$P = -\left(\frac{\partial A}{\partial V}\right)_T \tag{8.38}$$

The isothermal compressibility κ and the coefficient of thermal expansion α are defined by

$$\kappa = -\frac{1}{V}\left(\frac{\partial V}{\partial P}\right)_T \tag{8.39}$$

$$\alpha = \frac{1}{V}\left(\frac{\partial V}{\partial T}\right)_P \tag{8.40}$$

They are related to the Helmholtz free energy through the equations

$$\frac{1}{\kappa} = V\left(\frac{\partial^2 A}{\partial V^2}\right)_T \tag{8.41}$$

$$\frac{\alpha}{\kappa} = -\left[\frac{\partial}{\partial V}\left(\frac{\partial A}{\partial T}\right)_V\right]_T \tag{8.42}$$

In principle, the partition function is all that is needed to allow us to compute the thermodynamic properties of any system that can be represented by a canonical ensemble. We merely find the partition function for the system under consideration, use (8.31) for the Helmholtz free energy, and then apply the relation from thermodynamics appropriate to the property we wish to compute. This procedure is, of course, workable only for closed systems, that is, systems that do not exchange atoms or molecules with their environment, since this was one of the conditions on the canonical ensemble. More general methods are needed to deal with open systems.

In practice, there is another kind of problem that presents itself in the application of partition function theory. An examination of the preceding formulae shows that the derivatives can be divided into two groups: those that are taken at constant volume and those that are not. The former are much easier to handle than the latter, as is evident by thinking about what must be done to evaluate C_V and P. The temperature appears explicitly in the partition function, and if the energy levels E_i are known, a straight-forward differentiation according to (8.37) gives the heat capacity. In arriving at the pressure by means of (8.38), however, the temperature is held constant, and we must differentiate with respect to volume. But the volume enters into the partition function only implicitly, in that the energy levels are functions of the volume. This means that in addition to the energy levels themselves, we must know how they vary with the volume. This latter item of information is often not known with great precision, particularly in the theory of solids, and approximations of various kinds must be made. As a result, theories of the heat capacity can be developed with a certain degree of elegance. In constructing a theory of the equation of state of solids by using (8.38) to get a P-V-T relation, however, we must be satisfied with a lack of neatness coming from our imperfect knowledge of the effect of volume on the energy levels. In general, then, the "constant volume" properties are easier to describe by the canonical partition function than those that depend on volume derivatives.

Even with the limitations described above, the canonical ensemble is the most useful, and most widely used, ensemble, especially for elementary problems. This is so because we are very often interested in closed, simple systems and the "constant volume" properties are often sufficient to allow us to build a consistent theory of matter. Also, in certain important systems, such as the ideal gas and the free-electron gas, we do know how the energy levels vary with volume. In such cases, there are no difficulties.

1.9 THE GRAND CANONICAL ENSEMBLE

Now, let us suppose we want to describe a system in which fluctuations in the number of atoms or molecules as well as in the energy are permitted.

For simplicity of notation, we take the system to consist of two components, 1 and 2. We then form a grand canonical ensemble, in which the total number of molecules of both kinds and the total energy are constant. The member systems of the ensemble, however, can exchange molecules and energy. Let $n_j(N_1, N_2)$ denote the number of member systems that contain N_1 molecules of type 1, N_2 molecules of type 2 and are in the *j*th quantum state. There is, of course, a different spectrum of quantum states for each pair of values of N_1 and N_2 so that the possible energy levels are functions of the numbers of molecules and will therefore be labeled $E_j(N_1, N_2)$. The total number of systems in the ensemble is

$$X = \sum_{j,N_1,N_2} n_j(N_1, N_2) \tag{9.1}$$

where the summation is taken over all possible quantum states and all possible values of N_1 and N_2. The average number of molecules of each type per member system is given by

$$\bar{N}_1 = \frac{1}{X} \sum_{j,N_1,N_2} N_1 n_j(N_1, N_2) \tag{9.2}$$

$$\bar{N}_2 = \frac{1}{X} \sum_{j,N_1,N_2} N_2 n_j(N_1, N_2) \tag{9.3}$$

while the average energy per member system is

$$U = \frac{1}{X} \sum_{j,N_1,N_2} E_j(N_1, N_2) n_j(N_1, N_2) \tag{9.4}$$

We now follow a process similar to that used for the canonical ensemble. The number of complexions for a given distribution $n_j(N_1, N_2)$ is found, and this number is maximized, subject to the conditions of constant total energy and total number of particles. The combinatorial problem is similar to that for the canonical ensemble, and the number of complexions $W\{n_j(N_1, N_2)\}$ is given by

$$W\{n_j(N_1, N_2)\} = \frac{X!}{\displaystyle\prod_{j,N_1,N_2} n_j(N_1, N_2)!} \tag{9.5}$$

where the product is now taken over all quantum states and over all values of N_1 and N_2.

Taking the logarithm of (9.5), we get

$$\ln W\{n_j(N_1, N_2)\} = \ln X! - \sum_{j,N_1,N_2} \ln n_j(N_1, N_2)!$$

Now, take the variation of $\ln W\{n_j(N_1, N_2)\}$, $\bar{N}_1$, $\bar{N}_2$, U, and X with respect to variations in the $n_j(N_1, N_2)$, and set them equal to zero. The result is, using Stirling's approximation

$$\sum_{j,N_1,N_2} \delta n_j(N_1, N_2) \ln \frac{X}{n_j(N_1, N_2)} = 0 \tag{9.6}$$

$$\sum_{j,N_1,N_2} N_1 \, \delta n_j(N_1, N_2) = 0 \tag{9.7}$$

$$\sum_{j,N_1,N_2} N_2 \, \delta n_j(N_1, N_2) = 0 \tag{9.8}$$

$$\sum_{j,N_1,N_2} E_j(N_1, N_2) \, \delta n_j(N_1, N_2) = 0 \tag{9.9}$$

$$\sum_{j,N_1,N_2} \delta n_j(N_1, N_2) = 0 \tag{9.10}$$

Multiply (9.7) through (9.10) by the undetermined multipliers $-\gamma_1$, $-\gamma_2$, $-\beta$, and $-\alpha'$, respectively, add the results to (9.6) and set each term equal to zero. This gives

$$f_j(N_1, N_2) = \frac{n_j(N_1, N_2)}{X} = e^{-\alpha'} e^{-\beta E_j(N_1,N_2)} e^{-(\gamma_1 N_1 + \gamma_2 N_2)} \tag{9.11}$$

for the grand canonical distribution function, which is the probability that the system contains N_1 molecules of type 1, N_2 of type 2, and is in the jth quantum state with energy $E_j(N_1, N_2)$.

The connection with thermodynamics is easily made. Differentiating the equation for the energy, (9.4),

$$dU = \sum_{j,N_1,N_2} E_j(N_1, N_2) \, df_j(N_1, N_2) + \sum_{j,N_1,N_2} f_j(N_1, N_2) \, dE_j(N_1, N_2) \tag{9.12}$$

we conclude, as before, that the first term represents $T \, dS$ because the energy levels can change only by doing external work on the system. Therefore

$$T \, dS = \sum_{j,N_1,N_2} E_j(N_1, N_2) \, df_j(N_1, N_2) \tag{9.13}$$

Solving (9.11) for $E_j(N_1, N_2)$, substituting the result into (9.13), and integrating gives

$$S = -\frac{1}{\beta T} \sum_{j,N_1,N_2} f_j(N_1, N_2) \ln f_j(N_1, N_2) \tag{9.14}$$

It is easily shown that $\beta = \beta(T)$ by considering two different systems in equilibrium just as was done for the canonical ensemble, and that (βT) is not a function of $f_j \ln f_j$.

If the solution of $E_j(N_1, N_2)$ from (9.11) is substituted into (9.4), we get

$$U = -\frac{1}{\beta} \sum_{j,N_1,N_2} f_j(N_1, N_2) \ln f_j(N_1, N_2) - \frac{\alpha'}{\beta} - \frac{\gamma_1 \bar{N}_1}{\beta} - \frac{\gamma_2 \bar{N}_2}{\beta} \tag{9.15}$$

Compare (9.15) with the thermodynamic equation for open systems

$$U = TS - PV + \mu_1\bar{N}_1 + \mu_2\bar{N}_2 \qquad (9.16)$$

where μ_1 and μ_2 are the chemical potentials of species 1 and 2, respectively. We already have the identification of entropy in (9.14) so that a comparison of the last two equations gives

$$\frac{\alpha'}{\beta} = PV \qquad (9.17)$$

$$\frac{\gamma_1}{\beta} = -\mu_1 \qquad (9.18)$$

$$\frac{\gamma_2}{\beta} = -\mu_2 \qquad (9.19)$$

All that needs to be done now is to show that $\beta = 1/kT$. First, we define the grand partition function by

$$Q = e^{\alpha'} \qquad (9.20)$$

or, by summing (9.11) over all N_1, N_2, and j and remembering that the sum of the $f_j(N_1, N_2)$ must yield unity

$$Q = \sum_{j,N_1,N_2} e^{-\beta E_j(N_1,N_2)} e^{-(\gamma_1 N_1 + \gamma_2 N_2)} \qquad (9.21)$$

Using the definition (9.20), (9.11) is easily transformed into

$$\ln f_j(N_1, N_2) = -\ln Q - \beta E_j(N_1, N_2) - \gamma_1 N_1 - \gamma_2 N_2 \qquad (9.22)$$

Now, multiply all terms of this equation by $f_j(N_1, N_2)$ and sum over all N_1, N_2, and j. In view of (9.14) for the entropy and of the definition of the average energy U, and numbers of particles $\bar{N}_1$ and $\bar{N}_2$, we get

$$-\beta TS = -\ln Q - \beta U - \gamma_1\bar{N}_1 - \gamma_2\bar{N}_2$$

or, solving for the entropy,

$$S = \frac{1}{\beta T}\ln Q + \frac{U}{T} + \frac{\gamma_1}{\beta T}\bar{N}_1 + \frac{\gamma_2}{\beta T}\bar{N}_2 \qquad (9.23)$$

Since γ_1 and γ_2 are known in terms of the chemical potentials μ_1 and μ_2 from (9.18) and (9.19), we may write

$$S = \frac{1}{\beta T}\ln Q + \frac{U}{T} - \frac{\mu_1\bar{N}_1}{T} - \frac{\mu_2\bar{N}_2}{T} \qquad (9.24)$$

An alternative form for the entropy can be obtained by starting with the thermodynamic relation for open systems.

$$S = \left(\frac{\partial(PV)}{\partial T}\right)_{V, N_1, N_2} \tag{9.25}$$

which can be written as

$$S = -\frac{\alpha'}{\beta^2}\frac{d\beta}{dT} + \frac{1}{\beta}\left(\frac{\partial\alpha'}{\partial T'}\right)_{V, N_1, N_2} \tag{9.26}$$

where use has been made of (9.17) and β taken as a function of temperature $\beta(T)$. Since $\alpha' = \ln Q$, we get

$$\frac{\partial\alpha'}{\partial T} = \frac{1}{Q}\frac{\partial Q}{\partial T}$$

and from (9.21)

$$\frac{\partial Q}{\partial T} = -\sum_{j, N_1, N_2}\left(\frac{d\beta}{dT}E_j + \frac{d\gamma_1}{dT}N_1 + \frac{d\gamma_2}{dT}N_2\right)e^{-\beta E(N_1, N_2)}e^{-(\gamma_1 N_1 + \gamma_2 N_2)} \tag{9.27}$$

But from (9.18) and (9.19), $\gamma_1 = -\mu_1\beta$ and $\gamma_2 = -\mu_2\beta$; and since we are taking the derivatives at constant N_1 and N_2 we have

$$\frac{d\gamma_1}{dT} = -\mu_1\frac{d\beta}{dT} - \beta\frac{d\mu_1}{dT}$$

$$\frac{d\gamma_2}{dT} = -\mu_2\frac{d\beta}{dT} - \beta\frac{d\mu_2}{dT}$$

Therefore, (9.27) becomes

$$\frac{\partial Q}{\partial T} = -\frac{d\beta}{dT}\sum_{j, N_1, N_2}(E_j - \mu_1 N_1 - \mu_2 N_2)e^{-\beta E_j(N_1, N_2)}e^{-(\gamma_1 N_1 + \gamma_2 N_2)}$$

$$-\beta\sum_{j, N_1, N_2}\left[N_1\frac{d\mu_1}{dT} + N_2\frac{d\mu_2}{dT}\right]e^{-\beta E_j - \gamma_1 N_1 - \gamma_2 N_2} \tag{9.28}$$

Dividing this equation by Q gives

$$\frac{1}{Q}\frac{\partial Q}{\partial T} = -\frac{d\beta}{dT}U + \frac{d\beta}{dT}\mu_1\bar{N}_1 + \frac{d\beta}{dT}\mu_2\bar{N}_2 \tag{9.29}$$

where the terms containing the derivatives of the chemical potentials sum to zero because, from the Gibbs–Duhem equation

$$\left(\frac{\partial\mu_1}{\partial T}\right)_{N, V}\bar{N}_1 + \left(\frac{\partial\mu_2}{\partial T}\right)_{N, V}\bar{N}_2 = 0$$

The right-hand side of (9.28) is a result of the fact that $f_j(N_1 N_2)$ is

$$f_j(N_1, N_2) = \frac{e^{-\beta E_j(N_1, N_2)} e^{-(\gamma_1 N_1 + \gamma_2 N_2)}}{Q} \tag{9.30}$$

Finally, substituting (9.29) into (9.26) gives

$$S = \frac{1}{\beta} \frac{d\beta}{dT} \left[-\frac{\ln Q}{\beta} - U + \mu_1 \bar{N}_1 + \mu_2 \bar{N}_2 \right] \tag{9.31}$$

In deriving (9.31) from (9.26), α' was replaced by $\ln Q$ according to (9.20). Now, comparing (9.31) with (9.24) we see that

$$-\frac{1}{\beta} \frac{d\beta}{dT} = \frac{1}{T} \tag{9.32}$$

in complete agreement with (8.28). Therefore

$$\beta = \frac{1}{kT} \tag{9.33}$$

just as for the canonical ensemble. Clearly, the constant k in this equation must have the same value as that for the canonical ensemble. This can be seen by considering a grand canonical ensemble to be a collection of canonical ensembles, each canonical ensemble being formed by grouping together systems with the same N_1, N_2. The canonical ensembles are all in equilibrium with each other and all are at the same temperature. They must also all have the same β, as can be easily shown by the procedure in Section 8, leading to (8.9). In this way, k is shown to be a universal constant for all ensembles.

Let us summarize our results. For a grand canonical ensemble, the distribution function is

$$f_j(N_1, N_2) = \frac{1}{Q} \exp \left\{ -\frac{1}{kT} [E_j(N_1, N_2) - \mu_1 N_1 - \mu_2 N_2] \right\} \tag{9.34}$$

where the grand partition function Q is given by

$$Q = \sum_{j, N_1, N_2} \exp \left\{ -\frac{1}{kT} [E_j(N_1, N_2) - \mu_1 N_1 - \mu_2 N_2] \right\} \tag{9.35}$$

and Q is related to thermodynamics by

$$PV = kT \ln Q \tag{9.36}$$

The other thermodynamic functions can be obtained by using the appropriate formulae from the thermodynamics of open systems. For example,

$$\left(\frac{\partial(PV)}{\partial T}\right)_{V,\mu_1,\mu_2} = S \tag{9.37}$$

$$\left(\frac{\partial(PV)}{\partial \mu_1}\right)_{T,V,\mu_2} = \bar{N}_1 \tag{9.38}$$

$$A = \bar{N}_1\mu_1 + \bar{N}_2\mu_2 - PV \tag{9.39}$$

give the entropy, number of particles of type 1, and Helmholtz free energy, respectively.

The theory developed here is easily generalized in an obvious way if the system contains an arbitrary number of component species. The number of terms $N_j\mu_j$ is simply equal to the number of components.

1.10 THE PRESSURE ENSEMBLE

The theoretical analysis of the ensemble concept has certain broad similarities for all types of ensembles. We simply find the number of complexions of the ensemble for a fixed distribution of systems among the allowed values of those parameters that can fluctuate from one system to another. Then, we maximize the number of complexions with respect to a variation of the number of systems in each state, subject to the appropriate subsidiary conditions. Finally, we identify certain statistical mechanical equations with thermodynamic equations, thereby evaluating the undetermined multipliers.

In a canonical pressure ensemble, we keep the number of particles of all species fixed for all systems, but allow fluctuations in energy and volume. The resulting canonical pressure distribution function is

$$f_j(V) = \frac{1}{Z_P} \exp\left\{-\frac{1}{kT}\left[E_j(V) + PV\right]\right\} \tag{10.1}$$

where $f_j(V)$ is the probability that the system has a volume V and is in quantum state j with energy $E_j(V)$ and Z_P is the pressure partition function

$$Z_P \equiv \sum_{j,V} \exp\left\{-\frac{1}{kT}\left[E_j(V) + PV\right]\right\} \tag{10.2}$$

which is related to thermodynamics by

$$G = -kT \ln Z_P \tag{10.3}$$

where G is the Gibbs free energy.

1.11   FLUCTUATIONS IN ENERGY

We will not be much concerned with fluctuations in this book, but we should show that the statistical methods that have been introduced lead to fluctuations that are indeed small in macroscopic parameters of ordinary systems. This is a necessary requirement on statistical mechanics because real systems at equilibrium generally exhibit a remarkable constancy and no fluctuations can be determined by ordinary measurements. In this section a calculation of the fluctuations in energy to be expected in a system in thermal equilibrium will be given.

For a canonical ensemble, the average energy per system is U, and the deviation from this average by some system of the ensemble with energy E is $(E - U)$. We are interested in the mean fluctuations in energy, so that whatever measure of the fluctuation we take, it should be some average value. The mean deviation from the average, $\overline{(E - U)}$, where the bar indicates statistical mechanical averaging, is not suitable as a measure of fluctuations, since it is identically zero. To see this, we merely write

$$\overline{E - U} = \sum_j f_j(E_j - U) = \sum_j f_j E_j - U = U - U = 0 \qquad (11.1)$$

The mean square deviation, $\overline{(E - U)^2}$, is a measure of the spread of a statistical distribution and is, therefore, a suitable measure of the energy fluctuations. It can also be written as $\overline{E^2} - U^2$, since

$$\overline{(E - U)^2} = \overline{(E^2 - 2EU + U^2)} = \overline{E^2} - 2U^2 + U^2 = \overline{E^2} - U^2 \quad (11.2)$$

For the canonical distribution, the average energy is given by

$$U = \frac{\sum_j E_j e^{-E_j/kT}}{\sum_j e^{-E_j/kT}} \qquad (11.3)$$

Differentiating this, we get for the specific heat at constant volume

$$C_V = \left(\frac{\partial U}{\partial T}\right)_V = \frac{1}{kT^2}(E^2 - U^2) \qquad (11.4)$$

This follows immediately from the definitions of U and $\overline{E^2}$

$$\overline{E^2} = \frac{\sum_j E_j{}^2 e^{-E_j/kT}}{\sum_j e^{-E_j/kT}} \qquad (11.5)$$

The relative mean square deviation is therefore

$$\frac{\overline{E^2} - U^2}{U^2} = \frac{\overline{(E - U)^2}}{U^2} = \frac{C_V k T^2}{U^2} \tag{11.6}$$

and the magnitude of the expected energy fluctuation is the square root of this:

$$\left[\frac{\overline{(E - U)^2}}{U^2}\right]^{1/2} = \left[\frac{C_V k T^2}{U^2}\right]^{1/2} \tag{11.7}$$

In general, the energy U of a physical system is proportional to the number of particles N in the system. Therefore, the heat capacity is also proportional to N. For example, in an ideal gas, $U = 3NkT/2$, as we will see later, where N is the number of atoms in the gas. The heat capacity is then $3Nk/2$, so that for an ideal gas (11.7) gives

$$\left[\frac{\overline{(E - U)^2}}{U^2}\right]^{1/2} = \frac{2}{3}\frac{1}{\sqrt{N}} \tag{11.8}$$

for the magnitude of expected fluctuations in the energy. Since N is always very large for macroscopic systems ($\sim 10^{20}$ or greater), we see that the energy fluctuations are extremely small. Similar calculations, with similar results, can be carried out for other thermodynamic parameters.

In defining the various types of ensembles, we gave a description of the physical situation corresponding to each ensemble, but we also said that the choice of ensemble was generally conditioned by convenience in solving particular problems. The reason that any ensemble can be used in describing real systems is that fluctuations are generally negligible. The constraints placed on the various ensembles, therefore, have a negligible effect on the calculation of physical variables.

1.12 THE STATISTICAL ENTROPY

The identification of the entropy in terms of the probability distribution functions is one of the most important results of statistical mechanics. In fact, the theory of statistical mechanics can be worked out by defining the statistical entropy for a canonical ensemble by (8.30), postulating that the statistical entropy is identical to the thermodynamic entropy and then maximizing it, subject to the conditions of constant average energy and constant number of ensemble members. In this section we will investigate the properties of the statistical entropy function for a canonical ensemble

$$S = -k \sum_i f_i \ln f_i \tag{12.1}$$

In Section 8, the additive property of the entropy was used to perform an integration leading to the above equation. Conversely, it is easy to show that if we are given (12.1), then $S_{AB} = S_A + S_B$ for two systems A and B in thermal equilibrium. We will leave this as an exercise.

We will now consider the relation between entropy and the concept of "randomness." Let us start with a crystal at an extremely low temperature. As the crystal is heated, we know from experiment that its entropy increases. We also know that the entropy of the liquid is greater than that of the solid and that the entropy of the vapor is greater than that of the liquid. From our knowledge of the atomic theory of matter, it is evident that as the crystal is heated, melted, and vaporized, the atoms of which the material is composed undergo increasingly violent agitation. At low temperatures, the atoms in the crystal are localized in a small region around their lattice sites, but in the gas phase, each atom has a trajectory that can carry it to any point in the enclosing volume. Such considerations lead to the concept that a gas is a more highly randomized system than a solid. By this statement we mean that the precise positions of the atoms in a gas change radically with time in a chaotic, unpredictable manner. We conclude, then, that the entropy has something to do with a system's randomness.

These vague notions can be made precise by use of (12.1). In a non-random system, the probability that the system is in a particular state is vanishingly small except for some single state or some small group of states. In the extreme case, this means that all f_i are zero except one, f_m say, for which $f_m = 1$. But then (12.1) becomes $S = -kf_m \ln f_m = -k(1) \ln (1) = 0$. The statistical entropy then becomes zero, in agreement with our intuitive expectations. In a highly randomized system, however, a large number of f_i are not zero, and, furthermore, they are approximately equal. Since $f_i < 1$, there are, therefore, a great many terms in (12.1) containing logarithms of small numbers. These logarithms are negative and large in magnitude, so we have a high, positive value for the entropy, again in agreement with intuition.

An instructive interpretation of the statistical entropy can be obtained by substituting n_i/X for f_i in (12.1):

$$S = -k \sum_i \frac{n_i}{X} \ln \frac{n_i}{X} \tag{12.2}$$

Carrying out some algebraic reduction and remembering that $\sum n_i = X$, (12.2) becomes

$$XS = -k \sum_i n_i \ln n_i + kX \ln X \tag{12.3}$$

Now, use Stirling's approximation in the form

$$n_i \ln n_i = \ln n_i! + n_i$$
$$X \ln X = \ln X! + X \tag{12.4}$$

Equation 12.3 now reduces to

$$XS = k \ln \frac{X!}{\prod_i n_i!} \tag{12.5}$$

But inspection of (6.1) shows that the argument in the logarithm is just the number of complexions $W\{n_i\}$. Therefore

$$XS = k \ln W\{n_i\} \tag{12.6}$$

which shows that the entropy of the ensemble is proportional to the logarithm of the number of complexions. Of course, (12.6) refers to the equilibrium distribution of states in the ensemble because the identification of entropy with $\sum f_i \ln f_i$ was made for equilibrium systems. But if the statistical entropy is *defined* by (12.1) for any system, (12.6) shows that application of the method of undetermined multipliers to the entropy will always give the correct results.

Now, let us assume that entropy can be defined in nonequilibrium systems by (12.1) and let us investigate the change of the entropy with time. Differentiation of (12.1) gives, for the time derivative of S

$$\frac{dS}{dt} = -k \sum_i \frac{df_i}{dt} \ln f_i \tag{12.7}$$

We must now introduce the concept of conditional transition probabilities. On a large scale, a system away from equilibrium passes through a definite succession of states until finally equilibrium is reached. On a microscopic level, however, statistical mechanics can tell us only the probability that a system is in a given state. Furthermore, if it is in a given state at a particular time, it may be in any of a large number of states at some later time. To describe this, we define a conditional transition probability function Λ_{ij}, such that $\Lambda_{ij} \, dt$ gives the probability that a system known to be in the state i at time t is in the state j at time $t + dt$. This definition allows us to derive a useful expression for df_i/dt. To get the change in the probability that the system is in state i at time $t + dt$, just take the probability that some system in state j enters state i, subtract from it the probability that a system in state i leaves for some other state j, and then sum over all j, that is,

$$f_i(t + dt) - f_i(t) = \sum_{j \neq i} (f_j \Lambda_{ji} \, dt - f_i \Lambda_{ij} \, dt) \tag{12.8}$$

Dividing through by dt and taking the limit as $dt \to 0$ converts the left-hand side to a time derivative and we have

$$\frac{df_i}{dt} = \sum (f_j \Lambda_{ji} - f_i \Lambda_{ij}) \tag{12.9}$$

The requirement $j \neq i$ has been dropped in this equation, since nothing is changed by adding $f_i \Lambda_{ii}$ to the first term and subtracting it from the second term.

Now substitute (12.9) into (12.7):

$$\frac{dS}{dt} = -k \sum_{i,j} (f_j \Lambda_{ji} - f_i \Lambda_{ij}) \ln f_i \qquad (12.10)$$

The subscripts i and j are dummies over which summations are performed so that they can be interchanged without altering the value of the sum and we can write

$$\frac{dS}{dt} = -k \sum_{i,j} (f_i \Lambda_{ij} - f_j \Lambda_{ji}) \ln f_j \qquad (12.11)$$

Add this equation to the preceding one, divide by two, and rearrange terms a little to get

$$\frac{dS}{dt} = \frac{k}{2} \sum_{i,j} (\ln f_i - \ln f_j)(\Lambda_{ij} f_i - \Lambda_{ji} f_j) \qquad (12.12)$$

In order to proceed further, some assumption must be made about the Λ_{ij}. Remember that Λ_{ij} is a transition probability for a system going from state i to state j, given that it is initially in state i. The *total number* of transitions per unit time from i to j is proportional to $f_i \Lambda_{ij}$ and, therefore, in general the number of transitions from i to j is not equal to the number of transitions from j to i. However, we can postulate that *at equilibrium*, these numbers are indeed equal, that is,

$$f_i \Lambda_{ij} = f_j \Lambda_{ji} \qquad \text{(at equilibrium)} \qquad (12.13)$$

This is called the principle of detailed balance. Note that detailed balancing sets (12.12) equal to zero, so that it is sufficient to ensure equilibrium. Because fluctuations at equilibrium are so small, practically all of the states taking part in detailed balancing will be very close together in energy and, therefore, for most of the states, $f_i \approx f_j$. Equation 12.13, therefore, suggests that it is plausible to assume that

$$\Lambda_{ij} = \Lambda_{ji} \qquad (12.14)$$

This is called the principle of microscopic reversibility. The reasoning we have used to arrive at (12.14) merely suggests that $\Lambda_{ij} = \Lambda_{ji}$ is correct to a good approximation for a large number of physically important states. In the statistical study of kinetic phenomena, however, microscopic reversibility is often elevated to the rank of a postulate and used to deduce kinetic equations. If we accept (12.14), then (12.12) becomes

$$\frac{dS}{dt} = \frac{k}{2}\sum_{i,j}\Lambda_{ij}(\ln f_i - \ln f_j)(f_i - f_j) \tag{12.15}$$

The reason for putting the rate of change of entropy in this form is that every term in (12.15) is positive. If, for example, $f_i > f_j$, then $\ln f_i > \ln f_j$ and, conversely, if $f_i < f_j$, then $\ln f_i < \ln f_j$. The right-hand side of this equation is, therefore, always a positive quantity. This means that the entropy of a system not at equilibrium is always increasing. In other words, we have here a statistical expression for the second law of thermodynamics.

It is important to remember that in arriving at this result, it was necessary to introduce a new postulate, that of microscopic reversibility. This postulate is not needed, and not used, in the application of statistical mechanics to equilibrium systems.

1.13 PARTICLE STATISTICS

In the methods presented thus far, the statistics have always been applied to ensembles of macroscopic systems, and the distribution functions have always referred to the probability of a macroscopic system being in a particular state. Macroscopic systems can always be labeled and differentiated from one another and are, therefore, subject to the statistics of completely distinguishable entities. There are a great many physical applications, however, in which the physical system is composed of particles that interact with each other only very weakly. These applications can be conveniently handled by considering the statistics of the particles themselves and deriving equations for the distribution among particle states rather than total system states. There is, however, a close connection between the ensemble and particle approach to statistical problems. A physical system whose component particles are practically independent can itself be thought of as a canonical "ensemble," each particle being a member "system" of the "ensemble." This amounts to renaming the terms used in the theory of the macroscopic canonical ensemble, and the ensuing mathematical development is exactly the same. The ensemble distribution function f_j becomes the particle distribution function N_j/N:

$$\frac{N_j}{N} = \frac{e^{-\epsilon_j/kT}}{\sum_j e^{-\epsilon_j/kT}} \tag{13.1}$$

where N_j is the number of particles in the jth state, N is the total number of particles and ϵ_j is the energy of a molecule in the jth state. Equation 13.1

is known as the Maxwell-Boltzmann distribution law for nearly independent particles. Note that the particles have always been described as *nearly* independent. In order to apply the statistical methods to the particles, it is necessary that interactions among them be negligible. But it is also essential that there be *some* interactions, otherwise the particles could not exchange energy and come to thermal equilibrium.

An important point about the Maxwell-Boltzmann distribution, is that it describes *distinguishable* particles. But particle statistics differs fundamentally from ensemble statistics in that particles of atomic size are *not* distinguishable. This is a quantum mechanical result that merely asserts that no physically real effect can be detected in a system when two of its particles are interchanged. In quantum mechanical terms, this means that an interchange of two particles can alter the wave function of the system by at most a change in sign. For a system whose wave function is symmetric with respect to particle interchange, it is found that any number of particles can be in any given state. For a system with an antisymmetric wave function, not more than one particle can be in any given state.* The first case leads to the *Bose-Einstein* statistics and the second case leads to the *Fermi-Dirac* statistics.

Does all this mean that the Maxwell-Boltzmann distribution is useless? Not at all; for, as we shall see, the distribution functions derived by taking into account the symmetry of the wave functions reduce to a slightly corrected form of the Maxwell-Boltzmann distribution in the limit of high temperature.† In fact, the corrected Maxwell-Boltzmann distribution turns out to be quite accurate in the theory of the gaseous state down to quite low temperatures for all gases but helium. In the theory of metals, however, the conduction electrons are often treated as nearly independent particles, and in this case it is essential to use Fermi-Dirac statistics.

Before going on to a study of the quantum particle statistics, another connection between ensembles and particles must be pointed out. The entropy of an ensemble was shown to be

$$XS = k \ln W \qquad (13.2)$$

where W is the number of complexions of the entire ensemble. If each member system of the ensemble consists of independent particles, then the number of complexions of each system can be computed. That is, we can define w_j to be the number of ways of distributing all the particles in the jth

---

* This is just the Pauli exclusion principle which follows from writing the antisymmetric wave function for a system of independent particles as a Slater determinant of one-particle wave functions.

† The correction is needed to get rid of the particle distinguishability inherent in the Maxwell-Boltzmann statistics.

system in such a way that N_1 particles are in state 1, N_2 in state 2, and so on. Then W for the ensemble is just the product of all the w_j's for each member system, and we have

$$XS = k \sum_j \ln w_j \qquad (13.3)$$

Since entropy is additive, we therefore can write that the entropy of the jth system S_j is given by

$$S_j = k \ln w_j \qquad (13.4)$$

Now if we restrict our attention to a single system whose statistics we wish to investigate, we can drop the subscript and write

$$S = k \ln w \qquad (13.5)$$

This equation is valid regardless of the nature of the statistics to be used. The differences among the Maxwell-Boltzmann, Bose-Einstein, and Fermi-Dirac statistics are accounted for by different methods of computing w. All that we have done here is to give another instance of the universal validity of the relation between entropy and the number of complexions in statistical mechanics. This relation can be used in particle as well as ensemble statistics provided we know how to calculate w. All of the statistical mechanics of particles then follows, because maximizing the entropy is equivalent to maximizing the number of complexions. Working directly with the entropy is somewhat easier, since we already have its statistical identification.

To calculate w, consider a system of N identical, nearly independent particles in a fixed volume. For such a many particle system, the quantum energy levels are always very close together. Then the energy levels can always be arranged in groups such that the jth group has energy $\epsilon_j \pm \Delta\epsilon$ and there are ω_j particle states with energy in this range. Since the energy levels are closely spaced, $\Delta\epsilon$ can always be chosen in such a way that it is very much smaller than ϵ_j and ω_j is still very much greater than unity. Let N_j be the number of particles with energies in the range $\epsilon_j \pm \Delta\epsilon$. The system can now be characterized by the following scheme:

Group number	1,	2, ..., j, ...
Number of possible states in the group	ω_1,	$\omega_2, \ldots, \omega_j \cdots$
Number of particles in the group	N_1,	$N_2, \ldots, N_j \cdots$
Energy of a particle in the group	ϵ_1,	$\epsilon_2, \ldots, \epsilon_j \cdots$

Note that the subscript j refers to a group of states, all being considered as nearly identical, whereas in the ensemble statistics we were referring to individual states.

The statistical count of the number of complexions, w, is the number of ways that the particles can be arranged among all states, or levels, in all groups, since each such arrangement corresponds to a state of the entire system. Three cases must be considered.

A. The Fermi-Dirac Case

The particles are indistinguishable and not more than one particle can be in a given state.

The number of ways of putting N_j particles into ω_j levels so that there is no more than one particle per level is

$$\frac{\omega_j!}{N_j!(\omega_j - N_j)!} \tag{13.6}$$

The product of all terms similar to this is the number of ways of arranging the particles among the individual states; that is

$$w_{\text{FD}} = \prod_j \frac{\omega_j!}{N_j!(\omega_j - N_j)!} \tag{13.7}$$

The subscript FD has been put on the statistical count to identify it as the number of complexions for the Fermi-Dirac case.

B. The Bose-Einstein Case

The particles are indistinguishable and there is no limit to the number of particles that can be put in any level. The number of ways in which N_j particles can be distributed among ω_j levels in the jth group, with no restrictions on the number of particles per level, is

$$\frac{(\omega_j + N_j - 1)!}{(\omega_j - 1)! \, N_j!} \tag{13.8}$$

and the total number of ways of arranging all particles among all levels is the product

$$w_{\text{BE}} = \prod_j \frac{(\omega_j + N_j - 1)!}{(\omega_j - 1)! \, N_j!} \tag{13.9}$$

C. The Maxwell-Boltzmann Case

The particles are distinguishable and there is no limit to the number of particles that can be put in any level. First, we calculate the number of ways of arranging N particles in piles of N_j each. This is

$$\frac{N!}{\prod_j N_j!} \tag{13.10}$$

Now in each j-pile, we need the number of ways of arranging the N_j particles in the ω_j levels. This is $\omega_j^{N_j}$. Multiplying the product of this over all j by the expression (13.10) gives the total number of ways of arranging the particles among the levels. The number of complexions in the Maxwell-Boltzmann case is therefore

$$w_{\mathrm{MB}} = N! \prod_j \frac{\omega_j^{N_j}}{N_j!} \tag{13.11}$$

The probability density N_j/N, which gives the fraction of particles with energy ϵ_j, can be found for systems at equilibrium by maximizing the w's with respect to arbitrary variations in the N_j's, subject to the condition of constant total energy and constant number of particles. The procedure is completely analogous to that for the canonical ensemble, and is valid for the same reasons. Only the Fermi-Dirac case will be treated in detail here. The derivations for the other cases proceed in the same way.

From (13.7)

$$\ln w_{\mathrm{FD}} = \sum_j \ln \omega_j! - \sum_j \ln N_j! - \sum_j \ln (\omega_j - N_j)! \tag{13.12}$$

or, using Stirling's approximation,

$$\ln w_{\mathrm{FD}} = \sum_j \{\omega_j \ln \omega_j - N_j \ln N_j - (\omega_j - N_j) \ln (\omega_j - N_j)\} \tag{13.13}$$

From the condition that the total energy and the total number of particles must be constant, we have

$$N = \sum_j N_j \tag{13.14}$$

$$U = \sum_j N_j \epsilon_j \tag{13.15}$$

The maximization problem is therefore embodied in the three equations:

$$\delta \ln w_{\mathrm{FD}} = \sum_j \frac{\partial \ln w_{\mathrm{FD}}}{\partial N_j} \, \delta N_j = 0 \tag{13.16}$$

$$\delta N = \sum_j \delta N_j = 0 \tag{13.17}$$

$$\delta U = \sum_j \epsilon_i \, \delta N_j = 0 \tag{13.18}$$

Multiplying (13.17) and (13.18) by the parameters $-a$ and $-b$, respectively, and adding the results to (13.16) give a sum of terms that are now all independent and thus all equal to zero. Therefore

$$\frac{\partial \ln w_{FD}}{\partial N_j} - a - b\epsilon_j = 0 \qquad (13.19)$$

Evaluating the derivative and solving for N_j gives

$$N_j = \frac{\omega_j}{e^{a+b\epsilon_j} + 1} \qquad \text{(Fermi-Dirac)} \qquad (13.20)$$

A similar treatment for the Bose-Einstein and Maxwell-Boltzmann statistics give the following results:

$$N_j = \frac{\omega_j}{e^{a+b\epsilon_j} - 1} \qquad \text{(Bose-Einstein)} \qquad (13.21)$$

$$N_j = N\omega_j e^{-a-b\epsilon_j} \qquad \text{(Maxwell-Boltzmann)} \qquad (13.22)$$

Please take notice that if the exponential terms in the Fermi-Dirac and Bose-Einstein distributions are much greater than unity ($e^{a+b\epsilon_j} \gg 1$), then both distributions reduce to the same limiting form, namely

$$N_j = \omega_j e^{-a-b\epsilon_j} \qquad \text{(semiclassical)} \qquad (13.23)$$

In this limit, N_j/ω_j is very small because a very large term exists in the denominator of the distribution equations. The number of particles is a constant; therefore if $N_j/\omega_j \ll 1$, the available particles are sparsely spread out over a very large range of levels. But this is just the condition to be expected at high temperatures, because the high thermal energy can boost particles into a large number of high levels that they could not reach otherwise. Also notice that (13.23) is identical to the Maxwell-Boltzmann form except for the presence of N. The appearance of N in the Maxwell-Boltzmann distribution comes directly from treating the particles as distinguishable. If we correct for indistinguishability by dividing out the $N!$ in the expression for w_{MB}, and then carry out the Lagrangian method of undetermined multipliers, we will end up with precisely (13.23). This equation is then an equation that has been partially corrected for quantum mechanical effects. It partially takes indistinguishability into account, but not the symmetry properties of wave functions. It is useful because at high temperatures the levels are so sparsely occupied that there is not much chance of their being more than one particle per level. We will call the statistics leading to (13.23) semiclassical.

1.14 PARTICLE STATISTICS AND THERMODYNAMICS

Since we already have the formula for the statistical entropy, the contact between particle statistics and thermodynamics can be made in a very simple way. The starting point of each of the calculations for the particle distribution functions has been a variational equation of the form

$$\delta \ln w - b \delta U - a \delta N = 0 \qquad (14.1)$$

In computing the Fermi-Dirac distribution, an equation of this form was obtained by multiplying (13.17) and (13.18) by constants and adding the results to (13.16). Since (14.1) is valid for any arbitrary variations, it is also valid for variations connecting two equilibrium states, and in this case we can write it in terms of ordinary differentials as

$$d \ln w - b\, dU - a\, dN = 0 \qquad (14.2)$$

But $S = k \ln w$, so we have

$$dS = bk\, dU + ak\, dN \qquad (14.3)$$

and therefore, since our system has a fixed volume,

$$\left(\frac{\partial S}{\partial U}\right)_{V,N} = bk \qquad (14.4)$$

and

$$\left(\frac{\partial S}{\partial N}\right)_{V,U} = ak \qquad (14.5)$$

The thermodynamic formulas for these derivatives are

$$\left(\frac{\partial S}{\partial U}\right)_{V,N} = \frac{1}{T} \qquad (14.6)$$

$$\left(\frac{\partial S}{\partial N}\right)_{U,V} = \frac{\mu}{T} \qquad (14.7)$$

where μ is the chemical potential, or Gibbs free energy per particle. Comparing (14.6) with (14.4), and (14.7) with (14.5), we see that

$$bk = \frac{1}{T}$$

$$ak = -\frac{\mu}{T}$$

from which

$$b = \frac{1}{kT} \tag{14.8}$$

$$a = -\frac{\mu}{kT} \tag{14.9}$$

The parameter b has the same interpretation in the particle statistics as β has for ensemble statistics. However, whereas α was related to the Helmholtz free energy, a is given in terms of the Gibbs free energy.

Using (14.8) and (14.9), the particle distribution functions now become

$$N_j = \frac{\omega_j}{e^{(\epsilon_j-\mu)/kT} + 1} \quad \text{(Fermi-Dirac)} \tag{14.10}$$

$$N_j = \frac{\omega_j}{e^{(\epsilon_j-\mu)/kT} - 1} \quad \text{(Bose-Einstein)} \tag{14.11}$$

$$N_j = N\omega_j e^{(\mu-\epsilon_j)/kT} \quad \text{(Maxwell-Boltzmann)} \tag{14.12}$$

$$N_j = \omega_j e^{(\mu-\epsilon_j)/kT} \quad \text{(semiclassical)} \tag{14.13}$$

In what particular cases can these various distribution formulas be applied? Remember that the Fermi-Dirac statistics describe systems with an anti-symmetric wave function. Quantum theory tells us that such systems consist of particles with half-integral values of the spin. The Fermi-Dirac distribution is therefore required in treating collections of nearly independent electrons. The Bose-Einstein formulae refer to systems whose wave functions are antisymmetric and whose particles have integral spin values. The most important application of the Bose-Einstein statistics is to helium at low temperatures.

The semiclassical statistics are valid for any system of nearly independent particles provided that

$$e^{(\epsilon_j-\mu)/kT} \gg 1 \tag{14.14}$$

The accuracy of the semiclassical approximation depends therefore on the value of the chemical potential as well as on the temperature. Specific calculations of the validity of (14.14) can be made for individual systems. We will anticipate these calculations by pointing out that condition (14.14) is valid for all dilute gases at ordinary temperatures. In fact, quantum corrections become important only for hydrogen and helium at rather low temperatures.

The most important application of particle statistics in solid state physics is to the free-electron theory of metals. In this case, the semiclassical condition is not valid and it is essential to use the Fermi-Dirac statistics. A detailed treatment of Fermi-Dirac systems will be given later.

1.15 THE IDEAL GAS

We define an ideal gas as a many-particle system; all of the particles are identical, interact with each other to a negligible extent, and are described by semiclassical statistics. The particles are presumed to have no internal structure so that their energy levels are just those of the quantum mechanical particle-in-a-box. This definition leads to the usual thermodynamic and phenomenological equations for the ideal gas.

Now, let us investigate some useful statistical thermodynamic results for semiclassical statistics. By summing (14.13) over all j and solving for μ, we get

$$\mu = kT \ln N - kT \ln \sum_j \omega_j e^{-\epsilon_j/kT} \tag{15.1}$$

Because μ is the chemical potential per particle and the particles are independent, multiplication by N gives the Gibbs free energy of the system. Furthermore, let us define a particle partition function z in analogy with the canonical ensemble partition function

$$z \equiv \sum_j \omega_j e^{-\epsilon_j/kT} \tag{15.2}$$

Using this definition, multiplying (15.1) by N to get G, the Gibbs free energy, gives

$$G = NkT \ln N - NkT \ln z \tag{15.3}$$

The entropy S is obtained from the Gibbs free energy by the thermodynamic formula

$$S = -\left(\frac{\partial G}{\partial T}\right)_P \tag{15.4}$$

so that

$$S = -Nk(\ln N - \ln z) + NkT\left(\frac{\partial \ln z}{\partial T}\right)_P \tag{15.5}$$

We see that just as in ensemble statistics, the partition function has an essential part in the statistical thermodynamic equations. To evaluate the particle partition function for the ideal gas, we must draw on the results of quantum theory for the energy levels of a particle in a box. If the volume of the box is V and the mass of the particle is m, then its possible energy levels are

$$\epsilon_j = \frac{h^2}{8mV^{2/3}}(j_1{}^2 + j_2{}^2 + j_3{}^2) \tag{15.6}$$

where j_1, j_2, and j_3 are positive integers defining the jth state of the system. There is just one wave function for each set (j_1, j_2, j_3). The energy levels are so close together that the sum defining the partition function can be replaced by an integral

$$z = \int_0^\infty \omega(\epsilon)e^{-\epsilon/kT}\, d\epsilon \tag{15.7}$$

This equation is obtained from (15.2) by treating ϵ as a continuous variable and defining $\omega(\epsilon)\, d\epsilon$ to be the number of states (or energy levels) with energy in the range ϵ to $\epsilon + d\epsilon$.

To evaluate this integral, $\omega(\epsilon)$ must be known. Equation 15.6 for the energy levels can be used to get $\omega(\epsilon)$ by the following procedure. Construct a three-dimensional, simple cubic lattice with the edge of the unit cell being of unit length. To each lattice point there corresponds three integers giving its coordinates from one of the points taken as origin. We have, therefore, constructed a j-space containing a lattice in which each lattice point in the positive octant represents an energy level of our particle. If j is the magnitude of the vector connecting the origin to one of the lattice points, then

$$j^2 = j_1^2 + j_2^2 + j_3^2 \tag{15.8}$$

and (15.6) can be expressed as

$$\epsilon_j = \frac{h^2}{8mV^{2/3}} j^2 \tag{15.9}$$

Now, construct a spherical shell of thickness dj about the origin. Because the energy levels are closely spaced, we can choose the thickness dj in such a way that it is small compared to the radius of the shell but still contains a large number of lattice points. The volume of the shell is $4\pi j^2\, dj$, and since the unit cell of the lattice has unit volume, the volume of the shell is numerically equal to the number of points in it. Each point in the positive octant corresponds to a state of the particle. Therefore, the number of states contained in the shell is

$$\tfrac{1}{2}\pi j^2\, dj \tag{15.10}$$

All we need to convert this to the number of states in a given energy range $d\epsilon$ is a relation between $d\epsilon$ and dj. This is provided by (15.9) if we treat $\epsilon_j \to \epsilon$ as a continuous variable

$$d\epsilon = \frac{h^2}{8mV^{2/3}} 2j\, dj \tag{15.11}$$

Solving for dj and substituting into (15.10) gives $\omega(\epsilon)\, d\epsilon$, the number of states with energy in a range $d\epsilon$

$$\omega(\epsilon)\, d\epsilon = \frac{2\pi mV^{2/3}}{h^2} j\, d\epsilon \tag{15.12}$$

To get rid of the inconvenient j in this equation, just solve (15.9) for j (treating $\epsilon_j \to \epsilon$ as a continuous variable)

$$j = \frac{2\sqrt{2m}\, V^{1/3}}{h}\sqrt{\epsilon} \tag{15.13}$$

and substitute into (15.12) to get

$$\omega(\epsilon) = \frac{4\pi\sqrt{2}}{h^3} m^{3/2} V \sqrt{\epsilon} \tag{15.14}$$

This is the desired function $\omega(\epsilon)$. It is called the density-of-states function and is of prime importance in all theories of a large number of independent particles.

Now, we can get back to the business of evaluating the particle partition function (15.7). Substituting (15.4) into (15.7) gives

$$z = \tfrac{1}{2}CV\int_0^\infty \epsilon^{1/2}e^{-\epsilon/kT}\, d\epsilon \tag{15.15}$$

where we have defined C as the collection of constants

$$\tfrac{1}{2}C \equiv \frac{4\pi\sqrt{2}}{h^3} m^{3/2} \tag{15.16}$$

The integral is a standard form given by

$$\int_0^\infty \epsilon^{1/2}e^{-\epsilon/kT}\, d\epsilon = \frac{kT}{2}\sqrt{\pi kT} \tag{15.17}$$

and, therefore, after collecting constants, (15.15) reduces to

$$z = V\left(\frac{2\pi mkT}{h^2}\right)^{3/2} \tag{15.18}$$

The thermodynamic properties of the ideal gas now all fall out in an easy fashion. From (15.3), the Gibbs free energy is

$$G = NkT\left[\ln N - \ln V - \tfrac{3}{2}\ln\left(\frac{2\pi mkT}{h^2}\right)\right] \tag{15.19}$$

From (15.5), the entropy is

$$S = -Nk\left[\ln N - \ln V - T\left(\frac{\partial \ln V}{\partial T}\right)_P - \tfrac{3}{2} - \tfrac{3}{2}\ln\left(\frac{2\pi mkT}{h^2}\right)\right] \tag{15.20}$$

A simple form of the energy can be obtained by directly computing it from the distribution function. In the continuum notation, the number of particles with energy in the range $d\epsilon$ is

$$N(\epsilon)\, d\epsilon = \omega(\epsilon)e^{\mu/kT}e^{-\epsilon/kT}\, d\epsilon \tag{15.21}$$

The total energy of the gas is, therefore,

$$U = \int_0^\infty \epsilon N(\epsilon)\, d\epsilon = e^{\mu/kT}\int_0^\infty \omega(\epsilon)\epsilon e^{-\epsilon/kT}\, d\epsilon \tag{15.22}$$

or, using (15.4),

$$U = e^{\mu/kT}\frac{4\pi\sqrt{2}}{h^3}m^{3/2}V\int_0^\infty \epsilon^{3/2}e^{-\epsilon/kT}\, d\epsilon \tag{15.23}$$

The integral has the value $3\sqrt{\pi}\,(kT)^{5/2}/4$, so

$$U = e^{\mu/kT}\frac{3(m\pi)^{3/2}\sqrt{2}}{h^3}(kT)^{5/2}V \tag{15.24}$$

Dividing (15.19) by N gives μ, from which, taking exponentials of μ/kT, we get

$$e^{\mu/kT} = \frac{N}{V}\left(\frac{2\pi mkT}{h^2}\right)^{-3/2} \tag{15.25}$$

and combining this with (15.24), the energy reduces to

$$U = \tfrac{3}{2}NkT \tag{15.26}$$

To get the equation of state, we will use the energy in the form

$$U = \sum_j N_j\epsilon_j \tag{15.27}$$

from which

$$dU = \sum_j N_j\, d\epsilon_j + \sum_j \epsilon_j\, dN_j \tag{15.28}$$

Assuming that only pressure-volume work is allowed in the interaction of the system with its environment, the first law of thermodynamics is

$$dU = T\, dS - P\, dV \tag{15.29}$$

Just as we did when discussing ensemble theory, we identify the work terms in these last two equations by recognizing that the value of an energy level can only be changed by external work. Therefore

$$P\, dV = -\sum_j N_j\, d\epsilon_j \tag{15.30}$$

and

$$P = -\sum_j N_j\frac{\partial\epsilon_j}{\partial V} \tag{15.31}$$

Now, substitute (14.13) for N_j:

$$P = -e^{\mu/kT} \sum_j \omega_j e^{-\epsilon_j/kT} \frac{\partial \epsilon_j}{\partial V} \tag{15.32}$$

But, from the definition of the partition function,

$$\left(\frac{\partial z}{\partial V}\right)_T = -\frac{1}{kT} \sum_j \omega_j \frac{\partial \epsilon_j}{\partial V} e^{-\epsilon_j/kT} \tag{15.33}$$

and, therefore, (15.32) can be written as

$$P = kTe^{\mu/kT}\left(\frac{\partial z}{\partial V}\right)_T \tag{15.34}$$

Also, as we can see from either (15.1) or (15.3), $\mu = G/N$ is given in terms of z by

$$e^{\mu/kT} = \frac{N}{z} \tag{15.35}$$

and, finally, we have an equation for the pressure in terms of the partition function

$$P = NkT\left(\frac{\partial \ln z}{\partial V}\right)_T \tag{15.36}$$

The equation of state now follows directly from (15.18). Performing the differentiation indicated in (15.36), we get the equation of state for an ideal gas

$$PV = NkT \tag{15.37}$$

This is the ordinary form of the ideal gas law and identifies k as Boltzmann's constant or the gas constant per particle.

Now, we can get the enthalpy from (15.26) and (15.37). Since the enthalpy $H = U + PV$, we have

$$H = \tfrac{5}{2}NkT \tag{15.38}$$

Differentiating (15.26) and (15.38) with respect to temperature gives the heat capacity at constant volume and at constant pressure, respectively:

$$C_V = \frac{\partial U}{\partial T} = \tfrac{3}{2}Nk \tag{15.39}$$

$$C_P = \frac{\partial H}{\partial T} = \tfrac{5}{2}Nk \tag{15.40}$$

By using (15.37), the entropy can be put in terms of volume and temperature alone. Thus, (15.20) becomes

$$S = -Nk\left[\ln N - \ln V - \tfrac{5}{2} - \tfrac{3}{2}\ln T - \tfrac{3}{2}\ln\left(\frac{2\pi mk}{h^2}\right)\right] \quad (15.41)$$

and again using (15.37) to replace V by P in this equation, we get

$$S = -Nk\left\{\ln P - \tfrac{5}{2}\ln T - \tfrac{5}{2} - \tfrac{3}{2}\ln\left[k^{5/2}\left(\frac{2\pi m}{h^2}\right)\right]\right\} \quad (15.42)$$

for the entropy as a function of pressure and temperature.

This is as far as we will go with the theory of the ideal gas. In a later chapter, we will work out the theory of the Fermi-Dirac gas as applied to the free-electron theory of metals and semiconductors.

1.16   THE GRAND CANONICAL ENSEMBLE AND PARTICLE STATISTICS

The distribution functions for particle statistics were obtained in section 13 by treating each particle as a member system in a canonical ensemble. Because of this, it was necessary to arbitrarily divide the energy levels into groups with nearly equal energies, and then to apply combinatorial analysis to each group. The success of this procedure depends on the energy levels being very close together. This arbitrary element in the derivation can be avoided by the use of the grand canonical ensemble. In this section, we will derive the particle distribution functions using the grand canonical ensemble.

For a one-component system of identical, noninteracting particles, the grand canonical distribution function is

$$f_j(N) = \frac{1}{Q}\exp\left\{-\frac{1}{kT}[E_j(N) - \mu N]\right\} \quad (16.1)$$

where the energy of the system in the jth state containing N particles is

$$E_j(N) = \sum_k n_k \epsilon_k \quad (16.2)$$

n_k being the number of particles with energy ϵ_k, so that

$$\sum_k n_k = N \quad (16.3)$$

The summation in (16.2) is taken over all particle states.

The grand canonical partition function for our problem is

$$Q = \sum_{N,j}\exp\left\{-\frac{1}{kT}[E_j(N) - \mu N]\right\} \quad (16.4)$$

The state of our system is specified by the integers n_k, so that summation on (N, j) can be replaced by summation over all values of $(n_1, n_2, n_3 \cdots)$. Thus, using (16.2) and (16.3), (16.4) becomes

$$Q = \sum_{n_1, n_2 \cdots} \exp \left\{ -\frac{1}{kT} [n_1(\epsilon_1 - \mu) + n_2(\epsilon_2 - \mu) + \cdots] \right\} \quad (16.5)$$

and (16.1) becomes

$$f_{n_1 n_2 \cdots}(N) = \frac{1}{Q} \exp \left\{ -\frac{1}{kT} [n_1(\epsilon_1 - \mu) + n_2(\epsilon_2 - \mu) + \cdots] \right\} \quad (16.6)$$

The subscript j has been replaced by its equivalent $(n_1, n_2, \ldots)$. Actually, it is not necessary to specify the total number of particles, since this is determined when the n_k are specified.

The particle-distribution function defines the probability that a particle is in a given state. This can be obtained from the probability that n_i particles are in state i, which we will call $f(n_i)$. Since (16.5) is the probability that n_1 particles are in state 1, n_2 particles are in state 2, and so on; $f(n_i)$ is obtained by summing (16.5) over all values of every n_k except n_i. The result is

$$f(n_i) = \sum_{n_1, n_2 \cdots (\neq n_i)} f_{n_1, n_2} \cdots (N)$$

$$= \frac{1}{Q} \sum_{n_1, n_2 \cdots (\neq n_i)} \exp \left\{ -\frac{1}{kT} [n_1(\epsilon_1 - \mu) + n_2(\epsilon_2 - \mu) + \cdots] \right\} \quad (16.7)$$

But the sum in (16.7) is identical to that for Q as given by (16.5), except that the exponential containing n_i is not summed. Therefore, cancelling factors that appear in both sums, (16.7) simplifies to

$$f(n_i) = \frac{\exp [-n_i(\epsilon_i - \mu)/kT]}{\sum\limits_{n_i} \exp [-n_i(\epsilon_i - \mu)/kT]} \quad (16.8)$$

The mean number of particles in state i is obtained by multiplying $f(n_i)$ by n_i and summing over all possible values of n_i:

$$\langle n_i \rangle = \sum_{n_i} n_i f(n_i)$$

$$= \frac{\sum\limits_{n_i} n_i \exp [-n_i(\epsilon_i - \mu)/kT]}{\sum\limits_{n_i} \exp [-n_i(\epsilon_i - \mu)/kT]} \quad (16.9)$$

To complete the derivation, the sums in (16.9) must be evaluated. It is here that the difference between Fermi-Dirac and Bose-Einstein statistics appears. For Fermi-Dirac particles, a state can be empty, or hold one particle. Thus n_i can only take the values 0 and 1, so evaluation of the sums is trivial. The result is

$$\langle n_i \rangle_{\mathrm{FD}} = \frac{1}{e^{(\epsilon_i - \mu)/kT} + 1} \tag{16.10}$$

The case of Bose-Einstein statistics is only slightly more complex. For Bose-Einstein particles, any number of particles can be in a given state, so n_i can assume any positive integral value. The sums in (16.9) then have well-known values since they have the form

$$\sum_{j=0}^{\infty} x^j = \frac{1}{1 - x} \tag{16.11}$$

$$\sum_{j=0}^{\infty} j x^j = \frac{x}{(1 - x)^2} \tag{16.12}$$

Using these forms, (16.9) gives, for Bose-Einstein particles

$$\langle n_i \rangle_{\mathrm{BE}} = \frac{1}{e^{(\epsilon_i - \mu)/kT} - 1} \tag{16.13}$$

Note that if ω_i states are chosen that have energies clustered closely about ϵ_i, the number of particles in these states is $\omega_i \langle n_i \rangle \equiv N_i$, and (16.10) and (16.13) give results that are identical to (14.10) and (14.11), respectively.

1.17 THE SEMICLASSICAL APPROXIMATION

In a system whose particles obey the laws of classical mechanics, the state is completely determined by a specification of the coordinates and momentum of the particles. Thus, if the system has N particles, its state is defined by the $3N$ values of the particle coordinates $(q_1, q_2, \ldots, q_{3N})$ and the $3N$ values of the particle momenta $(p_1, p_2, \ldots, p_{3N})$. It is usual to think of these $6N$ variables as defining a $6N$-dimensional space, called the phase space, the values of the coordinates and momenta at any time being the *phase* of the system. The system can, therefore, be represented by the motion of a phase point $(q_1, q_2 \cdots q_{3N}, p_1, p_2 \cdots p_{3N})$ in phase space.

A statistical mechanics can, of course, be constructed from this classical picture and, in fact, this was done by Gibbs before the advent of the quantum theory. The classical statistical mechanics is a limiting case of quantum statistics and will not be developed in detail here. However, we will show how to express probability distribution functions and partition functions in terms of coordinates and momenta in the classical limit. This formulation is often useful when quantum effects can be neglected.

Let us rewrite the probability density and the partition function for a canonical ensemble (Section 8)

$$f_i = \frac{1}{Z} e^{-E_i/kT} \tag{17.1}$$

$$Z = \sum_i e^{-E_i/kT} \tag{17.2}$$

In these equations the index i refers to quantum states of the system. In a classical system, however, the state is determined by a point in phase space, and the energy is a function of the coordinates and momenta

$$E = E(p, q) \tag{17.3}$$

where (p, q) represents the set of all $3N$ momenta and $3N$ coordinates. It is clear, then, that the partition sum over quantum states in (17.2) can be replaced by an integral over coordinates and momenta in the classical limit, so we will write

$$Z_c = C \int e^{-E(p,q)/kT} \, dp \, dq \tag{17.4}$$

where

$$dp \, dq = dp_1 \, dp_2 \cdots dp_{3N} \, dq_1 \, dq_2 \cdots dq_{3N} \tag{17.5}$$

and the subscript c denotes the classical limit.

It is necessary to introduce the proportionality constant in (17.4) in order to keep the partition sum dimensionless. The differentials $dp \, dq$ have the dimensions of (action)^{3N} so that C will have the dimensions (action)$^{-3N}$.

Similarly, the classical limit of the probability distribution function is obtained from (17.1) by requiring it to represent that a phase point lies in the phase volume $dp \, dq$, so that the classical form of (17.1) is

$$f(p, q) \, dp \, dq = \frac{C}{Z_c} e^{-E(p,q)/kT} \, dp \, dq \tag{17.6}$$

The proportionality constant in (17.6) must be the same as that in (17.4) in order that the integral of the probability density over all p and q be unity.

The theory of the classical canonical ensemble can be developed from (17.4) and (17.6) in exactly the same way as for the quantum canonical ensemble. All statistical thermodynamic relations connecting thermodynamic quantities to the partition function remain unchanged. In particular, the Helmholtz free energy is given by

$$A = -kT \ln Z_c \tag{17.7}$$

so that all thermodynamic functions of a classical system can be computed if the classical partition function is known.

If the theory of the ideal gas is constructed from (16.4), all the results of Section 15 are recovered provided C is given by

$$C = h^{-3N} \tag{17.8}$$

The presence of Planck's constant can be understood physically by the following procedure: divide the classical phase space into cells of volume h^{3N}. According to the uncertainty principle, the coordinates and momenta can be measured with an accuracy limited by the relation

$$\delta p_j \, \delta q_j = h \tag{17.9}$$

where δp_j and δq_j are the limits of accuracy in a simultaneous determination of p_j and q_j. Therefore, all phase points in a phase cell of volume h^{3N} must be considered as representing the same state. Then, for a volume element in phase space given by

$$(\Delta p \, \Delta q)_k = (\Delta p_1 \, \Delta q_1 \, \Delta p_2 \, \Delta q_2 \cdots \Delta p_{3N} \, \Delta q_{3N})_k \tag{17.10}$$

the total number of possible distinct states in the range $\Delta p \, \Delta q$ is

$$\frac{(\Delta p \, \Delta q)_k}{h^{3N}} \tag{17.11}$$

The partition sum (16.2) can be rearranged by taking all the states that are in the range $(\Delta p \, \Delta q)_k$ and lumping them together so that

$$Z = \sum_k e^{-E_k/kT} \frac{(\Delta p \, \Delta q)_k}{h^{3N}} \tag{17.12}$$

The validity of this result depends on choosing $(\Delta p \, \Delta q)_k$ large enough so that it contains many possible states, and yet small enough so that the energy is nearly constant throughout the range $(\Delta p \, \Delta q)_k$. Then (17.12) can be approximated by an integral if $(\Delta p \, \Delta q)$ are replaced by differentials

$$Z_c = \frac{1}{h^{3N}} \int e^{-E(p,q)/kT} \, dp \, dq \tag{17.13}$$

which is the same as (17.4) with $C = h^{-3N}$

If the coordinates are expressed in a Cartesian reference system, the energy of the system is

$$E = \sum_{i=1}^{N} \frac{p_i^2}{2m_i} + \varphi(q) \tag{17.14}$$

where p_i and m_i are the momentum and mass of the ith particle, respectively, and $\varphi(q)$ is the potential energy of the system as a function of all coordinates.

Using (17.14) and (17.8), we rewrite (17.6) and (17.4) as

$$f(p, q) = \frac{\exp\left(-\sum_i p_i^2/2m_ikT\right)e^{-\varphi(q)/kT}}{\int \exp\left(-\sum_i p_i^2/2m_ikT\right)e^{-\varphi(q)/kT}\,dp\,dq} \qquad (17.15)$$

$$Z_c = \frac{1}{h^{3N}}\int \exp\left(-\sum_i p_i^2/2m_ikT\right)e^{-\varphi(q)/kT}\,dp\,dq \qquad (17.16)$$

The integrals in these equations are over all momenta and all coordinates. The momentum integrals can be factored out and performed so that

$$\int \exp\left(-\sum_i p_i^2/2m_ikT\right)\exp\left(-\varphi(q)/kT\right)\,dp\,dq$$

$$= \int \exp\left(-\sum_i p_i^2/2m_ikT\right)\,dp\int \exp\left(-\varphi(q)/kT\right)\,dq$$

$$= \int \exp\left(-p_i^2/2m_ikT\right)\,d\mathbf{p}_1\int \exp\left(-p_2^2/2m_2kT\right)\,d\mathbf{p}_2\cdots$$

$$\times \int \exp\left(-\varphi(q)/kT\right)\,dq$$

$$= \prod_i^N (2\pi m_ikT)^{3/2}\int \exp\left(-\varphi(q)/kT\right)\,dq \qquad (17.17)$$

The last step follows from the fact that

$$p_i^2 = p_{ix}^2 + p_{iy}^2 + p_{iz}^2$$

$$d\mathbf{p}_i = dp_x\,dp_y\,dp_z$$

and

$$\int_{-\infty}^{\infty} e^{-ax^2}\,dx = \left(\frac{\pi}{a}\right)^{1/2} \qquad (17.18)$$

Combining (17.16) with (17.17) gives

$$Z_c = \frac{1}{h^{3N}}\prod_{i=1}^N (2\pi m_ikT)^{3/2}\int e^{-\varphi(q)/kT}\,dq \qquad (17.19)$$

The integral in (17.19) is taken over all coordinates, and is called the configurational partition function

$$Z_q \equiv \int e^{-\varphi(q)/kT}\,dq \qquad (17.20)$$

If all the particles have the same mass, (17.19) becomes

$$Z_c = \left(\frac{2\pi m k T}{h^2}\right)^{3N/2} Z_q \qquad (17.21)$$

Equation 17.15 gives the probability density that the particles in the system have momenta $(p_1, p_2 \cdots)$ and coordinates $(q_1, q_2 \cdots)$. To get the probability density that the particles have coordinates $(q_1, q_2 \cdots)$ irrespective of their momenta, just integrate (17.15) over all momenta. The result is

$$f(q) = \int f(p, q)\, dp = \frac{e^{-\varphi(q)/kT}}{\int e^{-\varphi(q)/kT}\, dq} \qquad (17.22)$$

Similarly, the probability density that the particles have momenta $(p_1, p_2 \cdots)$ irrespective of the coordinates is

$$f(p) = \int f(p, q)\, dq$$

$$= \frac{\exp\left(-\sum_i p_i^2/2m_i kT\right)}{\int \exp\left(-\sum_i p_i^2/2m_i kT\right) dp}$$

$$= \frac{\exp\left(-\sum_i p_i^2/2m_i kT\right)}{\prod_i^N (2\pi m_i kT)^{3/2}} \qquad (17.23)$$

Although we have used the adjective "classical" to describe some of the steps in the above development, the results of this section are really semiclassical in the sense that they include the fact that momenta and coordinates cannot be simultaneously determined with infinite accuracy. A completely classical statistical mechanics would permit the state to be precisely specified by the p's and q's and this would lead to contradictions even in the limits where the above results hold. The presence of Planck's constant removes these contradictions and the semiclassical results are valid whenever the temperature is sufficiently high or the particle masses are sufficiently large.

chapter 2

STATISTICAL THERMO-
DYNAMICS OF SIMPLE
CRYSTALS

2.1 THE HARMONIC MODEL

A rigorous theory of crystals should start with the quantum mechanical wave equation that includes all the electrons and nuclei explicitly. After determining the energy levels from this rigorous equation, they would be used to evaluate the partition function, from which we could compute the Helmholtz free energy and all other properties of thermodynamic interest. Such a program presents insurmountable practical difficulties, and we must be content with a more modest approach based on two fundamental approximations.

In the first approximation, we recognize that the nuclei are much heavier than the electrons. If we give a nucleus a small displacement, the electrons will readjust their positions because of the forces of interaction between electrons and nuclei. Being very light, the electrons can perform this readjustment in a very short time.

The adiabatic approximation states that in a crystal, the readjustment time is so short that the time lag between the motion of the nuclei and the response of the electrons to this motion is negligible compared to the period of vibration of the nuclei. This means that the electrons are in a state determined by the instantaneous position of the nuclei, and if we solve a series of electronic wave equations, one for each of a set of fixed nuclear coordinates, we will get the right answer for the complete electrons-nuclei problem.

The adiabatic approximation allows us to separate the nuclear and electronic motions. If we focus attention on the nuclei, all the electronic information can be soaked up in a potential function that is a function of nuclear coordinates alone, and gives the potential energy of the crystal in terms of these coordinates.

Now we want to superimpose another approximation on this one. The atoms in a crystal vibrate around mean positions and when all atoms are actually at their mean positions, the potential energy is a minimum. If the amplitudes of the atomic vibrations are not too large, the potential energy can be expanded in a Taylor series of the displacements from equilibrium, retaining only terms that are quadratic in the displacements. This is called the harmonic approximation. It gives a somewhat simplified model of a crystal in which all the atoms interact in such a way that the crystal potential is quadratic in displacements from equilibrium.

The harmonic model accounts for a number of phenomena remarkably well, being especially successful in the theory of specific heats. But for some purposes, it requires extension and addition. To adequately describe thermal expansion, for example, third- and fourth-order terms must be included in the potential, while a description of metals at low temperatures requires that the harmonic model be supplemented by a theory for the highly mobile "free" electrons in a metal. On the whole, however, the model is an excellent one and serves as one of the basic pillars of solid state theory.

2.2 ONE-DIMENSIONAL WAVES AND NORMAL MODE ANALYSIS

In developing the statistical thermodynamics of crystals on the harmonic model, an appreciation of some of the facts concerning vibrations in crystals is necessary. These facts can be most easily introduced by considering a fictitious one-dimensional crystal composed of identical atoms of mass m arranged in a line. When the chain is in mechanical equilibrium, all the atoms are an equal distance, a, apart (Figure 2.1), and the potential energy has a minimum value. We will choose the energy scale so that this minimum potential energy is zero. The atoms will be labeled by an index j that runs from 0 to $(N - 1)$, where N is the total number of atoms, so the total length of the chain is Na. Also, the displacement of the jth atom will be denoted by u_j, and in keeping with the harmonic model, the potential energy of the chain V will be written as

$$V(u_0, u_1 \cdots u_{N-1}) = \tfrac{1}{2} \sum_{i,j} C_{ij} u_i u_j \qquad (2.1)$$

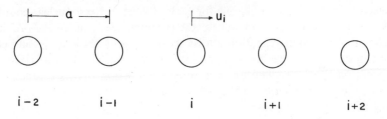

Fig. 2.1. Monatomic chain.

where the C_{ij} are defined by

$$C_{ij} = \left(\frac{\partial^2 V}{\partial u_i \, \partial u_j} \right)_0 \tag{2.2}$$

the derivative being evaluated at equilibrium where all $u_i = 0$. The C_{ij} are force constants coupling the ith and jth atoms and, therefore, depend only on the difference $(i - j)$. That is, the force between two atoms depends only on the distance between them and not on the absolute location of the pair in space, and two force constants C_{ij} and C_{mn} are equal if $|i - j| = |m - n|$.

Now let us find the equations of motion for the chain. This can be done by a simple application of Newton's second law, which can be written in the form

$$-\frac{\partial V}{\partial u_i} = m \frac{\partial^2 u_i}{\partial t^2} \tag{2.3}$$

t being the time, and the negative derivative of V being just the force on the ith atom. From (2.1) we get

$$-\frac{\partial V}{\partial u_i} = -\sum_j C_{ij} u_j \tag{2.4}$$

and combining this with (2.3) gives the equations of motion:

$$m \frac{\partial^2 u_i}{\partial t^2} + \sum_j C_{ij} u_j = 0 \tag{2.5}$$

What kind of a functional form would we expect for the u_i? Clearly the u_i can depend on only two variables; time and the position of the atom in the chain. The structure of (2.5) is very suggestive of at least a possible form for the time dependence. Equation 2.5 is very much like the simple harmonic oscillator equation, and would be identical to it if all terms in the summation were zero except for $j = i$. Physically, we know that at least a possible motion for an atom in the chain is simple harmonic. We will, therefore, try out a harmonic oscillator solution for (2.5) to see if it works, and write this trial solution as

$$u_j = v(j)e^{-i\omega t} \tag{2.6}$$

where ω is an angular frequency and $v(j)$ is a function of j only. The second derivative of (2.6) with respect to time is

$$\frac{\partial^2 u_i}{\partial t^2} = -\omega^2 v(j)e^{-i\omega t} \tag{2.7}$$

Substituting (2.6) and (2.7) into (2.5) gives

$$\sum_j C_{ij}v(j) - m\omega^2 v(i) = 0 \tag{2.8}$$

Thus far, all we have shown is that the equations of motion of the chain have simple harmonic oscillator solutions if a set of $v(i)$ is chosen that satisfies (2.8). We can go further by making use of the translational symmetry of the chain. If we replace i and j by $i+1$ and $j+1$, (2.8) becomes

$$\sum_j C_{i+1,j+1}v(j+1) - m\omega^2 v(i+1) = 0 \tag{2.9}$$

But from our previous discussion we know that $C_{i+1,j+1} = C_{ij}$; the solution of (2.8) and (2.9) can therefore differ by at most a constant factor. That is

$$v(j+1) = B(1)v(j) \tag{2.10}$$

$B(1)$ being a constant, the argument (1) identifying it as the constant that connects two sets of solutions related by a unit translation. Evidently, a similar relation holds for any translation, say by l units,

$$v(j+l) = B(l)v(j) \tag{2.11}$$

and clearly for two translations l and l'

$$B(l+l') = B(l)B(l') \tag{2.12}$$

Since B for a sum of two translations is the product of the B's for the two translations separately, it is much more convenient to define a parameter σ that allows us to write $B(l)$ in exponential form.

$$B(l) = e^{il\sigma} \tag{2.13}$$

This equation is to be taken as a definition of σ. Equation 2.12 is satisfied by writing $B(l)$ in this fashion since

$$e^{i(l+l')\sigma} = e^{il\sigma}e^{il'\sigma} \tag{2.14}$$

To learn more about the parameter σ, we must now consider the ends of the chain. If the chain is very long, we would not expect any assumptions about the ends to make much difference in any physical results concerning the chain as a whole. This statement can be proved mathematically, but we will accept it as physically reasonable and treat the ends on the basis of convenience. It is usual to treat the ends of the chain by postulating the periodic boundary conditions according to which the ends of the chain move in phase so that

$$u_0 = u_N \tag{2.15}$$

The physical picture corresponding to the periodic boundary conditions is easily constructed by imagining the chain to form a circle. If the number of atoms is very large, the properties of this circle are almost identical to those of the linear chain. With the periodic boundary condition (2.15), we obviously have

$$v(N) = v(0) \tag{2.16}$$

but since

$$v(N) = B(N)v(0) = e^{iN\sigma}v(0) \tag{2.17}$$

it follows that

$$e^{iN\sigma} = 1 \tag{2.18}$$

or

$$N\sigma = 2\pi n \tag{2.19}$$

where n is any integer $(0, 1, 2, \ldots)$. Choosing the periodic boundary conditions has allowed us to determine a set of σ's relating any $v(l)$ to $v(0)$:

$$v(l) = e^{il\sigma}v(0) \tag{2.20}$$

where σ is limited to the set of values

$$\sigma = 2\pi \frac{n}{N} \quad (n = \text{integer}) \tag{2.21}$$

We can now go back to (2.6) and write our proposed trial solution for the displacement as

$$u_l = v(0)e^{i(l\sigma - \omega t)} \tag{2.22}$$

$v(0)$ being a constant. Equation 2.22 has just the form of a plane wave. If we write $\sigma = aq$, where q is a new parameter,

$$u_l = v(0)e^{i(alq - \omega t)} \tag{2.23}$$

This is clearly the equation of a plane wave of frequency ω, al being the distance parameter and q being 2π times a wave number. From (2.21), the q's can only take the values

$$q = \frac{2\pi}{a}\frac{n}{N} \tag{2.24}$$

It should be noted that because of (2.18), the integer n in (2.23) produces a wave identical to that for the integer $n + N$. Therefore, n can be restricted to the range 0 to $N - 1$, thereby restricting q to be between 0 and $2\pi/a$. This gives us as many different trial solutions u_j, as there are atoms in the chain.

Because q has the interpretation of a wave number, we expect a close relationship between q and the frequency ω. This relationship can be found by substituting (2.20) into (2.8). If we use l and l' as running indices in (2.8) instead of i and j, the result is

$$\sum_l C_{l'l} e^{il\sigma} - m\omega^2 e^{il'\sigma} = 0 \qquad (2.25)$$

or

$$m\omega^2 = \sum_l C_{l'l} e^{i(l-l')\sigma} \qquad (2.26)$$

But it has already been pointed out that $C_{l'l}$ depends only on the difference $|l - l'|$; this means that the right-hand side of (2.26) is a summation over $(l - l')$, and when the summation is performed neither l nor l' will appear in the result. Equation 2.26 is, therefore, an equation for ω^2 as a function of σ or, as it is usually stated, a function of q. The ω's form a discrete set because the q's can only take the values determined by (2.24), and there are precisely N of them, one for each degree of freedom of the chain. From now on we will label ω to remind us of this fact, writing it as $\omega(q)$.

We have now arrived at a set of special solutions whose number is just equal to the number of degrees of freedom. To construct a general solution, all we need to do is take a linear combination of all the special solutions of the form of (2.23). That is

$$u_l = \sum_q A_q e^{i(alq - \omega(q)t)} \qquad (2.27)$$

where we have absorbed $v(0)$ into the constants A_q. First, note that we have succeeded in expressing the displacements as superpositions of plane waves. Second, (2.27) can be written as

$$u_l = \sum_q \eta_q e^{ialq} \qquad (2.28)$$

where η_q is defined by

$$\eta_q = A_q e^{-i\omega(q)t} \qquad (2.29)$$

and (2.28) can be regarded as connecting two different sets of coordinates u_l and η_q. The advantage in this is that the η_σ turn out to be dynamically separable whereas the u_l are not. To show how this comes about, substitute (2.28) into the equation of motion (2.5) to get

$$m \sum_q \ddot{\eta}_q e^{ialq} + \sum_{l'q} C_{ll'} \eta_q e^{ial'q} = 0 \qquad (2.30)$$

or

$$m \sum_q \ddot{\eta}_q + \sum_{l',q} C_{ll'} \eta_q e^{ia(l'-l)q} = 0 \qquad (2.31)$$

Because of (2.26), the double sum in (2.30) reduces to a single sum over q and we get

$$\sum_q [\ddot{\eta}_q + \omega^2(q)\eta_q] = 0 \qquad (2.32)$$

The η_q are a set of independent coordinates, and each term in (2.31) is independent. This means the sum can be zero only if the individual terms are zero and therefore

$$\ddot{\eta}_q + \omega^2(q)\eta_q = 0 \qquad (2.33)$$

This result could have been obtained directly by differentiating (2.29), but by using the method given here, we have shown that a coordinate transformation defined by (2.28) reduces the dynamical problem to that of a set of independent linear harmonic oscillators. This is an important result because it tells us how to compute the energy levels of our system. The energy levels do not depend on the coordinates used to describe the problem, so they can be taken to be the levels for a set of oscillators whose frequencies are determined by (2.26).

To illustrate the nature of the allowed frequencies $\omega(q)$ let us go back to the equation of motion (2.5). If only nearest neighbor forces are acting so that C_{ij} is zero, unless i and j label adjacent atoms, (2.5) becomes

$$m \frac{d^2 u_l}{dt^2} = c[u_{l+1} + u_{l-1} - 2u_l] \qquad (2.34)$$

when c is the force constant connecting nearest neighbors.

It has already been shown that (2.23) is a solution of (2.5) and, therefore, of (2.34) so that we can write

$$u_l = v(0)e^{-i\omega t}e^{ialq}$$

$$u_{l+1} = v(0)e^{-i\omega t}e^{ia(l+1)q}$$

$$u_{l-1} = v(0)e^{-i\omega t}e^{ia(l-1)q} \qquad (2.35)$$

$$\frac{d^2 u_l}{dt^2} = -v(0)\omega^2 e^{-i\omega t}e^{ialq}$$

Putting these in (2.34) gives

$$m\omega^2 = -2c[\cos qa - 1] \qquad (2.36)$$

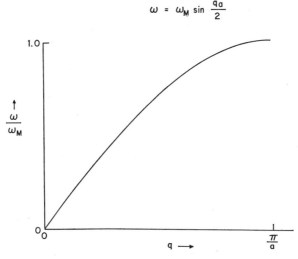

Fig. 2.2. Frequency-dispersion relation for monatomic chain-nearest:neighbor forces only; $\omega = \omega_M \sin (qa/2)$.

or, since $\cos qa = 1 - 2 \sin^2 (qa/2)$,

$$\omega = 2 \left(\frac{c}{m}\right)^{1/2} \sin \frac{qa}{2} \qquad (2.37)$$

Equation 2.37 shows how the frequency is related to the wave number q, and is called the frequency dispersion relation. Its form is shown in Figure 2.2.

Two important conclusions can be drawn from the above analysis that are valid for all discrete systems. The first is that a maximum frequency exists, so that waves higher than this maximum cannot be propagated through the lattice. For our chain, this frequency is $2(c/m)^{1/2}$. The second conclusion is that the wave velocity is not a constant but depends on frequency. This can be seen by remembering that the velocity v is defined by $v = \lambda \nu = \lambda \omega/2\pi$ where λ is the wavelength, which for the monatomic chain is $\lambda = 2\pi/q$.

A better appreciation of these results is obtained by comparing them to those for a continuous string. This can be done by recognizing that (2.34) is a difference equation that approaches a differential equation as the differences approach zero. Thus, if we make the substitution

$$\frac{u_{l+1} - u_l}{a} \rightarrow \frac{\partial u}{\partial x}$$

$$\frac{1}{a}\left[\frac{(u_{l+1} - u_l)}{a} - \frac{(u_l - u_{l-1})}{a}\right] \rightarrow \frac{\partial^2 u}{\partial x^2} \qquad (2.38)$$

thereby treating u as a continuous displacement in a continuous string, (2.34) becomes

$$m \frac{\partial^2 u}{\partial t^2} = ca^2 \frac{\partial^2 u}{\partial x^2} \qquad (2.39)$$

But this is just the equation of wave motion in one dimension

$$\frac{\partial^2 u}{\partial x^2} = \frac{1}{v_0{}^2} \frac{\partial^2 u}{\partial t^2} \qquad (2.40)$$

with the wave velocity v_0 given by

$$v_0 = \left(\frac{ca^2}{m}\right)^{1/2} \qquad (2.41)$$

It is readily verified by direct substitution that solutions of (2.40) exist that have the form

$$u(x, t) = A \exp\left[i\left(\frac{2\pi}{\lambda_0} x - \omega_0 t\right)\right] \qquad (2.42)$$

where A is a constant, and that these solutions are periodic in x and t with a frequency $\nu_0 = \omega_0/2\pi$ and wavelength λ_0 where

$$\nu_0 \lambda_0 = v_0 \qquad (2.43)$$

Since v_0 is a constant, (2.43) is just the ordinary frequency-wavelength-velocity relation for a nondispersive medium.

Comparing (2.42) with (2.23) we see that a correspondence exists between the continuous and discrete case such that al plays the role of x and q corresponds to $2\pi/\lambda_0$. In the continuous case, however, there is no upper limit on the frequency, and the frequency is not a function of wavelength.

Some further comment can now be made for the discrete chain. The maximum propagation frequency occurs at a value of $q = q_M$ that makes the sine unity in (2.37) so that

$$\frac{q_M a}{2} = \frac{\pi}{2} \qquad (2.44)$$

Since $q = 2\pi/\lambda$, the corresponding minimum wavelength λ_M is

$$\lambda_M = 2a \qquad (2.45)$$

This makes sense, since a wavelength less than the order of the lattice spacing has no meaning in a discrete structure.

Now let us find the frequencies in the chain when the wavelength is very large, that is, for small q, by expanding the sine in (2.37) and keeping only the first term to get

$$\omega = qa \left(\frac{c}{m}\right)^{1/2} \tag{2.46}$$

But $\omega = 2\pi\nu$ and $q = 2\pi/\lambda$ so this gives

$$\nu = \frac{1}{\lambda a}\left(\frac{c}{m}\right)^{1/2} \tag{2.47}$$

or, using (2.41),

$$\nu\lambda = v_0 \tag{2.48}$$

We see, therefore, that in the limit of large wavelengths, the chain behaves as a continuous string. This conclusion carries over to three dimensions, so that for long waves, a crystal can be treated as a continuous elastic solid.

This discussion has been restricted to a one-dimensional model, but it can be easily generalized to describe three-dimensional crystals. The wave number q becomes a wave-number vector $\mathbf{q}$ and instead of the condition (2.24) we get

$$\mathbf{q} = \frac{2\pi}{N^{1/3}} [n_1\mathbf{b}_1 + n_2\mathbf{b}_2 + n_3\mathbf{b}_3] \tag{2.49}$$

where n_1, n_2, and n_3 are integers ranging from 0 to N and $\mathbf{b}_1$, $\mathbf{b}_2$, and $\mathbf{b}_3$ are reciprocal lattice vectors defined by

$$\mathbf{a}_i \cdot \mathbf{b}_j = \delta_{ij} \qquad i, j = 1 \text{ to } 3 \tag{2.50}$$

where $\mathbf{a}_i$ are the basic lattice vectors defining the size and shape of the unit cell and δ_{ij} is the Kronecker delta. The displacement vector $\mathbf{u}_l$ of the lth atom in a three-dimensional crystal is related to a set of independent, harmonic oscillator type coordinates by equations similar to (2.26) but generalized to three dimensions.

It is a general result of mechanics that whenever the potential energy is a quadratic form in the coordinates, a set of coordinates exists whose equations of motion are those for simple harmonic oscillators. This can be proven in a direct manner without using the trial solutions we have introduced here. Such a set of coordinates are said to represent the normal modes of the motion, and what we have done with the linear chain is to analyze its motion into normal modes.

2.3 PARTITION FUNCTION AND FREE ENERGY OF THE HARMONIC CRYSTAL

Within the context of the harmonic model, the energy levels of a crystal are the same as those for a set of $3N$ independent simple harmonic oscillators. If ν_j is the frequency of the jth mode, then quantum theory tells us that the possible energy levels of the corresponding oscillator are

$$E_n = (n_j + \tfrac{1}{2})h\nu_j \qquad (3.1)$$

where n_j can take integral values from zero to infinity, and h is Planck's constant. The energy of the crystal is a sum of terms like (3.1) with $j = 1, 2, \ldots, 3N$. The state of the crystal is determined by all the n_j, so we will label the energy of a given crystal state with all the vibrational quantum numbers. If U_o is the potential energy of the crystal when all atoms are at their equilibrium positions, the energy of the crystal in a particular state determined by the set of integers $n_1, n_2, \ldots, n_{3N}$, is

$$E_{n_1, n_2, \cdots, n_{3N}} = U_o + \sum_{j=1}^{3N} (n_j + \tfrac{1}{2})h\nu_j \qquad (3.2)$$

It is easy to construct the canonical partition function of a crystal from this equation. All we need remember is that the partition function is a sum over *all* states, and that one crystal state is determined by a complete set of integers, one for each normal mode. We must therefore sum over all n_1, all n_2, all n_3, and so on and write the partition function as

$$Z = \sum_{n_1=0}^{\infty} \sum_{n_2=0}^{\infty} \cdots \sum_{n_{3N}=0}^{\infty} e^{-E_{n_1 n_2 \cdots n_{3N}}/kT} \qquad (3.3)$$

or, using (3.2),

$$Z = e^{-E_0/kT} \sum_{n_1=0}^{\infty} \sum_{n_2=0}^{\infty} \cdots \sum_{n_{3N}=0}^{\infty} e^{-1/kT \sum_{j=1}^{3N} n_j h\nu_j} \qquad (3.4)$$

where E_o is defined by

$$E_o = U_o + \sum_{j=1}^{3N} \tfrac{1}{2}h\nu_j \qquad (3.5)$$

The sum in (3.5) is the zero point vibrational energy.

Equation 3.4 looks complicated, but it is easily simplified. Because of the convenient properties of the exponential function, the multiple sum separates into a product of sums, all of which are alike except for the subscript j.

$$Z = e^{-E_0/kT} \sum_{n_1=0}^{\infty} e^{-n_1 h\nu_1/kT} \sum_{n_2=0}^{\infty} e^{-n_2 h\nu_2/kT} \sum_{n_{3N}=0}^{\infty} e^{-n_{3N} h\nu_{3N}/kT} \qquad (3.6)$$

We cannot do anything about the subscripts on the frequencies, but they certainly can be thrown away on the n's, so Z becomes

$$Z = e^{-E_0/kT} \prod_{j=1}^{3N} \sum_{n=0}^{\infty} e^{-n h\nu_j/kT} \qquad (3.7)$$

The infinite sum in this equation can be evaluated. If x_j is defined by

$$x_j = e^{-h v_j / kT} \tag{3.8}$$

then

$$\sum_{n=0}^{\infty} e^{-n h v_j / kT} = \sum_{n=0}^{\infty} x_j{}^n \tag{3.9}$$

which is nothing but a geometric series whose sum is $1/(1 - x_j)$. Therefore

$$\sum_{n=0}^{\infty} e^{-n h v_j / kT} = \frac{1}{1 - e^{-h v_j / kT}} \tag{3.10}$$

and substituting this in (3.7) gives

$$Z = e^{-E_0 / kT} \prod_{j=1}^{3N} \frac{1}{1 - e^{-h v_j / kT}} \tag{3.11}$$

and the Helmholtz free energy

$$A = -kT \ln Z \tag{3.12}$$

is

$$A = E_0 + kT \sum_{j=1}^{3N} \ln \left(1 - e^{-h v_j / kT}\right) \tag{3.13}$$

If only we knew the values of the v_j, we could now evaluate all the thermodynamic functions. This is, of course, the crux of the problem of building a statistical theory of vibrating crystals.

In principle, the frequencies can be determined by equations similar to (2.26), and this has actually been done for a number of crystals. The trouble with such an approach is twofold: first, the force constants for real crystals are generally not known to any degree of accuracy and, second, a separate numerical calculation is required for each case. The simplicity and coherence of being able to express the statistical thermodynamic results in a single set of equations is thereby lost.

Historically, a different approach was taken, based on simplified approximations to the frequency spectrum in solids. These approximations were suggested at about the same time as the dynamical method was developed, and they were so successful that the more rigorous dynamic calculations have been pursued vigorously only in fairly recent times.

It is extremely convenient to replace the sum in (3.12) by an integral. To do this, we define a frequency distribution function $g(v)$ such that $Ng(v)\,dv$ is the number of modes with frequencies between v and $v + dv$. There are $3N$ modes altogether, so the normalization condition for $g(v)$ is

$$\int_0^{\infty} g(v)\,dv = 3 \tag{3.14}$$

To get the Helmholtz free energy in integral form, just multiply the summand in (3.13) by $Ng(v)\,dv$, replace the summation by an integration, and drop the subscript j.*

$$A = E_0 + NkT \int_0^\infty g(v) \ln\left(1 - e^{-hv/kT}\right) dv \qquad (3.15)$$

If $g(v)$ were known, the entire problem would be solved. The approximations that were mentioned involve approximations to $g(v)$, and these will be considered later. But at this point, it is instructive to show how a knowledge of the frequencies as a function of the wave number allows a determination of $g(v)$ to be made. We will illustrate this for the one-dimensional model of the previous section. From (2.24), if q and n are treated as quasi-continuous variables

$$dn = \frac{Na}{2\pi}\,dq \qquad (3.16)$$

But the number of permissible values of n is just equal to the number of vibrational modes and therefore

$$dn = Ng(v)\,dv \qquad (3.17)$$

Combining (3.16) and (3.17) and solving for $g(v)$

$$g(v) = \frac{a}{2\pi}\frac{dq}{dv} = a\frac{dq}{d\omega} \qquad (3.18)$$

and it is clear that if the frequency is known as a function of wave number from a solution of the dynamical problem, $g(v)$ can be calculated. Similar considerations hold for the three-dimensional case, so that, in general, the frequency distribution function can be computed from the dispersion relation connecting frequency and wave number.

2.4 GENERAL HEAT-CAPACITY EQUATIONS

Of all the thermodynamic functions, the heat capacity is usually given special emphasis in any discussion of the statistical thermodynamics of

---

* As is evident from the definition of $g(v)$, all sums of functions of the frequencies can be replaced by integrals according to the following identity

$$\sum_j f(v_j) = N \int_0^\infty g(v) f(v)\,dv$$

where the v_j in $f(v_j)$ is replaced by the continuous variable v in $f(v)$. This identity will be used frequently in passing back and forth between the "discrete notation," in which the equations are expressed in terms of sums, and the "continuum notation," in which the sums are replaced by integrals.

crystals. It can be measured with considerable accuracy, and a reasonable quantity of data are available for comparing theory with experiment. Also, it is closely related to the energy, a function of central importance in statistical mechanics. Finally, heat capacity measurements have played a historic part in the growth of modern physics, since the experimental results were com- pletely inexplicable on the basis of classical physics and required the intro- duction of quantum energy levels. It is fortunate that heat capacity is related to the partition function by a constant volume temperature derivative. The theory can therefore be compared with experiment without any explicit knowledge of the variation of energy levels with volume.

The heat capacity formula can be obtained in two ways. The first is to use (8.37) of Chapter 1. The second is to calculate the average energy in terms of the partition function, and then differentiate the energy once with respect to temperature. The latter method will be used here. From the general definition of the statistical mechanical average, the energy of the crystal is

$$U = \frac{1}{Z} \sum_j E_j e^{-E_j/kT} \tag{4.1}$$

The partition function as usual is

$$Z = \sum_j e^{-E_j/kT} \tag{4.2}$$

If Z is differentiated with respect to the variable $1/kT$, the result is

$$\frac{\partial Z}{\partial \left(\frac{1}{kT}\right)} = - \sum_j E_j e^{-E_j/kT} \tag{4.3}$$

Combining (4.3) and (4.1) gives

$$U = - \frac{\partial \ln Z}{\partial \left(\frac{1}{kT}\right)} \tag{4.4}$$

This equation is perfectly general and is readily applied to the present problem. It is apparent from (3.12) and (3.15) that in terms of the frequency distribution function, the partition function is given by

$$\ln Z = - \frac{E_0}{kT} - N \int_0^\infty g(\nu) \ln (1 - e^{-h\nu/kT}) \, d\nu \tag{4.5}$$

Differentiating this according to (4.4) gives for the energy

$$U = E_0 + N \int_0^\infty g(\nu) \frac{h\nu e^{-h\nu/kT}}{1 - e^{-h\nu/kT}} \, d\nu \tag{4.6}$$

and the derivative of the energy with respect to T gives the heat capacity at constant volume

$$C_V = Nk \int_0^\infty g(\nu) \left(\frac{h\nu}{kT}\right)^2 \frac{e^{-h\nu/kT}}{(1 - e^{-h\nu/kT})^2} \, d\nu \qquad (4.7)$$

Some interesting comments can be made about the specific heat on the basis of this equation even without any specific knowledge of the frequency distribution function. First, it must be realized that the possible values of the frequency cannot be arbitrarily large. A high frequency means a short wavelength, but because of the discrete, atomistic structure of crystals, the shortest wavelength that has any physical meaning is of the order of magnitude of the lattice spacing. Therefore, some maximum frequency must exist. This is a consequence of the fact that the frequencies represent collective oscillations of groups of atoms in the crystal. As a result, a temperature T can always be chosen high enough that $h\nu/kT \ll 1$, and it is sensible to look for a high temperature limiting form of (4.7). Let us examine the function $E(x)$ defined by

$$E(x) = \frac{x^2 e^{-x}}{(1 - e^{-x})^2} \qquad (4.8)$$

A little algebra allows us to write $E(x)$ as

$$E(x) = \frac{x^2}{e^x + e^{-x} - 2} \qquad (4.9)$$

A series expansion can be found for this function by using the exponential expansions

$$e^x = \sum_{n=0}^\infty \frac{x^n}{n!} \qquad (4.10)$$

and

$$e^{-x} = \sum_{n=0}^\infty (-1)^n \frac{x^n}{n!} \qquad (4.11)$$

Substitution of these expansions into (4.9) gives

$$E(x) = \frac{x^2}{2} \left[\frac{x^2}{2!} + \frac{x^4}{4!} + \frac{x^6}{6!} + \frac{x^8}{8!} + \cdots\right]^{-1} \qquad (4.12)$$

or

$$E(x) = \frac{1}{2} \left[\frac{1}{2!} + \frac{x^2}{4!} + \frac{x^4}{6!} + \frac{x^6}{8!} + \cdots\right]^{-1} \qquad (4.13)$$

Inverting the series in the brackets gives $E(x)$ as

$$E(x) = 1 - \tfrac{1}{12} x^2 + \tfrac{1}{240} x^4 - \cdots \qquad (4.14)$$

If $h\nu/kT$ is identified with x, then (4.7) can be written in terms of $E(x)$ as

$$C_V = Nk \int_0^\infty g(\nu)E\left(\frac{h\nu}{kT}\right) d\nu \qquad (4.15)$$

Therefore, replacing x by $h\nu/kT$ in (4.14), and substituting the result in (4.15)

$$C_V = Nk \int_0^\infty g(\nu)\left[1 - \frac{1}{12}\left(\frac{h\nu}{kT}\right)^2 + \frac{1}{240}\left(\frac{h\nu}{kT}\right)^4\right] d\nu \qquad (4.16)$$

(high temperature)

The dots have been dropped and only the terms to the fourth power have been retained because this series approximation is useful only for high temperatures. Equation 4.16 can be integrated term by term if $g(\nu)$ is known. Actually, the first term can be integrated immediately because of the normalization condition of (3.14), so that

$$C_V = 3Nk - \frac{Nk}{12} \int_0^\infty g(\nu)\left(\frac{h\nu}{kT}\right)^2 d\nu + \frac{Nk}{240} \int_0^\infty g(\nu)\left(\frac{h\nu}{kT}\right)^4 d\nu \qquad (4.17)$$

(high temperature)

If the temperature is high enough, all terms but the first can be ignored and to a good approximation $C_v = 3Nk$. This is just the result a classical statistical treatment would give for all temperatures, and is independent of the nature of $g(\nu)$. In such a treatment, the partition function would be evaluated by taking the energy of an oscillator to be that given by classical mechanics, rather than the quantum levels used in the present treatment. The deviation of the heat capacity from the value of $3Nk$ is a direct quantum mechanical effect. Since the heat capacity is the derivative of the energy, the high temperature limiting form of the energy is $U = 3NkT$. There are $3N$ oscillators, so the energy per oscillator is kT. That is, the energy is equally divided among the oscillators. This is called the classical equipartition of vibrational energy. We have, of course, ignored the potential energy in this discussion, which does not contribute to the heat capacity in the harmonic model.

What happens at very low temperatures? In this case $h\nu/kT$ is very large and C_V approaches zero as $T \to 0$, as can be seen by the following operations on $E(x)$

$$\lim_{x \to \infty} E(x) = \lim_{x \to \infty} \frac{x^2 e^{-x}}{(1 - e^{-x})^2}$$

$$= \lim_{x \to \infty} \frac{x^2}{(e^x + e^{-x} - 2)} = 0 \qquad (4.18)$$

The form of $E(x)$ is such that it starts at $E(x) = 0$ at $T = 0$ and increases monotonically as T increases, asymptotically approaching unity as $T \to \infty$.

2.5   THE EINSTEIN APPROXIMATION

The simplest choice to make for the vibration spectrum in a crystal is to assume that all vibrational modes have the same frequency ν_E. In the discrete notation, this means that (3.13) for the Helmholtz free energy becomes

$$A = E_0 + 3NkT \ln (1 - e^{-h\nu_E/kT}) \tag{5.1}$$

In the continuum notation, $g(\nu)$ is zero for all frequencies except for $\nu = \nu_E$. Direct differentiation of (5.1) according to (8.37) of Chapter 1 gives the heat capacity:

$$C_V = 3Nk \left(\frac{h\nu_E}{kT}\right)^2 \frac{e^{-h\nu_E/kT}}{(1 - e^{-h\nu_E/kT})^2} \tag{5.2}$$

This is the Einstein formula for the heat capacity. The physical assumption on which it rests is that all atoms vibrate independently with the same frequency. Although this assumption is certainly a gross simplification of the situation in crystals, (5.2) gives a pretty fair representation of the experimental data. It is worthwhile to note parenthetically that the Einstein theory of specific heats is accurate for the vibrational contribution to the heat capacity of diatomic gases. In this case each molecule is an oscillator that is independent of all the others.

It is usual to define a parameter θ_E by the relation

$$\theta_E = \frac{h\nu_E}{k} \tag{5.3}$$

and rewrite (5.2) as

$$C_V = 3Nk \left(\frac{\theta_E}{T}\right)^2 \frac{e^{-\theta_E/T}}{(1 - e^{-\theta_E/T})^2} \tag{5.4}$$

θ_E is called the Einstein characteristic temperature and can be calculated from experiment by choosing the value of θ_E that gives the best fit of (5.4) to measured values of the heat capacity at various temperatures.

The form of the Einstein heat-capacity equation is shown in Figure (2.3), where $C_V/3Nk$ is plotted against T/θ_E. The figure shows that above $T = 2\theta_E$, the high temperature approximation, $C_V = 3Nk$, is fairly accurate.

An estimate of the magnitude of vibration frequencies in solids can be made from the fact that typical θ_E values are in the neighborhood of 300°K. Equation 5.3 then gives $\nu_E \approx 10^{13}$ sec^{-1}.

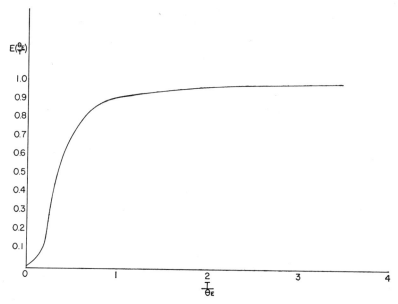

Fig. 2.3. The Einstein function: $E(\theta_E/T) = (\theta_E/T)^2 e^{-\theta_E/T}/(1 - e^{\theta/T})^2$.

2.6 THE SUPERPOSITION OF EINSTEIN OSCILLATORS

Examination of (4.7) shows that the most general form of the heat capacity is a superposition of Einstein functions, the contribution of Einstein oscillators of each frequency being determined by $g(\nu)$. This means that the general heat capacity-temperature curve is a sum of many curves similar to that in Figure 2.3. All the curves must start at the origin and approach the same value at high temperatures. Therefore, the deviations among theories with different assumptions about $g(\nu)$, and between these theories and experiment, will occur at low and intermediate temperatures. This is so because the Einstein function

$$E(x) = \frac{x^2 e^{-x}}{(1 - e^{-x})^2} \tag{6.1}$$

is relatively insensitive to changes in x at low x values.

Now let us consider the contribution of an Einstein oscillator to the heat capacity as a function of its frequency. If the temperature is very low, then the Einstein function $E(x)$ has a negligible value unless the frequency is also very low, as shown by (6.1), with $x = h\nu/kT$. That is, at low temperatures, only vibrations of low frequency, or long wavelength, contribute significantly to the heat capacity. High-frequency contributions are simply choked off by the form of the Einstein function. As the temperature is raised, the frequency values that can contribute to the heat capacity increase. The result is that not all of the vibrations are important to the heat capacity at all temperatures. In a sense, the equivalent physical statement to the quantum theory of heat capacity, is that at low temperatures, only low frequency vibrations are excited, at high temperatures, all vibrations are excited.

The Einstein function defined by (6.1) is tabulated in Appendix VII.

2.7 THE DEBYE MODEL

Debye assumed that the frequencies in an actual crystal are distributed as though the solid were an isotropic, elastic continuum rather than an aggregate of particles. With this assumption, the frequency distribution function can be obtained just by counting the number of possible sound waves with different frequencies that can exist in a given frequency range.

This procedure should work all right for long wavelengths. A low frequency vibration involves the cooperative motion of many atoms and in such a case the discrete nature of the medium is not important. This result is borne out by direct solution of the dynamical equations of motion. These solutions show that for long wavelengths, the dispersion relation between frequency and wave number is linear, corresponding to elastic waves being propagated with constant velocity. The Debye assumption cannot be right for high frequencies, however, because then the velocity of wave propagation does depend on frequency. Because of the discrete structure of crystals, there is the additional complication that only a finite number of vibrations exist and wavelengths shorter than the spacing between atoms has no meaning. This latter difficulty is easily handled by introducing a cutoff frequency above which no vibrations exist; but the dispersion problem at high frequencies is simply ignored in the Debye model.

The Debye model works quite well, and this is a surprise considering the physical assumptions it contains. Part of the reason for its success can be understood by referring to the results of the previous sections. In the limit of low temperature, it was shown that only low frequency vibrations are important, and the Debye model is adequate for low frequencies. In the limit of high temperatures, it was shown that the heat capacity is not sensitive to the form of $g(\nu)$, and if the temperature is high enough, it is even independent of any information concerning the vibrations at all. The Debye theory should therefore work very well at low and high temperatures, but some trouble might be expected at intermediate temperatures. It turns out that this is indeed the case, but even at intermediate temperatures, the troubles are not serious.

To derive an expression for $g(\nu)$ in the Debye approximation, suppose that the specimen is a cube of side L, and attach a coordinate system to one corner such that the Cartesian coordinates of any point in the specimen are in the range

$$0 \leq x \leq L, \qquad 0 \leq y \leq L, \qquad 0 \leq z \leq L \tag{7.1}$$

As an elastic continuum, any frequency can be impressed on the specimen. But the only waves that are of concern here are standing waves. All other waves would die out rapidly and cannot correspond to the thermal oscillations that exist in a crystal. The possible wavelengths are therefore determined by the dimensions of the specimen. Now consider a particular standing wave with wavelength λ_j, traveling in a direction determined by a unit vector $\mathbf{n}$, and let α_1, α_2, α_3 be the direction cosines between n and the x, y, and z axes. For a standing wave traveling parallel to the x axis, the half wavelength must be an integral submultiple of the length L of the box. For the wave traveling in the $\mathbf{n}$ direction, this condition must be fulfilled by the x component of the wave number vector $\mathbf{n}/\lambda$. That is

$$\frac{2}{\lambda_j}\alpha_1 = \frac{j_1}{L} \tag{7.2}$$

where j_1 is any integer from 1 to L. Similarly, for the projections of $\mathbf{n}/\lambda_j$ in the y and z directions

$$\frac{2}{\lambda_j}\alpha_2 = \frac{j_2}{L} \tag{7.3}$$

$$\frac{2}{\lambda_j}\alpha_3 = \frac{j_3}{L} \tag{7.4}$$

j_2 and j_3 also being any integers from 1 to L. Each possible wavelength is associated with a set of three integers (j_1, j_2, j_3), and we have a problem entirely analogous to that of determining the density of states of a gas of independent particles (Section 15, Chapter 1). If an integer space is constructed, and j defined by

$$j^2 = j_1{}^2 + j_2{}^2 + j_3{}^2 \tag{7.5}$$

then in quasi-continuum language, the number of possible wavelengths in a range $d\lambda$ corresponding to a range dj is

$$\tfrac{1}{2}\pi j^2 \, dj \tag{7.6}$$

(See 15.10 in Chapter I).

We do not yet have the number of possible vibrations in a given wavelength range. From the theory of elastic vibrations we know that three distinct waves are possible for each wavelength. One of these is longitudinal and two are transverse, so the number of vibrations in a given frequency range can be written as

$$3Ng(\nu)\,d\nu = Ng_l(\nu)\,d\nu + 2Ng_t(\nu)\,d\nu \tag{7.7}$$

$g_l(\nu)$ and $g_t(\nu)$ being the frequency distribution functions for longitudinal and transverse vibrations, respectively. Since there is just one longitudinal vibration per wavelength, (7.6) gives the number of longitudinal waves directly

$$Ng_l(\nu)\,d\nu = \tfrac{1}{2}\pi j^2\,dj \tag{7.8}$$

and all that is needed is a relation between ν and j. Addition of the squares of (7.2), (7.3), and (7.4) gives

$$\frac{4}{\lambda_j^2}(\alpha_1{}^2 + \alpha_2{}^2 + \alpha_3{}^2) = \frac{1}{L^2}(j_1{}^2 + j_2{}^2 + j_3{}^2) \tag{7.9}$$

or, recognizing that the sum of the squares of the direction cosines is unity, and going to the quasi-continuum language

$$\frac{4}{\lambda^2} = \frac{j^2}{L^2} \tag{7.10}$$

The wavelength is related to the frequency by $\lambda\nu = C$ where C is the wave velocity, and since the waves being considered are longitudinal

$$\frac{1}{\lambda} = \frac{\nu}{C_l} \tag{7.11}$$

C_l being the longitudinal velocity of sound. Equation 7.10, therefore, becomes

$$\frac{4\nu^2}{C_l{}^2} = \frac{j^2}{L^2} \tag{7.12}$$

so that

$$j^2 = \frac{4L^2}{C_l{}^2}\nu^2 \tag{7.13}$$

and

$$dj = \frac{2L}{C_l}\,d\nu \tag{7.14}$$

Using (7.13) and (7.14), (7.8) becomes

$$Ng_l(\nu) = \frac{4\pi V}{C_l{}^3}\nu^2 \tag{7.15}$$

where L^3 has been replaced by the volume V. The calculation of $g_t(\nu)$ goes along in the same way with the result

$$2Ng_t(\nu) = \frac{8\pi V}{C_t{}^3}\nu^2 \tag{7.16}$$

C_t being the transverse velocity of sound. Combining (7.15) and (7.16) according to (7.7) gives the frequency distribution function.

$$g(v) = 4\pi \frac{V}{N} \left[\frac{1}{C_l^3} + \frac{2}{C_t^3} \right] v^2 \tag{7.17}$$

Notice that (7.17) contains constants that can be experimentally measured independently of heat capacity data (density V/N and sound velocities C_l and C_t). The frequencies will be integrated out in the statistical thermodynamic equations, and the value of the coefficient of v^2 as given by (7.17) can be compared with the value obtained from thermodynamic data.

It is convenient to lump the constants in (7.17) into a single symbol B, and write

$$g(v) = Bv^2 \tag{7.18}$$

The constant B can be related to the maximum cutoff frequency required by the discrete structure of the solid by the normalization condition. Calling this frequency v_D and using (3.13) gives

$$\int_0^{v_D} g(v) \, dv = B \int_0^{v_D} v^2 \, dv = 3 \tag{7.19}$$

so that

$$B = \frac{9}{v_D^3} \tag{7.20}$$

Later, a characteristic Debye temperature θ_D, analogous to the Einstein temperature will be needed, defined by

$$\frac{h v_D}{k} = \theta_D \tag{7.21}$$

In terms of θ_D

$$B = \frac{9h^3}{k^3 \theta_D^3} \tag{7.22}$$

By using experimental data for density and the velocity of sound, an estimate can be made of the v_D and θ_D values to be expected in the Debye theory. Combining (7.20) and the original definition of B as the coefficient of v^2 in (7.17), and solving for v_D gives

$$v_D = \left(\frac{9N}{4\pi V} \right)^{1/3} \left[\frac{1}{C_l^3} + \frac{2}{C_t^3} \right]^{-1/3} \tag{7.23}$$

If we take the velocity of sound to be of the order of 10^5 cm/sec (ignoring the difference between C_l and C_t for the purposes of this calculation), and assume that the density N/V is 10^{23} atoms/cm³, then v_D is of the order of

10^{12}–10^{13} vibrations per second and, as calculated from (7.21), θ_D is in the neighborhood of 50 to 500°K.

2.8   THE DEBYE ENERGY AND HEAT CAPACITY

Upon substitution of the Debye frequency distribution (7.18) into the general equations for the energy and heat capacity given by (4.6) and (4.7), we get

$$(U - E_0) = 9NkT \left(\frac{h}{k\theta_D}\right)^3 \int_0^{v_D} \frac{h v}{kT} \frac{v^2}{e^{h v/kT} - 1} \, dv \tag{8.1}$$

and

$$C_V = 9NkT \left(\frac{h}{k\theta_D}\right)^3 \int_0^{v_D} v^2 \left(\frac{h v}{kT}\right)^2 \frac{e^{-h v/kT}}{(1 - e^{-h v/kT})^2} \, dv \tag{8.2}$$

where the cutoff frequency of Debye has been introduced as the upper limit in the integral, and the constant B was used in the form given by (7.22). $(U - E_0)$ is the thermal energy, and is important not only in the theory of the heat capacity, but also in the theory of the equation of state.

Define a parameter x by

$$x \equiv \frac{h v}{kT} \tag{8.3}$$

and transform to this parameter in (8.1) and (8.2). The results are

$$U - E_0 = 9Nk\theta_D \left(\frac{T}{\theta_D}\right)^4 \int_0^{\theta_D/T} \frac{x^3 \, dx}{e^x - 1} \tag{8.4}$$

$$C_V = 9Nk \left(\frac{T}{\theta_D}\right)^3 \int_0^{\theta_D/T} \frac{x^4 e^{-x}}{(1 - e^{-x})^2} \, dx \tag{8.5}$$

These equations show that, in the Debye theory, the heat capacity is a universal function of a single parameter θ_D/T, and that the thermal energy is another universal function of θ_D/T, multiplied by the Debye temperature. These formulae can be put in compact form by making the following definitions:

$$D_E(x_D) = \frac{3}{x_D^4} \int_0^{x_D} \frac{x^3 \, dx}{e^x - 1} \tag{8.6}$$

and

$$D(x_D) = \frac{3}{x_D^3} \int_0^{x_D} \frac{x^4 e^{-x}}{(1 - e^{-x})^2} \, dx \tag{8.7}$$

Then, if we also define

$$x_D \equiv \frac{\theta_D}{T} \tag{8.8}$$

(8.4) and (8.5) can be written as

$$U - E_0 = 3Nk\theta_D D_E(x_D) \tag{8.9}$$

$$C_V = 3NkD(x_D) \tag{8.10}$$

$D_E(x_D)$ and $D(x_D)$ are called the Debye energy function and the Debye heat-capacity function respectively. They cannot be evaluated in closed form, but their properties have been thoroughly studied. They are given numerically as functions of x_D in Appendix VIII, which can be used in numerical analyses. Note that $D_E(x_D)$ is the thermal energy per oscillator in units of $k\theta_D$, and $D(x_D)$ is the heat capacity per oscillator in units of k.

Analytic approximations to the Debye energy and heat capacity can be obtained in the limits of low and high temperatures that are useful for the study of crystal properties. For the high-temperature limits, the procedure is to expand the integrands in (8.6) and (8.7) in a power series about $x = 0$, since for high T, x is small. To do this for $D_E(x_D)$, define the function

$$f(x) = \frac{x}{e^x - 1} \tag{8.11}$$

and expand it in a Taylor series about $x = 0$ to the second order in x;

$$f(x) = f(0) + f'(0)x + \tfrac{1}{2}f''(0)x^2 \tag{8.12}$$

The derivatives of (8.11) are

$$f'(x) = \frac{1}{e^x - 1} - \frac{xe^x}{(e^x - 1)^2}$$

or

$$f'(x) = \frac{e^x(1 - x) - 1}{(e^x - 1)^2} \tag{8.13}$$

$$f''(x) = \frac{e^x(1 - x) - e^x}{(e^x - 1)^2} - 2\frac{e^{2x}(1 - x) - e^x}{(e^x - 1)^3} \tag{8.14}$$

Using the series expansion for the exponential

$$e^x = 1 + x + \frac{x^2}{2!} + \cdots \tag{8.15}$$

in (8.11), (8.13), and (8.14), and taking the limits as $x \to 0$ gives

$$f(0) = 1$$
$$f'(0) = -\tfrac{1}{2}$$
$$f''(0) = \tfrac{1}{6} \tag{8.16}$$

Thus (8.12) becomes

$$f(x) = 1 - \tfrac{1}{2}x + \tfrac{1}{12}x^2 \tag{8.17}$$

Multiplying this by x^2 gives the integrand we are after, which can be substituted into (8.6) to yield

$$D_E(x_D) = \frac{3}{x_D{}^4} \int_0^{x_D} \left[x^2 - \frac{x^3}{2} + \frac{x^4}{12} \right] dx \tag{8.18}$$

Now perform the integration to get

$$D_E(x_D) = \frac{1}{x_D} - \frac{3}{8} + \frac{1}{20} x_D \qquad (x_D \ll 1) \tag{8.19}$$

The high-temperature limit ($x_D \ll 1$) for the Debye heat-capacity function is obtained even more simply. The integrand of (8.7) is just x^2 times the Einstein function, so that from (4.14), the high-temperature limit of the integrand is

$$x^2[1 - \tfrac{1}{12}x^2 + \tfrac{1}{240}x^4]$$

Putting this in (8.7) and performing the integration gives

$$D(x_D) = 1 - \tfrac{1}{20}x_D{}^2 \qquad (x_D \ll 1) \tag{8.20}$$

where only the first two terms have been retained.

Substitution of (8.19) and (8.20) into (8.9) and (8.10) and using the definition (8.8) of x_D, gives the high-temperature limit for the energy and the heat capacity as functions of temperature.

$$U - E_0 = 3NkT \left[1 - \frac{3}{8}\frac{\theta_D}{T} + \frac{1}{20}\left(\frac{\theta_D}{T}\right)^2 \right] \qquad (T \gg \theta_D) \tag{8.21}$$

$$C_V = 3Nk \left[1 - \frac{1}{20}\left(\frac{\theta_D}{T}\right)^2 \right] \qquad (T \gg \theta_D) \tag{8.22}$$

These equations show that at high temperatures the Debye theory approaches the classical limit in which each oscillator has an energy kT and a heat capacity k.

The low-temperature approximation to the Debye energy function can be obtained by writing the integral in (8.6) as a difference of two integrals:

$$\int_0^{x_D} \frac{x^3\,dx}{e^x - 1} = \int_0^{\infty} \frac{x^3\,dx}{e^x - 1} - \int_{x_D}^{\infty} \frac{x^3\,dx}{e^x - 1} \tag{8.23}$$

The first term in this equation is a definite integral whose value is known to be $\pi^4/15$. For low T, x_D is large, so the second integral can be approximated by neglecting unity in the numerator of the integrand

$$\int_{x_D}^{\infty} \frac{x^3 \, dx}{e^x - 1} \simeq \int_{x_D}^{\infty} x^3 e^{-x} \, dx = e^{-x_D}(x_D^3 - 3x_D^2 - 6x_D - 6) \quad (8.24)$$

Putting these results in (8.23) and multiplying by $3/x_D^4$ gives the low-temperature approximation to (8.6).

$$D_E(x_D) = \frac{3}{x_D^4}\left[\frac{\pi^4}{15} - e^{-x_D}(x_D^3 - 3x_D^2 - 6x_D - 6)\right] \quad (x_D \gg 1) \quad (8.25)$$

A sufficiently accurate form of (8.25) is

$$D_E(x_D) = \frac{3}{x_D^4}\left[\frac{\pi^4}{15} - x_D^3 e^{-x_D}\right] \quad (x_D \gg 1) \quad (8.26)$$

The Debye heat capacity at low temperatures is computed in a similar fashion. The integral in (8.7) is written as

$$\int_0^{x_D} \frac{x^4 e^{-x}}{(1 - e^{-x})^2} \, dx = \int_0^{\infty} \frac{x^4 e^{-x}}{(1 - e^{-x})^2} \, dx - \int_{x_D}^{\infty} \frac{x^4 e^{-x}}{(1 - e^{-x})^2} \, dx \quad (8.27)$$

The first term is a definite integral whose value is $4\pi^4/15$. In the second integral e^{-x} can be neglected relative to unity in the denominator of the integrand because x is large. The integral is then easily evaluated. Equation 8.26 then becomes

$$\int_0^{x_D} \frac{x^4 e^{-x}}{(1 - e^{-x})^2} \, dx \simeq \frac{4\pi^4}{15} - x_D^4 e^{-x_D} \quad (8.28)$$

In this approximation, powers of x_D lower than the fourth were neglected in the preexponential term. Multiplying this by $3/x_D^3$ gives the low temperature Debye heat-capacity function

$$D(x_D) = \frac{4\pi^4}{5x_D^3} - 3x_D e^{-x_D} \quad (x_D \gg 1) \quad (8.29)$$

The low-temperature approximations to the energy and heat capacity are now readily obtained from (8.9) and (8.10) as

$$U - E_0 = 3NkT\left[\frac{\pi^4}{5}\left(\frac{T}{\theta_D}\right)^3 - 3e^{-\theta_D/T}\right] \quad (T \ll \theta_D) \quad (8.30)$$

$$C_V = 3Nk\left[\frac{4\pi^4}{5}\left(\frac{T}{\theta_D}\right)^3 - 3\left(\frac{\theta_D}{T}\right)e^{-\theta_D/T}\right] \quad (T \ll \theta_D) \quad (8.31)$$

In the limit of low temperature, the Debye theory predicts that the heat capacity at constant volume varies as the cube of the temperature according to (8.31).

The above results are, of course, valid only when the lattice vibrations are the sole agents responsible for the variation of energy with temperature. When other factors are present, such as phase transformations or "free" electrons in metals, additional contributions to the heat capacity must be considered.

2.9 RELATION BETWEEN EINSTEIN AND DEBYE CHARACTERISTIC TEMPERATURES

The Einstein model can be thought of as describing the vibrational crystal phenomena by replacing the various frequencies by a single average frequency. Given the Debye model, it is worthwhile to ask how an equivalent Einstein frequency can be computed from it. One possibility, of course, is simply to equate the Einstein frequency to the average frequency in the Debye model. This average is

$$\bar{\nu} = \frac{\int_0^{\nu_D} \nu g(\nu)\, d\nu}{\int_0^{\nu_D} g(\nu)\, d\nu} = \frac{3}{\nu_D^3} \int_0^{\nu_D} \nu^3\, d\nu \tag{9.1}$$

or

$$\bar{\nu} = \tfrac{3}{4}\nu_D$$

so that if the average of the Debye spectrum is identified with the Einstein frequency, we have

$$\left. \begin{aligned} \nu_E &= 0.75\nu_D \\ \theta_E &= 0.75\theta_D \end{aligned} \right\} \quad \text{(average)} \tag{9.2}$$

An alternative method of relating the Einstein and Debye frequencies is to require that the heat-capacity curves given by the two theories match in some range of temperature. Analysis of the Einstein heat capacity shows that at low temperatures it varies as $e^{-\theta_E/T}(\theta_E/T)^2$. Since the Debye heat capacity varies as $(T/\theta_D)^3$, no comparison can be made. At high temperatures, however, (4.17) shows that the largest term that contains any information about the frequency spectrum is essentially an average of the *square* of the frequency. That is, at high temperatures we have from (4.17)

$$C_V = 3Nk - \frac{Nk}{4}\left(\frac{h}{kT}\right)^2 \overline{\nu^2}(\text{high } T) \tag{9.3}$$

where

$$\overline{\nu^2} = \frac{\displaystyle\int_0^\infty \nu^2 g(\nu)\,d\nu}{\displaystyle\int_0^\infty g(\nu)\,d\nu}$$

or

$$\overline{\nu^2} = \tfrac{1}{3}\int_0^\infty \overline{\nu}^2 g(\nu)\,d\nu \tag{9.4}$$

Therefore, to make the Einstein and Debye theories match in the region of high temperature, it is only necessary to identify the Einstein frequency with the root mean square of the Debye spectrum.

$$\nu_E = (\overline{\nu^2})^{1/2} \tag{9.5}$$

but

$$\overline{\nu^2} = \frac{1}{3}\frac{9}{\nu_D{}^3}\int_0^\infty \nu^4\,d\nu = \tfrac{3}{5}\nu_D{}^2 \tag{9.6}$$

so

$$\left.\begin{aligned}\nu_E &= 0.775\nu_D\\[4pt]\theta_E &= 0.775\theta_D\end{aligned}\right\}\quad\text{(root mean square)} \tag{9.7}$$

Use of (9.7) gives a fairly precise match between the Einstein and Debye representation of heat-capacity data at high temperatures. For an overall

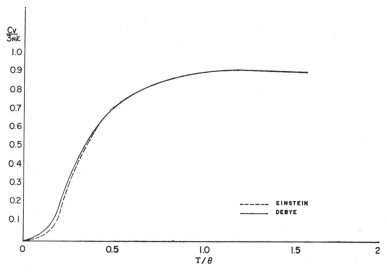

Fig. 2.4. Einstein and Debye heat-capacity curves for $\theta_E = 0.75\theta_D$.

fit at all temperatures, the straight average of (9.2) gives a better correspondence between the two theories. Figure 2.4 shows a comparison of the heat capacity per oscillator in the Einstein and Debye theories when $\theta_E = 0.75\theta_D$. It is evident that the two theories give significantly different results only below $T \simeq (1/2)\theta_D$.

2.10 COMPARISON OF DEBYE THEORY WITH EXPERIMENT

The most obvious way to compare the Debye theory to experiment is to try to fit heat-capacity data to (8.5), treating θ_D as a disposable parameter to be computed from the data. This has been done for a large variety of crystals with the result that a value of θ_D can always be chosen so that deviation of the heat-capacity data from that predicted from the Debye theory is quite small. We can therefore conclude that the Debye theory gives an excellent representation of the heat capacity. θ_D values calculated from heat-capacity data are given in Table 2.1.

Comparison with heat-capacity data, however, is certainly not the most sensitive way of testing the Debye theory. The Debye theory and the Einstein theory are based on extremely different frequency distribution models, yet we saw that the heat capacities predicted by the two theories are very much alike. We can only conclude that the heat capacity is rather insensitive to the form of the frequency-distribution function. This is comforting in that we can use the Debye theory with an assurance that is not really justified by the fundamental assumptions, but it is not much help if we want to investigate the validity of these assumptions.

Another way of testing the theory is to utilize the definition of the frequency-distribution function in the Debye model. From (7.17), (7.18), and (7.22), it is easy to see that

$$\theta_D = \frac{h}{k}\left(\frac{9N}{4\pi V}\right)^{1/3}\left[\frac{1}{C_l^3} + \frac{2}{C_t^3}\right]^{-1/3} \tag{10.1}$$

so that if the velocities of sound are known, θ_D can be computed. But the velocities of sound can either be measured directly or calculated from elastic data. This then gives us a method of calculating θ_D based directly on the core of the model, and this value can be compared with that computed from heat-capacity data. Such a comparison is given in Table 2.2, which shows that in many cases θ_D calculated from elastic data differs from that computed from heat-capacity data by 5% or less. Some crystals, especially anisotropic ones, show considerably larger deviations.

A third method of comparing Debye theory with experiment is to test the T^3 law at low temperatures. If Debye theory is correct, then at sufficiently

Table 2.1. Debye Characteristic Temperatures from Heat-Capacity Data

Element	θ_D K	Element	θ_D K
A	85	Mg	318
Ag	215	Mn	400
Aℓ	394	Mo	380
As	285	Na	150
Au	170	Ne	63
B	1250	Ni	375
Be	1000	Pb	88
Bi	120	Pd	275
C(diamond)	1860	Pr	74
Ca	230	Pt	230
Cd	120	Sb	200
Co	385	Si	625
Cr	460	Sn(gray)	260
Cu	315	Sn(white)	170
Fe	420	Ta	225
Ga	240	Th	100
Ge	360	Ti	380
Gd	152	Tℓ	96
Hg	100	V	390
In	129	W	310
K	100	Zn	234
Li	400	Zr	250
La	132		

From a compilation by J. De Launay, *Solid State Physics*, Vol. 2, F. Seitz and D. Turnbull, Eds., Academic Press, New York, p. 233.

Table 2.2. Comparison of Debye Temperatures Based on Heat Capacity and Velocities of Sound

Element	θ_D (Sound Velocity)	θ_D (Heat Capacity)
Al	399 °K	394 °K
Fe	467	420
Cu	329	315
Ag	212	215
Cd	168	120
Pt	226	230
Pb	72	88

Taken from a compilation by J. C. Slater, *Introduction to Chemical Physics*, McGraw-Hill Book Company, New York, 1939, p. 237.

low temperature a plot of C_V against T^3 should be linear, and the Debye temperature θ_D can be calculated from the slope. The evidence is very good that for simple solids the T^3 law actually holds if the temperatures are low enough. It was previously thought that $\theta_D/12$ was about the upper limit for which the T^3 law could hold with any accuracy. More recent work shows, however, that for a number of solids the temperature must be below $\theta_D/50$ for the T^3 law to be valid.

Another test can be made by using precision heat-capacity data to compute θ_D values at various temperatures from (8.5). If the Debye theory is correct, the θ_D values should be the same for all temperatures. Any deviation from a constant in a θ_D-T plot reflects an inaccuracy in the Debye theory. This method is a sensitive one and shows that for many crystals θ_D is not constant with temperature. However, deviations from constancy are generally smaller than 20% and often less than 10%. Crystals with the diamond and hexagonal structures show greater deviations than crystals with face-centered or body-centered cubic structures.

The general conclusion to be drawn from the comparison with experiment is that the Debye theory is extremely useful and astonishingly accurate considering its doubtful axiomatic ancestry. Detailed comparisons clearly show that the theory contains some defects that can be removed only by going to an actual lattice dynamics calculation of the frequency-distribution function.

2.11   EQUATION OF STATE

The equation of state is contained in the thermodynamic equation (8.38) of Chapter 1, that is,

$$P = -\left(\frac{\partial A}{\partial V}\right)_T \qquad (11.1)$$

All that is needed is to differentiate a statistical mechanical expression for the free energy. It was pointed out in Chapter 1, Section 8, that difficulties sometimes arise in developing a theory containing volume derivatives. In the present case, the troubles can be circumvented just by taking the frequencies to be functions of volume. This is a reasonable thing to do, because if a pressure is exerted on a crystal so that its volume decreases, the atoms get closer together. Equation 2.26 for the linear chain shows that the larger the force constants coupling the atoms, the greater the frequency. The force constants are derivatives of interatomic force functions, and since the slope of the force curve increases with decreasing distance, the frequency should be increased by a decrease in volume. Conversely, the frequency is decreased by an increase in volume. Differentiating (3.13) with respect to volume gives for the pressure

$$P = -\left(\frac{\partial E_0}{\partial V}\right)_T - kT \sum_{j=1}^{3N} \frac{e^{-h\nu_j/kT}}{(1 - e^{-h\nu_j/kT})}\left[\frac{h}{kT}\right]\frac{\partial \nu_j}{\partial V} \qquad (11.2)$$

The derivatives of the frequencies will continually appear in this discussion, but there is no easy method of obtaining them from first principles. To deal with the variation of frequency with volume, Gruneisen assumed that

$$\frac{d \ln \nu_j}{d \ln V} = -\gamma \qquad (11.3)$$

where γ is a positive constant, the same for all normal modes. Adopting (11.3) implies that the frequencies vary with volume according to the formula

$$\frac{\nu_j}{\nu_j{}^0} = \left(\frac{V_0}{V}\right)^{\gamma} \qquad (11.4)$$

ν_j being the frequency when the volume is V and $\nu_j{}^0$ being the frequency when the volume is V_0.

The Gruneisen assumption will be adopted here, and γ will be treated as a parameter to be determined by experiment. It follows directly from (11.3) that

$$\frac{d\nu_j}{dV} = -\frac{\gamma \nu_j}{V} \qquad (11.5)$$

$$\frac{d^2\nu_j}{dV^2} = (\gamma^2 + \gamma)\frac{\nu_j}{V^2} \qquad (11.6)$$

By means of these relations, it is not only possible to write statistical mechanical formulas for the volume-dependent properties, but also, several interesting connections among the various thermodynamic properties can be made. In making these connections, the energy and heat-capacity equations are needed. For present purposes, convert (4.6) and (4.7) to the discrete notation and rewrite them as

$$kT \sum_j \frac{h\nu_j}{kT} \frac{e^{-h\nu_j/kT}}{1 - e^{-h\nu_j/kT}} = U - E_0 \tag{11.7}$$

$$k \sum_j \left(\frac{h\nu_j}{kT}\right)^2 \frac{e^{-h\nu_j/kT}}{(1 - e^{-h\nu_j/kT})^2} = C_V \tag{11.8}$$

Now use (11.5) to rewrite (11.2) for the pressure as

$$P = -\left(\frac{\partial E_0}{\partial V}\right)_T + \gamma \frac{kT}{V} \sum_j \left(\frac{h\nu_j}{kT}\right) \frac{e^{-h\nu_j/kT}}{(1 - e^{-h\nu_j/kT})} \tag{11.9}$$

or, replacing the sum by $(U - E_0)$ according to (11.7).

$$P = -\left(\frac{\partial E_0}{\partial V}\right)_T + \frac{\gamma}{V}(U - E_0) \tag{11.10}$$

This is the Mie-Gruneisen equation of state. The pressure is displayed as a sum of two terms: the first is the negative derivative of E_0 and does not contain the temperature explicitly; the second is proportional to the thermal energy $(U - E_0)$, which is an explicit function of temperature.

The compressibility and the thermal expansion are readily obtained from the thermodynamic formulas given in Section 8 of Chapter 1, that is,

$$\frac{1}{\kappa} = V\left(\frac{\partial^2 A}{\partial V^2}\right)_T = -V\left(\frac{\partial P}{\partial V}\right)_T \tag{11.11}$$

$$\frac{\alpha}{\kappa} = -\left[\frac{\partial}{\partial V}\left(\frac{\partial A}{\partial T}\right)_V\right]_T \tag{11.12}$$

From (11.9) and (11.11)

$$\frac{1}{\kappa} = V\left(\frac{\partial^2 E_0}{\partial V^2}\right)_T + \frac{(\gamma^2 + \gamma)}{V} kT \sum_j \left(\frac{h\nu_j}{kT}\right) \frac{e^{-h\nu_j/kT}}{1 - e^{-h\nu_j/kT}}$$
$$- \frac{\gamma^2 kT}{V} \sum_j \left(\frac{h\nu_j}{kT}\right)^2 \frac{e^{h\nu_j/kT}}{(e^{h\nu_j/kT} - 1)^2} \tag{11.13}$$

where use has been made of (11.5) and (11.6). The form of (11.13) is considerably simplified if (11.7) and (11.8) are used, that is,

$$\frac{1}{\kappa} = V \left(\frac{\partial^2 E_0}{\partial V^2} \right)_T + \frac{(\gamma^2 + \gamma)}{V} (U - E_0) - \frac{\gamma^2 C_V T}{V} \tag{11.14}$$

Differentiating the free-energy equation (3.12) first with respect to temperature and then with respect to volume, we get from (11.12),

$$\frac{\alpha}{\kappa} = \frac{k\gamma}{V} \sum_j \left(\frac{h\nu_j}{kT} \right)^2 \frac{e^{-h\nu_j/kT}}{(1 - e^{-h\nu_j/kT})^2} \tag{11.15}$$

where the volume derivatives of the frequencies have been reduced by the Gruneisen assumption. Combining (11.15) with (11.8)

$$\frac{\alpha}{\kappa} = \frac{\gamma C_V}{V} \tag{11.16}$$

This is known as Gruneisen's equation. It can be used to compute γ, since all other quantities in it are measurable. Values obtained in this way are given in Table 2.3.

Equations 11.10, 11.14, and 11.16 show that the introduction of Gruneisen's constant allows the equation of state and its accompanying parameters (κ and α) to be expressed in terms of thermal properties; that is, the heat

Table 2.3. Gruneisen Constants for Some Elements

Element	γ
Si	0.44
Ge	0.72
Cu	2.00
Ag	2.4
Au	3.0
Na	1.14
Al	2.34

From a compilation by W. B. Daniels, *Lattice Dynamics*, Proceedings of Copenhagen Conference, August 5–9, 1963, R. F. Wallis, Ed., Pergamon Press, Elmsford, N.Y.

capacity C_V and the thermal energy $(U - E_0)$. The theoretical development for these thermal properties can therefore be carried over bodily into equation of state theory. In terms of the Debye theory, the heat capacity and thermal energy are given by (8.10) and (8.9), respectively. Therefore, in the Debye theory, the equation-of-state relations given by (11.10), (11.14), and (11.16) become

$$P = -\left(\frac{\partial E_0}{\partial V}\right)_T + \frac{3Nk\theta_D\gamma}{V} D_E(x_D) \tag{11.17}$$

$$\frac{1}{\kappa} = V\left(\frac{\partial^2 E_0}{\partial V^2}\right)_T + \frac{(\gamma^2 + \gamma)}{V} 3Nk\theta_D D_E(x_D) - \frac{3NkT\gamma^2}{V} D(x_D) \tag{11.18}$$

$$\frac{\alpha}{\kappa} = \frac{3Nk\gamma}{V} D(x_D) \tag{11.19}$$

The limiting cases for high and low temperatures are readily obtained using (8.18), (8.19), (8.25), and (8.28) in the above. The results are, for $T \gg \theta_D$,

$$P = -\left(\frac{\partial E_0}{\partial V}\right)_T + \frac{3NkT\gamma}{V}\left[1 - 3\frac{\theta_D}{T} + \frac{1}{20}\left(\frac{\theta_D}{T}\right)^2\right] \tag{11.20}$$

$$\frac{1}{\kappa} = V\left(\frac{\partial^2 E_0}{\partial V^2}\right)_T + \frac{3NkT\gamma}{V}\left[1 - 3\frac{(\gamma + 1)}{8}\frac{\theta_D}{T} + \frac{(2\gamma + 1)}{20}\left(\frac{\theta_D}{T}\right)^2\right] \tag{11.21}$$

$$\frac{\alpha}{\kappa} = \frac{3Nk\gamma}{V}\left[1 - \frac{1}{20}\left(\frac{\theta_D}{T}\right)^2\right] \tag{11.22}$$

while for $T \ll \theta_D$

$$P = -\left(\frac{\partial E_0}{\partial V}\right)_T + \frac{3Nk\theta_D\gamma}{V}\left(\frac{T}{\theta_D}\right)^4\left[\frac{\pi^4}{5} - 3\left(\frac{\theta_D}{T}\right)^3 e^{-\theta_D/T}\right] \tag{11.23}$$

$$\frac{1}{\kappa} = V\left(\frac{\partial^2 E_0}{\partial V^2}\right)_T - \frac{3Nk\theta_D\gamma}{V}\left(\frac{T}{\theta_D}\right)^4\left[\frac{(3\gamma - 1)\pi^4}{5} - \frac{\theta_D^5}{5} e^{-\theta_D/T}\right] \tag{11.24}$$

$$\frac{\alpha}{\kappa} = \frac{3Nk\gamma}{V}\left(\frac{T}{\theta_D}\right)^3\left[\frac{4\pi^4}{5} - 3\left(\frac{\theta_D}{T}\right)^4 e^{-\theta_D/T}\right] \tag{11.25}$$

To discuss the implications of these results, it is necessary to make some comment on E_0 and its derivatives. Let us recall that, from (3.5), E_0 is given by

$$E_0 = U_0 + \sum_{j=1}^{3N} \tfrac{1}{2}h\nu_j \tag{11.26}$$

where U_0 is the potential energy of the crystal when all atoms are at rest at their mean positions. The second term is the zero point vibrational energy, and is readily evaluated in the Debye theory by converting the sum to an integral and using the Debye distribution function.

$$\sum_{j=1}^{3N} \tfrac{1}{2}h\nu_j \rightarrow \frac{9\nu_D^4}{2}N \int_0^{\nu_D} \nu^3 \, d\nu = \tfrac{9}{8}Nh\nu_D = \tfrac{9}{8}Nk\theta_D \qquad (11.27)$$

Thus in the Debye theory, we have

$$E_0 = U_0 + \tfrac{9}{8}Nk\theta_D \qquad (11.28)$$

Since the Gruneisen assumption applies to ν_D as well as to all other vibration frequencies, and since ν_D is proportional to the Debye temperature, we have from (11.5) and (11.6)

$$\frac{d\theta_D}{dV} = -\frac{\gamma\theta_D}{V} \qquad (11.29)$$

$$\frac{d^2\theta_D}{dV^2} = (\gamma^2 + \gamma)\frac{\theta_D}{V^2} \qquad (11.30)$$

Therefore, differentiating (11.28) gives

$$\frac{\partial E_0}{\partial V} = \frac{\partial U_0}{\partial V} - \frac{9Nk\gamma\theta_D}{8V} \qquad (11.31)$$

$$\frac{\partial^2 E_0}{\partial V^2} = \frac{\partial^2 U_0}{\partial V^2} + \frac{9N(\gamma^2 + \gamma)k\theta_D}{8V^2} \qquad (11.32)$$

Equation 11.32 shows that the contribution of the vibrational zero point energy to the derivatives of E_0 that appear in the equations for the pressure and the bulk modulus vary with volume. From (11.29), $\theta_D \propto V^{-\gamma}$, so the last term in (11.32) varies as $V^{-2-\gamma}$. But the volume increases with temperature because of thermal expansion and, therefore, the zero point energy contributes a term to the bulk modulus that increases with temperature.

The derivatives of U_0 with respect to volume are also functions of temperature because the potential energy of the crystal changes as the mean atomic positions change relative to each other, and the volume increases with temperature. The first term in (11.32) also contributes to the temperature dependence of the bulk modulus.

If we start at absolute zero, the potential energy is a minimum and, as we see from (11.25), the thermal-expansion coefficient vanishes. At low temperatures, therefore, U_0 and its derivatives cannot depend very strongly on temperature. Also, since $\alpha \rightarrow 0$ as $T \rightarrow 0$, the second term in (11.32) has only a weak temperature dependence at low temperatures. As $T \rightarrow 0$, therefore, the temperature dependence of the bulk modulus should be given primarily by the T^4 factor in the second term of (11.24).

At high temperatures, however, the thermal-expansion coefficient is appreciable. Therefore, U_0 and its derivatives will vary with temperature, and the temperature dependence of the bulk modulus is contained in both of the terms in (11.21). In this equation, we see that the thermal-energy contribution (second term) to the bulk modulus increases with temperature. Since the atoms get further apart as the temperature increases, we would expect that the contribution of the potential energy to the bulk modulus would decrease with temperature. That is, as the temperature goes up, the vibrations make the crystal harder, while the spreading apart of the atoms makes it softer. In fact, experiment shows that the latter effect predominates and the bulk modulus continually decreases with increasing temperature.

At temperatures so low that the temperature dependence of the first term in (11.24) can be neglected, and only the T^4 term in the thermal energy contribution need be retained, we have

$$\frac{1}{\kappa} = \left(\frac{1}{\kappa}\right)_0 - \frac{9Nk\theta_D\gamma(\gamma - 1)\pi^4}{5V}\left(\frac{T}{\theta_D}\right)^4 \tag{11.33}$$

where

$$\left(\frac{1}{\kappa}\right)_0 \equiv V\left(\frac{\partial^2 E_0}{\partial V^2}\right)_{T=0} \tag{11.34}$$

is the bulk modulus at absolute zero. The T^4 dependence displayed in (11.33) holds in a semiquantitative way for many materials, the validity of this result increasing with decreasing temperature.

2.12 HEAT CAPACITY AT CONSTANT PRESSURE

The canonical ensemble leads to the Helmholtz free energy as the thermodynamic potential most naturally related to the energy levels of the crystal. This means that the theory gives the heat capacity at constant volume rather than constant pressure. But experimentally, it is the constant-pressure heat capacity that is measured. From thermodynamics, the relation between the constant pressure and constant-volume heat capacities is

$$C_P = C_V + TV\frac{\alpha^2}{\kappa} \tag{12.1}$$

Substitution of (11.16) into this relation gives

$$C_P = C_V(1 + \gamma\alpha T) \tag{12.2}$$

As shown in Section 11, the thermal-expansion coefficient α vanishes at low temperatures and increases with increasing temperature. At low temperatures, therefore, C_P and C_V are nearly equal. Even at high temperatures, the difference is not large. If we take $\gamma = 2$ and $\alpha = 10^{-5}$ as typical values, then at $1000°K$, $\gamma \alpha T = 2 \times 10^{-2}$. The two heat capacities therefore differ by only several percent.

For a harmonic crystal, $\gamma = 0$ and $C_P = C_V$. The difference between the two heat capacities is therefore an anharmonic effect along with the thermal expansion, the temperature dependence of the elastic constants and the temperature dependence of the equation of state.

2.13 DEBYE THEORY AND THE GRUNEISEN ASSUMPTION

In the Debye theory, the ν_j are possible sound wave frequencies and they depend on the volume according to equations similar to (7.13). If we stay in the discrete notation and solve for ν_j, this equation gives

$$\nu_j = 2jCV^{-1/3} \tag{13.1}$$

where $C = C_l$ for longitudinal waves and $C = C_t$ for transverse waves. This can be rewritten as

$$\ln \nu_j = \ln C - \tfrac{1}{3} \ln V + \ln (2j) \tag{13.2}$$

so that

$$\frac{d \ln \nu_j}{d \ln V} = -\frac{1}{3} + \frac{d \ln C}{d \ln V} \tag{13.3}$$

From the theory of elasticity, it is known that the transverse and longitudinal velocities of sound in an isotropic body are related to the elastic properties by

$$C_t = \sqrt{\frac{\mu}{\rho}} \tag{13.4}$$

$$C_l = \sqrt{\frac{\lambda + 2\mu}{\rho}} \tag{13.5}$$

where ρ is the density and μ and λ are the Lame constants. The Lame constants are related to the compressibility κ and to Poisson's ratio σ by the equations

$$\frac{1}{\kappa} = \lambda + \tfrac{3}{2}\mu \tag{13.6}$$

$$\sigma = \frac{\lambda}{2(\lambda + \mu)} \tag{13.7}$$

It is clear from (13.3) and the existence of two sound velocities, that a Debye solid requires two Gruneisen constants given by

$$\gamma_t = \frac{1}{3} - \frac{d \ln C_t}{d \ln V} \tag{13.8}$$

$$\gamma_l = \frac{1}{3} - \frac{d \ln C_l}{d \ln V} \tag{13.9}$$

Using (13.4) and (13.5), and the fact that the density ρ is inversely proportional to the volume, (13.8) and (13.9) give

$$\gamma_t = -\frac{1}{2} \frac{d \ln \mu}{d \ln V} - \frac{1}{6} \tag{13.10}$$

$$\gamma_l = -\frac{1}{2} \frac{d \ln (\lambda + 2\mu)}{d \ln V} - \frac{1}{6} \tag{13.11}$$

Strictly speaking, therefore, the Debye theory requires that two Gruneisen constants be introduced in the theory of the equation of state, rather than just one. If this were done, for example, in the calculation of the pressure, instead of (11.9), the result would be

$$P = -\frac{\partial E_0}{\partial V} + \frac{kT\gamma_l}{V} \sum_l \left(\frac{h\nu_j}{kT}\right) \frac{1}{(e^{h\nu_j/kT} - 1)}$$
$$+ \frac{kT\gamma_t}{V} \sum_t \left(\frac{h\nu_j}{kT}\right) \frac{1}{(e^{h\nu_j/kT} - 1)} \tag{13.12}$$

where the first sum on the right is over all longitudinal modes and the second sum is over all transverse modes. Following this procedure would destroy the simplicity of the Mie-Gruneisen theory and a more complicated analysis would be necessary.

In the continuum notation, (3.13) for the free energy is

$$A = E_0 + 3NkT \int_0^\infty g(\nu) \ln (1 - e^{-h\nu/kT}) \, d\nu \tag{13.13}$$

and upon introducing the Debye distribution function

$$A = E_0 + \frac{27NkT}{\nu_D{}^3} \int_0^{\nu_D} \nu^2 \ln (1 - e^{-h\nu/kT}) \, d\nu \tag{13.14}$$

This equation gives the free energy in terms of only one volume dependent frequency ν_D. It appears, therefore, that only one Gruneisen constant is needed in the Debye theory. But this is a direct result of assuming that only a single cutoff frequency ν_D is needed for both longitudinal and transverse modes. If two such frequencies were introduced, one for each type of mode, then two Gruneisen constants would be required as shown above.

A single Gruneisen constant is generally adequate for the description of equation-of-state data to a fair degree of accuracy.

2.14   VIBRATIONAL ANHARMONICITY

If the harmonic assumption were correct, then the thermal expansion coefficient would be zero and the compressibility would be independent of temperature. This can be seen by looking at (2.26), which gives the normal mode frequencies in terms of the force constants $C_{l'l}$. For a harmonic crystal, in which the potential energy is truly a quadratic function of the atomic displacements, the force constants are independent of the atomic positions and therefore independent of volume. Equation 2.26 shows that in this case the frequencies are also independent of volume and therefore Gruneisen's constant is zero. It is evident that thermal expansion and temperature variation of the elastic constants are anharmonic effects; that is, they must depend on higher-order terms in the potential energy that have been neglected in writing (2.1).

By introducing a nonzero Gruneisen constant in the equation-of-state theory, we have in fact violated the harmonic assumption and introduced the anharmonic effects in an arbitrary way. The relationship between the Gruneisen method and anharmonicity can be found if we start with an anharmonic potential for the monatomic chain:

$$V = \tfrac{1}{2} \sum_{i,j} C_{ij} u_i u_j + \tfrac{1}{6} \sum_{i,j,k} C_{ijk} u_i u_j u_k + \tfrac{1}{24} \sum_{i,j,k,l} C_{ijkl} u_i u_j u_k u_l \qquad (14.1)$$

where u_i are the displacements from the mean atomic positions. The last two terms are anharmonic terms that were neglected in (2.1). The constants C_{ijk} and C_{ijkl} are third- and fourth-order derivatives evaluated at $u_i = 0$.

From (14.1)

$$\frac{\partial^2 V}{\partial u_i \partial u_j} = C_{ij} + \sum_k C_{ijk} u_k + \sum_{k,l} C_{ijkl} u_k u_l \qquad (14.2)$$

That is, when anharmonicity is included, the second derivatives of the potential are not constants, but depend on atomic displacements and therefore on the length of the chain. (For the three-dimensional case, the second derivatives would be functions of volume.) Now what if we want to retain the simplicity of the harmonic assumption and write the potential energy as a quadratic form even if we know it is not entirely accurate? Equation 14.2 shows that it would be more accurate to take the second-order force "constants" to be functions of length (volume in the three-dimensional case) than to treat them as bona-fide constants. This is precisely what is done in the Gruneisen theory. It amounts to shifting the minimum of potential energy as the volume changes.

This discussion also shows that the Gruneisen method only accounts for part of the effect of anharmonicity, since it does not truly reflect the existence of the anharmonic terms, but tries to make volume-dependent second-order derivatives wholly responsible for higher-order terms. It is therefore not surprising that the Gruneisen theory only gives a semiquantitive description of the volume-dependent properties of crystals.

2.15 THE PHONON GAS

In the harmonic approximation, the statistical thermodynamic functions are written in terms of simple harmonic oscillator frequencies. This is possible because a coordinate transformation exists that transforms the atomic coordinates to normal mode coordinates, thereby converting the dynamical equations of motion to a set of equations each involving only one normal mode. This procedure in effect reduces the many-body problem for the crystal to a set of one-body problems.

The equivalence of the harmonic approximation to an independent particle model can be shown by considering the statistics of a collection of particles with the following properties:

1. Each particle can exist in any of a set of energy levels $\epsilon_j = h\nu_j$.

2. There is no restriction on the number of particles that can occupy a given level.

3. There is no restriction on the total number of particles in the system; that is, particles are not conserved.

If the ν_j are the possible frequencies of electromagnetic radiation in a heated cavity, then the particles are photons, and the theory of blackbody radiation can be developed from these three properties. If the ν_j are the possible frequencies of a vibrating solid, then the "particles" are called phonons and, as we shall see, the statistical thermodynamics worked out in the previous sections of this chapter is equivalent to the theory of the phonon gas.

The second property listed above requires that phonons be described by Bose-Einstein statistics. Because of the third property, however, the probability function is not quite the same as for material Bose-Einstein particles. Phonons can be created and destroyed, so that in solving the variational problem for the statistical count, we no longer have the restriction $\delta N = 0$ (see Chapter 1, Section 13). The Lagrangian multiplier, a, therefore does not appear in the distribution function, and for phonons $\mu = 0$. Instead of

(14.11 Chapter 1), we therefore have for the number of phonons with energy in the range $\epsilon_j \pm \Delta\epsilon$,

$$N_j = \frac{\omega_j}{e^{\epsilon_j/kT} - 1} \tag{15.1}$$

or, in continuum language,

$$N(\epsilon)\, d\epsilon = \frac{\omega(\epsilon)\, d\epsilon}{e^{\epsilon/kT} - 1} \tag{15.2}$$

To write this in terms of the frequencies, we recall that $Ng(\nu)\, d\nu$ is the number of modes in a frequency range ν to $\nu + d\nu$, which we now identify with the number of phonon states in the same frequency range. This amounts to choosing a density of states for phonons that agrees with the frequency distribution in crystals. Equation 15.2 can now be written as

$$N(\nu)\, d\nu = \frac{Ng(\nu)\, d\nu}{e^{h\nu/kT} - 1} \tag{15.3}$$

where $N(\nu)\, d\nu$ is the number of phonons with frequencies in the range ν to $\nu + d\nu$.

The energy of the system of phonons is now readily obtained:

$$U_p = \int_0^\infty h\nu N(\nu)\, d\nu = N \int_0^\infty \frac{g(\nu)h\nu\, d\nu}{e^{h\nu/kT} - 1} \tag{15.4}$$

The heat capacity at constant volume of the phonons is

$$C_V = \left(\frac{\partial U_p}{\partial T}\right)_V = Nk \int_0^\infty g(\nu) \left(\frac{h\nu}{kT}\right)^2 \frac{e^{h\nu/kT}}{(e^{h\nu/kT} - 1)^2}\, d\nu \tag{15.5}$$

The entropy of phonons is given by

$$S = \int_0^T \frac{C_V}{T}\, dT = Nk \int_0^\infty g(\nu) \int_0^T \frac{1}{T}\left(\frac{h\nu}{kT}\right)^2 \frac{e^{h\nu/kT}}{(e^{h\nu/kT} - 1)^2}\, dT\, d\nu \tag{15.6}$$

The inner integral can be evaluated by means of the transformation

$$y \equiv e^{h\nu/kT} \tag{15.7}$$

which gives

$$\int_0^T \frac{1}{T}\left(\frac{h\nu}{kT}\right)^2 \frac{e^{h\nu/kT}}{(e^{h\nu/kT} - 1)^2}\, dT = -\int_\infty^{e^{h\nu/kT}} \frac{\ln y}{(y - 1)^2}\, dy \tag{15.8}$$

To compute the integral on the right, integrate by parts. The result is

$$\int_0^T \frac{1}{T}\left(\frac{h\nu}{kT}\right)^2 \frac{e^{h\nu/kT}}{(e^{h\nu/kT} - 1)^2}\, dT = \left.\frac{\ln y}{(y - 1)}\right|_\infty^{e^{h\nu/kT}} - \int_\infty^{e^{h\nu/kT}} \frac{dy}{y(y - 1)}$$

$$= \frac{h\nu}{kT}(e^{h\nu/kT} - 1)^{-1} - \ln(1 - e^{-h\nu/kT}) \tag{15.9}$$

Substituting this into (15.6) gives the entropy

$$S = \frac{N}{T} \int_0^\infty \frac{g(\nu)h\nu}{e^{h\nu/kT} - 1} \, d\nu - Nk \int_0^\infty g(\nu) \ln (1 - e^{-h\nu/kT}) \, d\nu \quad (15.10)$$

Since the Helmholtz free energy is $A = U - TS$, multiplying (15.10) by T and subtracting it from (15.4) give the phonon free energy as

$$A_p = NkT \int_0^\infty g(\nu) \ln (1 - e^{-h\nu/kT}) \, d\nu \quad (15.11)$$

Now compare (15.4), (15.5), and (15.11) to (4.6), (4.7), and (3.15). The two sets of equations are identical except that equations (4.6) and (3.15) contain a zero point energy term E_0. This term would have appeared in the phonon analysis also, if we had defined the phonons as moving in a constant potential U_0, rather than in a zero potential, and if $h\nu_j/2$ were added to the definition of ϵ_j.

The above analysis shows that as far as the vibrations are concerned, the crystal can be treated as a phonon gas. This concept is widely used in the study of solids.

chapter 3

FREE-ELECTRON THEORY

3.1 FREE ELECTRONS IN METALS

A metal differs from other types of solids in that its properties, such as high electrical and thermal conductivity, imply that it contains a number of highly mobile electrons. In fact, a wealth of experimental data and theoretical analysis strongly suggest that metals contain electrons that are not bound to any particular atom but belong to the crystal as a whole. In the language of quantum theory, this means that the wave function of a given electron is spread out throughout the crystal rather than being localized near a particular atom. In particle language, the electrons are free to wander about the entire crystal rather than being confined to a restricted, localized orbit.

One of the earliest successful approximations to the theory of metals is the free-electron theory. In this approximation, the metal is taken to be a collection of ion cores, each core consisting of a nucleus and a shell of tightly bound electrons, and a number of electrons that are essentially independent. These electrons are treated as being completely free, the metal acting simply as a box to contain them.

We want to develop the Fermi-Dirac theory of particles because of its application to the theory of electrons in metals and, therefore, we have a right to ask some questions about the validity of the free-electron approxi-mation. For example, we want to know how many of the electrons in a metal are to be considered free and how many are tied up in ion cores. Usually, the number of valence electrons is taken to be the number of free electrons per atom. This whole question is not quite as important as it sounds, because when we come to apply our theory to actual metals, we can take the density of electrons to be a parameter determined by experiment.

A much more basic question is involved in trying to understand how it is that the free-electron theory of metals is successful at all. We know that none of the electrons in a metal can be independent. They interact with each other and with the ion cores through strong Coulomb forces. However, there are several factors that permit the valence electrons to be treated as free. First, on the average over a large volume, the ionic charges just cancel the electronic charges. The electrons move, so to speak, in a background of positive charge that tends to neutralize them, so that in the free-electron model, we think of the three-dimensional periodic potential of the ion cores to be smoothed out to some average value. Second, if we solve the electrostatic problem of a charged electron in a "sea" of mobile charged particles, we find that the electrical potential of an electron is screened by the other electrons. This has the effect of greatly reducing the range in which the interaction between two particles is appreciable. All of this accounts for the fact that to a first approximation the electron-electron interactions can be neglected.

Finally, we have a remarkable result from quantum mechanics that states that a perfectly periodic potential offers no resistance to electron motion. The actual resistivity observed in metals arises from imperfections in the periodic structure rather than from electron-ion core interactions. These imperfections include atomic vibrations, vacant sites, dislocations, impurity atoms, grain boundaries or anything else that interrupts the regular crystalline array of ion cores. This does not mean that the regular ion-core structure has no influence at all on the electron motion. In fact it does, but this influence manifests itself in the inertial properties of the electron. It is often sufficient to assign some fictitious "effective mass" m^* to the electron rather than the actual electronic mass m. The electrons then move through the crystal just like free Fermi-Dirac particles with mass m^*.

Therefore, there is some justification in theory and certainly in practice for the applicability of free-electron theory to metals. The theory works best for the alkali metals and reasonably well for the noble metals. It is interesting to note that in the case of the alkali metals, the electrons-in-a-box model is not far from the truth. A calculation based on experimental values of ionic and metallic radii shows that the ion cores in sodium take up only a bit more than one-tenth of the volume of the metal. Nine-tenths of it is empty except for the free electrons.

3.2 CHOICE OF STATISTICS FOR THE ELECTRON GAS

Our first job is to show that the semiclassical statistics are not adequate for free electrons in metals, and that quantum statistics must be used. It is clear from our previous results, (14.10 and 14.13, Chapter 1) that the quantum statistics reduce to the semiclassical statistics if

$$e^{(\epsilon_j-\mu)/kT} \gg 1 \tag{2.1}$$

for all ϵ_j. We can write this inequality as $e^{-\mu/kT} \gg e^{-\epsilon_j/kT}$, and since ϵ_j is always positive, we can only strengthen the condition (2.1) for the applicability of semiclassical statistics if we write

$$e^{-\mu/kT} \gg 1 \tag{2.2}$$

The theory of the ideal gas has been worked out already on the assumption that (2.2) is correct. Also, we obtained a formula for μ in terms of particle mass, particle density, and temperature (15.25, Chapter 1). This formula gives

$$e^{-\mu/kT} = \left(\frac{2\pi mkT}{h^2}\right)^{3/2} \frac{V}{N} \tag{2.3}$$

so that we can test the validity of (2.2) for various types of particles. If (2.2) is indeed fulfilled, the use of the semiclassical statistics is justified. If we find otherwise, quantum statistics are necessary.

From (2.3), we see that high-particle mass, high temperature, and low-particle density contribute to the accuracy of the semiclassical formula. Let us evaluate (2.3) for a collection of hydrogen atoms. Inserting the values of k, h, and m, we have

$$e^{-\mu/kT} = 1.89 \times 10^{20} T^{3/2} \frac{V}{N} \tag{2.4}$$

for hydrogen atoms. Let us assume that the particle density is $N/V = 1.89 \times 10^{20}$ cm^{-3}. This corresponds to a pressure of about seven atmospheres at 0°C. Then (2.4) becomes $e^{-\mu/kT} = T^{3/2}$, and for a temperature as low as 10°K, we have $e^{-\mu/kT} \simeq 31.6$. This means that even at quite low temperatures, a collection of hydrogen atoms is described rather accurately by semiclassical statistics. Quantum statistics are not needed for collections of atoms or molecules, except at extremely low temperatures.

On the other hand, a similar calculation for electrons gives the opposite result. If we use the electronic mass in (2.3), we get

$$e^{-\mu/kT} = 2.42 \times 10^{15} T^{3/2} \frac{V}{N} \tag{2.5}$$

Let us take $N/V = 2.42 \times 10^{22}$ cm^{-3}. This is roughly the magnitude of the density of free electrons in metals. Therefore

$$e^{-\mu/kT} = 10^{-7} T^{3/2} \tag{2.6}$$

and we see that $e^{-\mu/kT} \ll 1$ at all but very high temperatures. In fact even at $T = 10,000$°K, $e^{-\mu/kT} = 0.1$, so that for all temperatures of actual interest, it is necessary to use quantum statistics for electrons in metals; we actually have the opposite condition to that given by (2.2).

In semiconductors, the density of free electrons ranges from 10^9 to 10^{17} cm^{-3}, so that in these systems, the semiclassical statistics are often valid.

3.3   THE DISTRIBUTION OF FREE ELECTRONS

We will start our discussion of the statistics of free electrons by taking advantage of the fact that the possible energy levels are very close together. Equation 14.10 of Chapter 1 can then be written in the continuum form

$$N(\epsilon) = \frac{\omega(\epsilon)}{e^{(\epsilon-\mu)/kT} + 1} \tag{3.1}$$

where we now interpret $N(\epsilon)$ to be a function such that $N(\epsilon)\,d\epsilon$ is the number of electrons that have an energy in the range ϵ to $\epsilon + d\epsilon$. The energy levels are those of a particle-in-a-box and are given by (15.6, Chapter 1), just as in the case of an ideal gas. The density of states $\omega(\epsilon)$, which determines the number of states in an energy range $d\epsilon$, is therefore given by a formula just like (15.14, Chapter 1), except for one minor modification. Because every electron can exist in two possible states of spin, there are two electrons that can have an energy determined by a particular set of integers j_1, j_2, and j_3, rather than just one. For electrons, therefore, the density-of-states function is just twice the value for an ideal gas, and we have

$$\omega(\epsilon) = \frac{8\pi\sqrt{2}}{h^3}\, m^{3/2} V \sqrt{\epsilon} \tag{3.2}$$

or, using the constant C defined in Chapter 1 (15.16)

$$\omega(\epsilon) = CV\sqrt{\epsilon} \tag{3.3}$$

Using this equation, let us write (3.1) as

$$N(\epsilon) = CV\sqrt{\epsilon} f(\epsilon) \tag{3.4}$$

where $f(\epsilon)$, defined by

$$f(\epsilon) = \frac{1}{e^{(\epsilon-\mu)/kT} + 1} \tag{3.5}$$

is called the Fermi function.

The properties of the Fermi function are of basic importance to subsequent discussions, and should be thoroughly understood. The Fermi function has the physical significance that it gives the probability that a particular state of energy ϵ is occupied by an electron, because when it is multiplied by the number of states with given energy, it yields the number of electrons with that energy.

From our previous calculations, (2.5) and (2.6), we know that μ is positive and much greater than kT for ordinary temperatures, and $f(\epsilon) < 1$ since the exponential is always positive. The form of the function is such that it is practically unity at low energies and decreases to zero as ϵ approaches the maximum electron energy. How wide is the energy range over which $f(\epsilon)$ differs significantly from zero or unity, and how rapidly does $f(\epsilon)$ drop from unity to zero over this range? If $\epsilon = \mu$, then $f(\epsilon) = f(\mu) = \frac{1}{2}$; if $\epsilon < \mu$, then the greater the difference $(\mu - \epsilon)$, the closer $f(\epsilon)$ is to unity. For an energy difference as small as $(\mu - \epsilon) = 3kT$, (3.5) tells us that $f(\epsilon)$ differs from unity by only about 5%. If $\epsilon > \mu$, then the greater the difference $(\epsilon - \mu)$, the closer $f(\epsilon)$ is to zero. For $(\epsilon - \mu) = 3kT$, we find from (3.5) that $f(\epsilon)$ differs from zero by only about 5%. We therefore conclude that the width of the energy range in which $f(\epsilon)$ differs significantly from zero or unity is five or six times the thermal energy kT and that the Fermi function has the shape shown in Figure 3.1. The lower the temperature, the more rapid is the drop of $f(\epsilon)$ from unity to zero, and in the limit of zero temperature, $f(\epsilon)$ is a step function whose value is unity for $0 \leq \epsilon \leq \mu$ and zero for $\epsilon > \mu$, as shown in the figure.

In analyzing the Fermi function, we have incidentally arrived at a physical interpretation of the chemical potential μ for free electrons. It is the energy of that state for which the probability of being occupied by an electron is equal to one-half. In solid state work, μ is usually called the Fermi energy. Because $\mu \gg kT$ for ordinary temperatures, the width of the transition

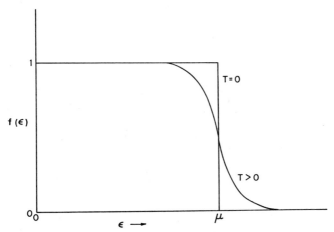

Fig. 3.1. The Fermi function $f = [e^{(\epsilon-\mu)/kT} + 1]^{-1}$.

region from $f(\epsilon) \simeq 1$ to $f(\epsilon) \simeq 0$ is very narrow compared to the Fermi energy of the electrons. Actually, at room temperature $kT \simeq (\frac{1}{40})$ eV while μ as determined from experiment for metals is of the order of 2 to 6 eV. This means that the Fermi function is almost a step function at temperatures that are not too high. That is, room temperature is a low temperature for electrons, and low-temperature approximations are adequate for the free-electron theory in metals. In the limit as $T \to 0$, $f(\epsilon) \to 1$ for $\epsilon < \mu$ and $f(\epsilon) \to 0$ for $\epsilon > \mu$, so (3.4) becomes

$$
\begin{aligned}
N(\epsilon) &= CV\sqrt{\epsilon} \qquad \epsilon < \mu_0 \\
&= 0 \qquad\qquad \epsilon > \mu_0
\end{aligned}
\tag{3.6}
$$

where μ_0 has been written for the Fermi energy at absolute zero. An expression for μ_0 can be obtained by integrating (3.6) from 0 to μ_0 since μ_0 is the maximum energy an electron can have at $T = 0$;

$$
\int_0^{\mu_0} N(\epsilon)\, d\epsilon = CV \int_0^{\mu_0} \epsilon^{1/2}\, d\epsilon
\tag{3.7}
$$

The left-hand side of this equation is just the total number of electrons N. Performing the integration on the right-hand side and solving for μ_0 gives

$$
\mu_0 = \left(\frac{3N}{2CV}\right)^{2/3}
\tag{3.8}
$$

or using the definition of C

$$
C = \frac{8\pi\sqrt{2}}{h^3} m^{3/2}
\tag{3.9}
$$

(3.8) becomes

$$
\mu_0 = \frac{h^2}{8m}\left(\frac{3N}{\pi V}\right)^{2/3}
\tag{3.10}
$$

The magnitude of the Fermi energy to be expected for electrons in metals can be obtained from this equation by inserting values for h, m, and N/V. It turns out to be several electron volts.

The average electron energy is also easily computed. From the definition of the average

$$
\bar{\epsilon} = \frac{1}{N} \int \epsilon N(\epsilon)\, d\epsilon
\tag{3.11}
$$

and since at absolute zero $N(\epsilon) = CV\sqrt{\epsilon}$, we have for $\bar{\epsilon}_0$, the average energy at absolute zero:

$$
\bar{\epsilon}_0 = \frac{CV}{N} \int_0^{\mu_0} \epsilon^{3/2}\, d\epsilon
\tag{3.12}
$$

Integration gives

$$\bar{\epsilon}_0 = \frac{2}{5}\frac{CV}{N}\mu_0^{5/2} \tag{3.13}$$

Solving (3.8) for CV/N and substituting the result in (3.13) we get

$$\bar{\epsilon}_0 = \tfrac{3}{5}\mu_0 \tag{3.14}$$

that is, the average energy of an electron is $\tfrac{3}{5}$ of the maximum energy.

It is sometimes convenient, and always instructive, to express scientific results in alternative forms. Thus far, we have discussed the properties of independent electrons in essentially particle language. Now let us use the wave particle duality of electrons as given by quantum mechanics to recast our results in wave language. The particles move in a box of zero potential and their energy is all kinetic. Therefore, in particle language

$$\epsilon = \frac{\mathbf{p}^2}{2m} = \tfrac{1}{2}mv^2 \tag{3.15}$$

where $\mathbf{p}$ is the electron momentum and v is the velocity. But from the de Broglie relation, the wave and particle aspects of an electron are connected by

$$\mathbf{p} = \frac{h}{\lambda}\mathbf{e} \tag{3.16}$$

where $\mathbf{e}$ is a unit vector in the direction of electron motion and λ is the electron wavelength. We now define a vector $\mathbf{k}$ by

$$\mathbf{k} = \frac{2\pi}{\lambda}\mathbf{e} \tag{3.17}$$

This is called the wave number vector. It determines the wavelength and direction of motion of the electron, and when combined with (3.16) it determines the momentum.

$$\mathbf{p} = \hbar\mathbf{k} \tag{3.18}$$

where $\hbar = h/2\pi$. (The factor 2π is introduced in defining $\mathbf{k}$ because it turns out to be convenient in the quantum theory of crystals.)

Using (3.18), the energy (3.15) in wave language becomes

$$\epsilon = \frac{\hbar^2\mathbf{k}^2}{2m} \tag{3.19}$$

At absolute zero, the k vectors with maximum magnitude are determined by replacing ϵ by μ_0 in (3.19); that is,

$$\mathbf{k}_M^{0^2} = \frac{2m}{\hbar^2}\mu_0 \tag{3.20}$$

The wave-number vectors may be treated as quasi-continuous parameters determining the energy, but in fact they form a discrete set defined by the energy levels of a particle-in-a-box. If we combine these energy levels (15.6, Chapter 1) with (3.19), we see that the **k** can only take on values that satisfy the following equation:

$$\mathbf{k}^2 = \frac{\pi^2}{V^{2/3}}(j_1{}^2 + j_2{}^2 + j_3{}^2) \qquad (3.21)$$

If we define k_x, k_y, and k_z to be the x, y, and z components of **k**

$$\mathbf{k} = k_x\mathbf{i} + k_y\mathbf{j} + k_z\mathbf{k} \qquad (3.22)$$

i, **j**, and **k** being the unit vectors in the three orthogonal Cartesian directions, then from the definition of the scalar product $\mathbf{k} \cdot \mathbf{k} = \mathbf{k}^2$, (3.21) and (3.22) tell us that the wave number vector is

$$\mathbf{k} = \frac{\pi}{V^{1/3}}(j_1\mathbf{i} + j_2\mathbf{j} + j_3\mathbf{k}) \qquad (3.23)$$

The energy is a quadratic function of the magnitude of **k**, so that in a space in which the coordinates are k_x, k_y, and k_z, it is a spherically symmetric function. The vectors $\mathbf{k}_M^0$ of (3.20) map a sphere in this **k**-space whose surface defines the maximum energy of a free electron. All the possible wave-number vectors, as given by (3.23), start at the origin of **k**-space and terminate somewhere within this sphere. The sphere itself is called the Fermi sphere, and its surface the Fermi surface for free electrons.

It is clear that the Fermi-Dirac distribution function can be expressed in terms of velocities, momenta, or wave-number vectors as well as energies. It is only necessary to find an expression for the density of states $\omega(\epsilon)$ in the new variable and to multiply by the Fermi function. If we pick out a set of contiguous states, their number remains the same whether we label them with velocities, momenta, energies, or **k**-values. That is,

$$\omega(\epsilon)\,d\epsilon = \omega(v)\,dv = \omega(p)\,dp = \omega(k)\,dk \qquad (3.24)$$

where $\omega(v)\,dv$ is the number of states in which electron velocities have magnitudes in the range v to $v + dv$. $\omega(p)\,dp$ and $\omega(k)\,dk$ are similarly defined. We rewrite (3.24) as

$$\omega(k) = \omega(\epsilon)\frac{d\epsilon}{dk}$$

$$\omega(p) = \omega(\epsilon)\frac{d\epsilon}{dp} \qquad (3.25)$$

$$\omega(v) = \omega(\epsilon)\frac{d\epsilon}{dv}$$

Using (3.15) and (3.19) to get the derivatives, and (3.2) for $\omega(\epsilon)$, (3.25) give

$$\omega(k) = V\frac{k^2}{\pi^2} \tag{3.26}$$

$$\omega(p) = V\frac{p^2}{\pi^2\hbar^3} \tag{3.27}$$

$$\omega(v) = V\frac{m^3v^2}{\hbar^3} \tag{3.28}$$

Multiplying these equations by the Fermi function gives the distribution of electrons in terms of k, p, and v, respectively:

$$N(k) = V\frac{k^2}{\pi^2}f \tag{3.29}$$

$$N(p) = V\frac{p^2}{\pi^2\hbar^3}f \tag{3.30}$$

$$N(v) = V\frac{m^3v^2}{\hbar^3}f \tag{3.31}$$

It is often necessary to count the number of electrons, with momenta in a given range, moving in a particular direction. To do this, construct a spherical polar coordinate system in momentum space with radial coordinate p, colatitude θ, and azimuthal angle φ, and define a function $\omega(p, \theta, \varphi)$ such that $\omega(p, \theta, \varphi)\,d\Omega_p$ is the number of states for which the momentum vectors terminate in an element $d\Omega_p$, given in spherical coordinates as

$$d\Omega_p = p^2 \sin\theta \, d\theta \, d\varphi \, dp \tag{3.32}$$

The function $\omega(p)\,dp$ is just the integral of $\omega(p, \theta, \varphi)\,d\Omega_p$ over θ and φ, so that from (3.27)

$$\omega(p)\,dp = \frac{Vp^2}{\pi^2\hbar^3}\,dp = \int_0^{2\pi}\int_0^{\pi}\omega(p, \theta, \varphi)\sin\theta\,d\theta\,d\varphi p^2\,dp \tag{3.33}$$

The distribution of electrons in momentum is isotropic. That is, for a given p, the number of states is the same regardless of the direction of the momentum (see previous k space discussion), so that $\omega(p, \theta, \varphi)$ is the same for all (θ, φ). The density of states function can, therefore, be taken out of the integral sign on the right-hand side of (3.33); therefore

$$\omega(p, \theta, \varphi)\int_0^{2\pi}\int_0^{\pi}\sin\theta\,d\theta\,d\varphi = \frac{V}{\pi^2\hbar^3} \tag{3.34}$$

Evaluating the integral and solving for $\omega(p, \theta, \varphi)$ we get

$$\omega(p, \theta, \varphi) = \frac{2V}{h^3} \tag{3.35}$$

The number of states in a volume element in momentum space does not depend on the choice of coordinate systems so that we can write

$$
\begin{aligned}
d\Omega_p &= dp_x \, dp_y \, dp_z \\
\omega(p, \theta, \varphi) \, d\Omega_p &= \omega(p_x, p_y, p_z) \, dp_x \, dp_y \, dp_z
\end{aligned}
\tag{3.36}
$$

where p_x, p_y, and p_z are the components of momentum in Cartesian coordinates. Combining (3.35) and (3.36)

$$\omega(p_x, p_y, p_z) = \frac{2V}{h^3} \tag{3.37}$$

Multiplying by the Fermi function gives the momentum distribution function as

$$N(p_x, p_y, p_z) = \frac{2V}{h^3} f \tag{3.38}$$

where $N(p_x, p_y, p_z) \, dp_x \, dp_y \, dp_z$ is the number of electrons with momentum components in the range p_x to $p_x + dp_x$, p_y to $p_y + dp_y$, and p_z to $p_z + dp_z$. In a similar way, it is easy to show that the distribution of electrons with velocity components in the range v_x to $v_x + dv_x$, v_y to $v_y + dv_y$, and v_z to $v_z + dv_z$ is

$$N(v_x, v_y, v_z) = \frac{2Vm^3}{h^3} f \tag{3.39}$$

and the corresponding distribution in terms of wave-number vector components is

$$N(k_x, k_y, k_z) = \frac{V}{4\pi^3} f \tag{3.40}$$

3.4 THERMODYNAMIC PROPERTIES OF THE FERMI GAS

The statistical thermodynamic formulae for a free-electron gas can be obtained if the Fermi energy and the internal energy are known as functions of temperature. The equation for the Fermi energy is obtained from the obvious requirement that the total number of electrons be given by the integral of the distribution function over all energies:

$$N = \int_0^\infty N(\epsilon) \, d\epsilon \tag{4.1}$$

The total energy of the electrons is just

$$U = N\bar{\epsilon} = \int_0^\infty \epsilon N(\epsilon)\, d\epsilon \tag{4.2}$$

If we insert the expression for the distribution function, these equations become

$$N = CV \int_0^\infty \frac{\sqrt{\epsilon}}{e^{(\epsilon - \mu)/kT} + 1}\, d\epsilon \tag{4.3}$$

$$U = CV \int_0^\infty \frac{\epsilon^{3/2}}{e^{(\epsilon - \mu)/kT} + 1}\, d\epsilon \tag{4.4}$$

All that is needed to get μ and U as functions of temperature is to evaluate the integrals. They cannot be found in closed form, but a method has been developed to compute integrals of this type in terms of rapidly convergent series. This method depends on the fact that the derivative of the Fermi function is practically zero everywhere except in the vicinity of the Fermi energy, where it is large, as can be seen by direct inspection of Figure 2.1.

The integrals in (4.3) and (4.4) are examples of the Fermi integral, which is defined by the general form

$$I = \int_0^\infty g(\epsilon) f(\epsilon)\, d\epsilon \tag{4.5}$$

where $f(\epsilon)$ is the Fermi function defined by (3.5) and $g(\epsilon)$ is any monotonically increasing function whose value is zero when $\epsilon = 0$. If (4.5) is integrated by parts, the result is

$$I = F(\epsilon) f(\epsilon)\big|_0^\infty - \int_0^\infty F(\epsilon) \frac{\partial f}{\partial \epsilon}\, d\epsilon \tag{4.6}$$

where

$$F(\epsilon) = \int_0^\epsilon g(x)\, dx \tag{4.7}$$

But $F(\epsilon)$ clearly is zero at $\epsilon = 0$, and $f(\epsilon)$ is zero at $\epsilon \to \infty$. Thus the first term on the right of (4.6) vanishes and

$$I = -\int_0^\infty F(\epsilon) \frac{\partial f}{\partial \epsilon}\, d\epsilon \tag{4.8}$$

In Appendix V it is shown that a rapidly converging series expression can be obtained for the Fermi integral. An excellent approximation to I is obtained by retaining only the first three terms of the series:

$$I = F_0(\mu) + \frac{(2\pi kT)^2}{24} F_2(\mu) + 1.22 \times 10^{-3}(2\pi kT)^4 F_4(\mu) \tag{4.9}$$

where
$$F_0(\mu) = F(\mu)$$
$$F_2(\mu) = \left(\frac{d^2F}{d\epsilon^2}\right)_{\epsilon=\mu} \tag{4.10}$$
$$F_4(\mu) = \left(\frac{d^4F}{d\epsilon^4}\right)_{\epsilon=\mu}$$

Even the truncated form (4.9) is more accurate than is usually required, and we will see later that taking only the first two terms is almost always adequate.

It is now easy to develop the formulae for the thermodynamic functions. To get the Fermi energy μ, write (4.3) as

$$\frac{N}{CV} = \int_0^\infty \epsilon^{1/2} f(\epsilon)\, d\epsilon \tag{4.11}$$

Solving (3.8) for N/CV and substituting the result into the left-hand side of (4.11) gives

$$\tfrac{2}{3}\mu_0^{3/2} = \int_0^\infty \epsilon^{1/2} f(\epsilon)\, d\epsilon \tag{4.12}$$

The right-hand side of this equation is just a Fermi integral of the form of (4.5) with $g(\epsilon)$ defined as

$$g(\epsilon) \equiv \epsilon^{1/2} \tag{4.13}$$

and

$$F(\epsilon) \equiv \int_0^\epsilon \epsilon^{1/2}\, d\epsilon = \tfrac{2}{3}\epsilon^{3/2} \tag{4.14}$$

The first four derivatives of $F(\epsilon)$ are:
$$F_1(\epsilon) = \epsilon^{1/2}$$
$$F_2(\epsilon) = \tfrac{1}{2}\epsilon^{-1/2}$$
$$F_3(\epsilon) = -\tfrac{1}{4}\epsilon^{-3/2} \tag{4.15}$$
$$F_4(\epsilon) = \tfrac{3}{8}\epsilon^{-5/2}$$

To apply (4.9), we need $F_0(\mu)$, $F_2(\mu)$, and $F_4(\mu)$. For our present problem these are
$$F_0(\mu) = \tfrac{2}{3}\mu^{3/2}$$
$$F_2(\mu) = \tfrac{1}{2}\mu^{-1/2} \tag{4.16}$$
$$F_4(\mu) = \tfrac{3}{8}\mu^{-5/2}$$

Substitution of these into the right-hand side of (4.9) gives the Fermi integral in (4.12) so that

$$\tfrac{2}{3}\mu_0^{3/2} = \tfrac{2}{3}\mu^{3/2} + \frac{(2\pi kT)^2}{48}\mu^{-1/2} + \frac{3(1.22)\times 10^{-3}}{8}(2\pi kT)^4\mu^{-5/2} \tag{4.17}$$

from which we get

$$\mu_0 = \mu\left[1 + \frac{1}{8}\left(\frac{\pi kT}{\mu}\right)^2 + 27.4 \times 10^{-4}\left(\frac{\pi kT}{\mu}\right)^4\right]^{2/3} \tag{4.18}$$

The rapidity of convergence of the series expansion for the Fermi integral can be shown by substituting reasonable values of the temperature and Fermi energy into this equation, and comparing the second two terms in the brackets to unity. The higher the temperature and the lower the Fermi level, the more slowly our series converges. Let us therefore be conservative and take kT to be about one tenth of an electron volt and μ to be one electron volt. This corresponds to a metal with a low Fermi level at a temperature of about $1200°K$. Even so, the second term in the brackets of (4.18) amounts to only about 10^{-2}, while the third term is a trivial 2×10^{-7}. With practically no loss of accuracy at all temperatures of experimental interest, the third term can be dropped. Furthermore, the second term itself amounts to little more than a correction term, and μ changes so slowly with temperature that it can be replaced by μ_0 in the bracketed terms. We therefore write (4.18) as

$$\mu_0 = \mu\left[1 + \frac{1}{8}\left(\frac{\pi kT}{\mu_0}\right)^2\right]^{2/3} \tag{4.19}$$

Solving for μ gives

$$\mu = \mu_0\left[1 + \frac{1}{8}\left(\frac{\pi kT}{\mu_0}\right)^2\right]^{-2/3} \tag{4.20}$$

and, remembering that the second term on the right is small, we can expand the right hand side in a series and keep only the first two terms to get

$$\mu = \mu_0\left[1 - \frac{1}{12}\left(\frac{\pi kT}{\mu_0}\right)^2\right] \tag{4.21}$$

which shows that the Fermi level decreases with increasing temperature.

A similar calculation can be performed to obtain a formula for the energy. From (4.4)

$$\frac{U}{CV} = \int_0^\infty \epsilon^{3/2} f(\epsilon)\, d\epsilon \tag{4.22}$$

This is an integral of the form of (4.5) with $g(\epsilon) = \epsilon^{3/2}$ and $F(\epsilon) = 2\epsilon^{5/2}/5$. Application of (4.9) to this integral gives

$$\frac{U}{CV} = \tfrac{2}{5}\mu^{5/2} + \frac{(\pi kT)^2}{4}\mu^{1/2} \tag{4.23}$$

To get the energy in terms of μ_0, simply substitute (4.21) into (4.23).

$$\frac{U}{CV} = \tfrac{2}{5}\mu_0^{5/2}\left[1 - \frac{1}{12}\left(\frac{\pi kT}{\mu_0}\right)^2\right]^{5/2} + \frac{(\pi kT)^2}{4}\mu_0^{1/2}\left[1 - \frac{1}{12}\left(\frac{\pi kT}{\mu_0}\right)^2\right]^{1/2} \qquad (4.24)$$

But from (3.8), CV is given by

$$CV = \frac{3N}{2}\mu_0^{-3/2} \qquad (4.25)$$

so that putting this into (4.24) gives for the energy per particle

$$\frac{U}{N} \equiv \bar{\epsilon} = \tfrac{3}{5}\mu_0\left[1 - \frac{1}{12}\left(\frac{\pi kT}{\mu_0}\right)^2\right]^{5/2} + \frac{3(\pi kT)^2}{8\mu_0}\left[1 - \frac{1}{12}\left(\frac{\pi kT}{\mu_0}\right)^2\right]^{1/2} \qquad (4.26)$$

Using the binomial theorem to expand the powers of the brackets and retaining terms only up to $(kT/\mu_0)^2$ gives

$$\frac{U}{N} = \tfrac{3}{5}\mu_0\left[1 + \frac{5}{12}\left(\frac{\pi kT}{\mu_0}\right)^2\right] \qquad (4.27)$$

and we see that the energy increases with temperature.

The heat capacity at constant volume is just

$$C_V = \left(\frac{\partial U}{\partial T}\right)_V \qquad (4.28)$$

so that differentiating U from (4.27) gives

$$C_V = \frac{\pi^2 Nk}{2}\left(\frac{kT}{\mu_0}\right) \qquad (4.29)$$

and the heat capacity is a linear function of temperature.

The entropy is easily computed from the thermodynamic equation

$$S = \int_0^T \frac{C_V}{T}\,dT \qquad (4.30)$$

which becomes, on using (4.29),

$$S = \frac{\pi^2 Nk}{2}\left(\frac{kT}{\mu_0}\right) \qquad (4.31)$$

That is, the entropy of a Fermi-Dirac gas is equal to its heat capacity.

Since the Helmholtz free energy is given by $A = U - TS$, we get from (4.27) and (4.31)

$$A = \tfrac{3}{5}N\mu_0\left[1 - \frac{5}{12}\left(\frac{\pi kT}{\mu_0}\right)^2\right] \qquad (4.32)$$

This completes our collection of basic equations for the thermodynamic properties in terms of the Fermi level at absolute zero.

3.5   THE ELECTRONIC HEAT CAPACITY IN METALS

Before the advent of quantum theory and the development of Fermi-Dirac statistics, there was a great inconsistency in the interpretation of experiments with metals. The data on electrical and thermal conductivity could only be understood by postulating the existence of free electrons. If such electrons were present, however, classical statistical mechanics predicted that they should contribute an amount to the heat capacity equal to $3k$ per particle. As a result, the specific heat of electrical conductors should be twice that of insulators. In fact, it is found that conductors and insulators have about the same specific heat. This is easily understood in terms of the quantum statistic result given by (4.29). In the neighborhood of 1200°K, $kT/\mu_0 \approx 0.05$ if μ_0 is 2 eV, while it decreases to 0.01 if μ_0 is 10 eV. These values of μ_0 are chosen because the Fermi energies of most metals lie in this range. With these figures, (4.29) shows that the electronic heat capacity at high temperatures can be expected to be between $0.25k$ and $0.05k$ per particle. This means that the electronic heat capacity can contribute only something of the order of 2 to 8 % of the DuLong-Petit value at high temperatures, and this contribution decreases with temperature.

In spite of its small value, the electronic heat capacity can be separated from that arising from the crystal vibrations. Although the electronic heat capacity decreases linearly with decreasing temperature, at low temperatures the lattice heat capacity decreases as T^3, as described by the Debye law. Therefore at low enough temperatures, the electronic contribution is a large fraction of the lattice contribution. At low temperatures, the total heat capacity is

$$C_V(\text{total}) = A_1 T + A_2 T^3 \tag{5.1}$$

where the first term on the right is the electronic contribution and the second term is the lattice contribution. The constant A_1 is given by the coefficient of T in (4.29)

$$A_1 = \frac{\pi^2 N k^2}{2\mu_0} \tag{5.2}$$

or, using (3.10),

$$A_1 = \frac{4m\pi^2 k^2 N}{h^2} \left(\frac{\pi V}{3N}\right)^{2/3} \tag{5.3}$$

while A_2 is given by the Debye theory as

$$A_2 = \frac{12\pi^4 k N}{5\theta_D^3} \tag{5.4}$$

If the number of free electrons in the metal is known, A_1 can, of course, be computed from (5.3). It is quite reasonable to take N to be the number of

Table 3.1. Effective Mass and Fermi Level for Some Metals

Metal	m*/m	μ(e.v.)
Li	1.19	4.72
Na	1.0	3.12
K	0.99	2.14
Rb	0.97	1.82
Cs	0.98	1.53
Cu	0.99	7.04
Ag	1.01	5.51
Au	1.01	5.51

The Fermi levels were computed from free-electron theory. The effective masses of the alkali metals were obtained theoretically from pseudopotentical theory. See M. L. Cohen and V. Heine, *Solid State Physics*, Vol. 24, H. Ehrenreich, F. Seitz, and D. Turnbull, Eds., Academic Press, New York, 1970, p. 67. The effective masses of the noble metals were computed by K. Kambe, *Phys. Rev.*, **99**, 419 (1955).

valence electrons. The computed value of A_1 is not always in good agreement with the experimental value if the actual electronic mass is used in the calculation. The reason is to be found in the quantum theory of electrons moving in the three dimensionally periodic potential of the crystal. The effect of the periodic potential is to alter the inertial properties of the electron, so that while electrons in a metal can still be treated by free-electron theory with reasonable accuracy, the electronic mass m must be replaced by an "effective mass" m^*. The ratio m^*/m is a measure of the influence of the lattice potential on the quasi-free electrons. This ratio, as computed from heat capacity data, is shown for several metals in Table 3.1 along with values of the Fermi energy.

3.6 EQUATION OF STATE OF THE FREE-ELECTRON GAS

The equation of state of the electron gas is derived by taking the negative derivative of the free energy with respect to volume. To do this, the free energy must first be expressed as a function of volume and temperature by substituting (3.8) into (4.32).

$$A = \frac{3N}{5}\left(\frac{3N}{2C}\right)^{2/3} V^{-2/3} - \frac{N(\pi kT)^2}{4}\left(\frac{2C}{3N}\right)^{2/3} V^{2/3} \qquad (6.1)$$

Evaluation of the pressure according to the equation

$$P = -\left(\frac{\partial A}{\partial V}\right)_T \tag{6.2}$$

gives

$$P = \frac{2N}{5}\left(\frac{3N}{2C}\right)^{2/3}V^{-5/3} + \frac{N}{6}(\pi kT)^2\left(\frac{2C}{3N}\right)^{2/3}V^{-1/3} \tag{6.3}$$

This is the free-electron equation of state. It can be expressed in terms of μ_0 by using (3.8).

$$P = \frac{2N\mu_0}{5V}\left[1 + \frac{5}{12}\left(\frac{\pi kT}{\mu_0}\right)^2\right] \tag{6.4}$$

For an electron density of 10^{22} cm^{-3} and a Fermi energy of 5 eV, the leading term in (6.4) gives a pressure of about 3×10^4 atmospheres. This shows that, as far as the free electrons in a metal are concerned, the metal should fly apart into atoms. The metal is held together, of course, by the interaction between the electrons and the ions. It is clear that any theory of cohesion in metals must take the free electrons into account.

If (6.4) is compared to (4.27) for the energy, we see that

$$PV = \tfrac{2}{3}U \tag{6.5}$$

This relationship between PV and the energy is also correct for the ideal gas, as can be seen by combining (15.26) and (15.37) of Chapter 1.

The bulk modulus, $1/\kappa$, is given by

$$\frac{1}{\kappa} = -V\left(\frac{\partial P}{\partial V}\right)_T \tag{6.6}$$

Differentiating (6.3) for the pressure and using this equation, we get

$$\frac{1}{\kappa} = \frac{2N}{3V}\left(\frac{3N}{2CV}\right)^{2/3} + \frac{N(\pi kT)^2}{18V}\left(\frac{2CV}{3N}\right)^{2/3} \tag{6.7}$$

or, using (3.8),

$$\frac{1}{\kappa} = \frac{2N\mu_0}{3V}\left[1 + \frac{1}{12}\left(\frac{\pi kT}{\mu_0}\right)^2\right] \tag{6.8}$$

Again the leading term is sufficient to give an estimate of the magnitude of $1/\kappa$. Taking $N/V = 10^{22}$ and $\mu_0 = 5$ eV gives $1/\kappa \simeq 5 \times 10^{10}$ dynes/cm^2. This value is of the order of the measured bulk modulus in metals. The free electrons therefore contribute an appreciable amount to the elastic properties of metals.

The equations in this section are evidently not applicable to electrons in metals just as they stand. Ignoring the electron-lattice interaction gives results that are fine for an idealized Fermi-Dirac gas, but are just too far off for free electrons in metals. The reason for this is that the electron-lattice interaction energy is a function of the volume and therefore must be included in equation-of-state considerations. This lattice effect is rather insensitive to temperature, so that we were able to ignore it in the derivation of the heat capacity, but it must be considered in any discussion of the equation of state. This can be done in at least a formal way by recognizing that the potential energy of an electron in a solid is not zero, as has been assumed up till now. The potential energy is actually a periodic function of position within the metal. To a first approximation, we can treat the periodic potential of the crystal as being smeared out to some average value and merely take an electron at rest to have some energy $-u^0$. The zero of energy is taken to be that of an electron at rest outside the metal. On this scale all electronic energies are negative. The energies in the equations derived from the Fermi-Dirac statistics need not be changed at all since they are all kinetic energies referred to the ground state (zero kinetic energy) of the electrons and are therefore positive.

To incorporate $-u^0$ into the theory, just add it to the free energy of (4.32) to get

$$A_m = \tfrac{3}{5} N \mu_0 \left[1 - \frac{5}{12} \left(\frac{\pi k T}{\mu_0} \right)^2 \right] - N u^0(V) \tag{6.9}$$

The pressure and the bulk modulus are obtained by adding the contributions arising from u^0 to (6.4) and (6.8)

$$P_m = \frac{2N\mu_0}{5V} \left[1 + \frac{5}{12} \left(\frac{\pi k T}{\mu_0} \right)^2 \right] + \frac{du^0}{dV} \tag{6.10}$$

$$\frac{1}{\kappa_m} = \frac{2N\mu_0}{3V} \left[1 + \frac{1}{12} \left(\frac{\pi k T}{\mu_0} \right)^2 \right] - NV \frac{d^2 u^0}{dV^2} \tag{6.11}$$

The subscript m on the free energy, pressure, and compressibility in the last three equations are meant to show that we are now dealing with a model for a metal and not just an idealized Fermi gas. Of course, we still have only the electronic contributions in these equations. Contributions whose source is in ion-ion interactions and crystal vibrations are not included.

Equations 6.10 and 6.11 show that both the pressure and the bulk modulus are decreased by the ion-electron interaction. The ion-electron interaction helps hold the crystal together, whereas the Fermi gas contribution tends to disrupt it.

It is of interest to note here the functional dependence of u^0 on volume. From a quantum theoretic analysis, Frohlich has shown that for good metals, this dependence can be approximated by*

$$-u^0 = -\frac{3}{r_s} + \frac{r_0^2}{r_s^3} \tag{6.12}$$

where r_0 is a constant, and r_s is the Wigner-Seitz radius, which is related to the volume by

$$\tfrac{4}{3}\pi r_s^3 = \frac{V}{N} \tag{6.13}$$

The energy in (6.12) is given in rydbergs. This formulation includes the ion-electron interactions, but still neglects the electron-electron interactions.

3.7 THOMAS-FERMI THEORY: FIELD OF A POINT CHARGE

The free-electron theory presumes that the potential energy and electron density are constants. Of course, this is not so in a metal, and some simplified method of estimating the spatial dependence of the density and the potential is desirable. To construct such a method, we start with the electron distribution function in terms of momentum given by (3.30). All thermal effects will be neglected by taking $T = 0$. Since the Fermi function is unity for p less than p_F, where p_F is the momentum at $\epsilon = \mu$, and zero above $p = p_F$, (3.30) becomes

$$N(p) = \frac{8\pi V p^2}{h^3} \qquad p < p_F$$
$$N(p) = 0 \qquad p > p_F \tag{7.1}$$

If this equation is integrated over all p, the result is equal to the total number of electrons

$$N = \int_0^\infty N(p)\,dp = \int_0^{p_F} \frac{8\pi V}{h^3} p^2\,dp \tag{7.2}$$

so that the electron density $n = N/V$ is given by

$$n = \frac{8\pi}{3h^3} p_F^3 \tag{7.3}$$

The basic assumption of Thomas-Fermi theory is that (7.3) is valid even when the electron density depends on position. Adopting this assumption, we write

$$n(\mathbf{r}) = \frac{8\pi}{3h^3} [p_F(\mathbf{r})]^3 \tag{7.4}$$

* Frohlich, H. *Proc. Roy. Soc.* **A239**, 311 (1937).

where **r** is the position vector. The variations in electron density arise because of the space-dependent potential acting on the electrons. The connection between the potential and the electron density is given from electrostatics by Poisson's equation

$$\nabla^2 \Phi = 4\pi e n \tag{7.5}$$

where Φ is the electrostatic potential field, $-e\Phi$ is the potential energy of an electron in this field, $-en$ is the charge density, and $-e$ is the charge of an electron. Since the various parts of our system must be in equilibrium with one another, the chemical potential (i.e., the Fermi energy μ_0) must be constant. From the law of conservation of energy, therefore,

$$\mu_0 = \frac{p_F{}^2(\mathbf{r})}{2m} - e\Phi(\mathbf{r}) \tag{7.6}$$

where μ_0 is independent of position. Solving (7.6) for $p_F{}^3$ and substituting into (7.4) gives

$$n(\mathbf{r}) = \frac{8\pi}{3h^3} [2m(\mu_0 + e\Phi)]^{3/2} \tag{7.7}$$

This equation allows Poisson's equation to be written in terms of Φ only, thereby giving the Thomas-Fermi equation.

$$\nabla^2 \Phi = \frac{32\pi^2 e}{3h^3} [2m(\mu_0 + e\Phi)]^{3/2} \tag{7.8}$$

Once the boundary conditions on Φ are given, (7.8) can be solved for $\Phi(\mathbf{r})$. This can then be substituted in (7.7) to get $n(\mathbf{r})$.

The methods of solution generally require numerical analysis however, and it is desirable to have a form that can be solved analytically. Let us now restrict ourselves to systems in which $e\Phi \ll \mu_0$. This is not too restrictive an assumption because μ_0 is usually large. Then write (7.7) in the form

$$n(\mathbf{r}) = \frac{8\pi}{3h^3} (2m\mu_0)^{3/2} \left(1 + \frac{e\Phi}{\mu_0}\right)^{3/2} \tag{7.9}$$

and treat $e\Phi/\mu_0$ as a small quantity. Then to a good approximation

$$\left(1 + \frac{e\Phi}{\mu_0}\right)^{3/2} = 1 + \frac{3e\Phi}{2\mu_0} \tag{7.10}$$

Therefore

$$n(\mathbf{r}) = \frac{8\pi}{3h^3} (2m\mu_0)^{3/2} \left[1 + \frac{3}{2} \frac{e\Phi}{\mu_0}\right] \tag{7.11}$$

Using (3.8) and (3.9), this equation can be simplified to read

$$n(\mathbf{r}) = n_0\left(1 + \frac{3}{2}\frac{e\Phi}{\mu_0}\right) \tag{7.12}$$

where n_0 is the number of electrons per unit volume that would be present if $\Phi = 0$. Substituting (7.12) into (7.5) gives

$$\nabla^2\Phi = 4\pi n_0 e\left(1 + \frac{3}{2}\frac{e\Phi}{\mu_0}\right) \tag{7.13}$$

This is the *linearized* Thomas-Fermi equation.

The Thomas-Fermi theory developed so far holds only in regions where there is no positive charge. The right-hand side of Poisson's equation should contain -4π times the *total* charge density so that in regions where a positive charge density exists, the appropriate term must be added. In the application of free-electron theory to metals, the positive ions are taken to be smeared out into a continuous distribution so that the positive and negative charge densities are equal. Poisson's equation then reduces to Laplace's equation $\nabla^2\Phi = 0$. What happens if an excess point charge is inserted into the system? This can be done, for example, by replacing one of the atoms in the metal by another of different valence, or by removing an atom to form a vacancy. In the first case, the excess charge is the difference between the valencies of the host and impurity atom. In the second case, the electrical equivalent of removing a positive ion is to introduce a negative charge of the same magnitude. To be specific, a cadmium impurity atom (Cd^{++}) in silver (Ag^+) would introduce a unit positive charge; a vacancy in silver introduces a unit negative charge.

The introduction of such a point charge induces a charge redistribution in the electron gas since electrons are repelled or attracted by the point charge depending on its sign. The redistributed electron density is related to the potential Φ arising from the point charge by (7.11). The positive charge density, everywhere except at the excess point charge, is en_0 where n_0 is the original electron density. The total charge density is therefore

$$en_0 - en_0\left(1 + \frac{3}{2}\frac{e\Phi}{\mu_0}\right) = -\tfrac{3}{2}n_0\frac{e^2\Phi}{\mu_0}$$

Multiplying this by -4π and substituting into Poisson's equation

$$\nabla^2\Phi = -4\pi \text{ (charge density)}$$

gives the linearized Thomas-Fermi equation for the potential around an excess point charge:

$$\nabla^2\Phi = \frac{6\pi e^2 n_0\Phi}{\mu_0} \tag{7.14}$$

To solve this equation for Φ, first express the left-hand side in terms of spherical coordinates centered on the excess charge. The system is spherically symmetric about the origin so that Φ depends only on the distance r from the charge and not on the angular coordinates. Therefore

$$\nabla^2\Phi = \frac{d^2\Phi}{dr^2} + \frac{2}{r}\frac{d\Phi}{dr} \tag{7.15}$$

and (7.14) becomes

$$\frac{d^2\Phi}{dr^2} + \frac{2}{r}\frac{d\Phi}{dr} = \frac{\Phi}{\lambda^2} \tag{7.16}$$

where λ is defined by

$$\lambda^2 = \frac{\mu_0}{6\pi e^2 n_0} \tag{7.17}$$

The differential equation (7.16) must satisfy two boundary conditions: at very small distances, Φ must approach the coulomb potential field of the excess charge q, because the effects of the electron gas must be negligible very close to the origin; at very large distances, the effect of the excess charge must be negligible, so that Φ must approach zero. That is

$$\lim_{r \to 0} \Phi(r) = \frac{q}{r}$$
$$\lim_{r \to \infty} \Phi(r) = 0 \tag{7.18}$$

Now define a function f by

$$f(r) = r\Phi(r) \tag{7.19}$$

from which we find that

$$\frac{d^2\Phi}{dr^2} + \frac{2}{r}\frac{d\Phi}{dr} = \frac{1}{r}\frac{d^2f}{dr^2} \tag{7.20}$$

or, equating the right-hand sides of (7.20) and (7.16) and then using (7.19),

$$\frac{d^2f}{dr^2} - \lambda^{-2}f = 0 \tag{7.21}$$

This is just a second-order linear differential equation with constant coefficients whose auxiliary equation is $m^2 - \lambda^{-2} = 0$. The roots of the auxiliary equation are $\pm\lambda^{-1}$ and the general solution of (7.21) is therefore

$$f = C_1 e^{r/\lambda} + C_2 e^{-r/\lambda}$$

or, combining this with (7.19),

$$\Phi(r) = \frac{C_1}{r} e^{r/\lambda} + \frac{C_2}{r} e^{-r/\lambda} \tag{7.22}$$

From the second of the boundary conditions (7.19), C_1 must be zero, while from the first boundary condition, $C_2 = q$. Equation 7.22 therefore reduces to

$$\Phi(r) = \frac{q}{r} e^{-r/\lambda} \qquad (7.23)$$

The potential field around an excess charge in a free-electron gas therefore has the form of a coulomb field modified by a decreasing exponential function. Physically, the free electrons screen the interaction between an electron and the point charge, by piling up around the origin (if q is positive) or leaving a hole around it (if q is negative).

The distribution of electrons around the point charge is obtained by substituting (7.23) into (7.12)

$$n(r) = n_0 \left(1 + \frac{q}{\mu_0 r} e^{-r/\lambda} \right) \qquad (7.24)$$

For distances much greater than λ, the effect of the excess charge is small; λ is a measure of the effectiveness of the screening of the excess charge by the free electrons, and is accordingly called the screening distance. The screening distance can be calculated from the Fermi energy and electron density by use of (7.17), and it turns out to be of the order of 1 Å°.

It is obvious that the Thomas-Fermi theory gives only a rough approximation to the actual potential and electron density around impurities and defects in real metals. However, it has been used profitably in both alloy and defect theory to give qualitative and semiquantitative results. The physical significance of the above development goes beyond its applicability to such problems. What we have done here amounts to at least a partial justification of the use of free-electron theory in metals. The screening distance is short; this means that the interaction between any two charges is negligible unless these charges are close together. Ignoring all electron-electron interactions in the metal is therefore not as bad an approximation as it sounds. Of course, the situation is more complex than can be presented here, but more sophisticated investigations bear out these conclusions.

3.8 BAND THEORY AND ELECTRONS IN SOLIDS

At the beginning of this chapter, reasons were presented for believing that free-electron theory is a fair approximation for describing electrons, at least in some solids. With the material presented in the preceding sections, we are now in a position to examine this question in a little more detail. In doing so, we shall draw from the results of the band theory of crystals and show

how it strongly suggests that the form of Fermi-Dirac theory is applicable to a number of solids.

Quantum theoretic analysis of the motion of electrons in periodic potentials shows that an electron in a crystal can take on any of an infinite number of energy values. These energy levels are determined by the chemical constitution and the structure of the crystal, and although there are an infinite number of them, not all energy values can be assumed by an electron. In general, electrons can exist in certain bands of energy, consisting of a set of semicontiguous energy states, but there are other energy ranges in which they are not allowed to exist. The general energy spectrum for an electron in a crystal consists of a series of allowed bands separated by forbidden energy gaps. The gaps between the allowed bands of energy vary in width from one substance to another and may even vanish with consequent overlapping of adjacent bands.

To acquire some idea of the origin of bands, consider a large number N, of identical atoms all far away from each other, but occupying the relative positions they would have in a crystal. Each of the atoms in this "expanded lattice" is independent of the others, so all the electrons are in free-atom states, and their energies correspond to the sharp lines of atomic spectra. Now start shrinking the lattice. As the atoms get closer together, they interact with one another. An electron on a particular atom now experiences a potential from other atoms as well as its own free atom potential. The result of this is that the original sharp atomic level spreads out into a band of levels as shown in Figure 3.2. Several important points that arise naturally from this description should be noted. First, since each atomic level gives rise to one band, there are an infinite number of bands. Second, since each

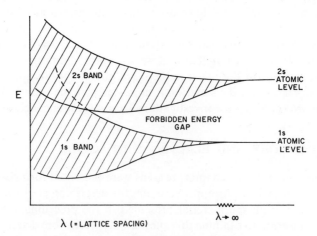

Fig. 3.2. The formation of bands from atomic levels.

atomic level can hold two electrons, and each atom contributes one level to each band, the total number of electrons that can be accommodated in a band is just twice the number of atoms in the crystal. Finally, it is clear that the gap width between bands, the band width itself, and the occurrence or non-occurrence of overlapping bands depends on the details of the interaction between the electrons and atoms of the crystal.

The most obvious success of the band model is that it accounts for the existence of conductors, semiconductors, and insulators. Consider a monovalent metal crystal. A free atom of this crystal would have one electron in an outer energy level. The energy band corresponding to this level can accommodate two electrons per atom, so it is only half full. With this in mind, apply an electric field to the crystal. In order for electrons to move under the influence of the field and thereby contribute to an electric current, they must absorb some energy from the field. Electrons near the bottom of the band cannot do this because if they absorb some energy this will put them in a level slightly above their normal level. But this upper level already holds two electrons, and according to the Pauli exclusion principle it can accept no more, so our electron has to stay where it is. Electrons just at the top of the occupied portion of the band, however, can contribute to electrical conduction. They just absorb energy from the electric field and jump into the empty levels in the top part of the band. Furthermore, when it jumps to any empty level, the electron leaves an empty level behind it. This empty state is called a "hole," and it too can contribute to conductivity since electrons near it in energy can jump into it under the action of the applied fields. The situation described here is characteristic of metals. The high mobility of the free electrons in a metal comes about because they exist in a partially filled band. A partially filled band is called a "conduction band."

If the free atoms from which the crystal is made contain two electrons in the outermost energy level, then the corresponding band in the crystal contains two electrons per atom and is completely filled. Assume that there is no overlapping of bands, so that the band above the filled one is completely empty, and the two bands are separated by a forbidden gap. Such a filled band is called "a valence band." It is clear from the above discussion that, provided only that the gap width is larger than the energy that can be supplied by the electric field, the material will be an insulator. Of course, if the two bands overlap, then a set of $4N$ contiguous states exists containing only $2N$ electrons, and the material is a metal.

All of this is at absolute zero. If we raise the temperature of an insulator, there is a finite probability that a thermal fluctuation will occur that supplies enough energy to an electron at the top of the filled band to kick it into the upper, empty band. If this happens, a current can flow because the electron has empty states available to it. The greater the energy gap, the larger the energy needed for the kick and the more improbable the required thermal fluctuation. If the gap is large relative to kT, very few electrons will be excited to the conduction band, and the material remains an insulator. If the gap is not too large relative to kT, an appreciable number of electrons reach the conduction band, and the material is a semiconductor. Energy-band diagrams for insulators, metals, and semiconductors are shown in Figure 3.3.

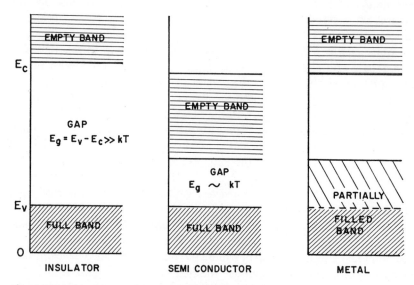

Fig. 3.3. Energy bands in insulators, metals, and semiconductors.

One of the most important results of band theory is that in many cases the energy of an electron near the bottom of a band is given, at least approximately, by

$$\epsilon = E_B + \gamma \mathbf{k}^2 \tag{8.1}$$

where E_B is the energy at the bottom of the band, $\mathbf{k}$ is a wave-number vector, and γ is a constant that depends on the interaction of the electron with the rest of the crystal. If we define a constant m^* by

$$m^* = \frac{\hbar^2}{2\gamma} \tag{8.2}$$

then we can write (8.1) as

$$\epsilon = E_B + \frac{\hbar^2 \mathbf{k}^2}{2m^*} \tag{8.3}$$

Comparing (8.2) and (3.19), we see that if (8.1) is correct for an electron in a crystal, then the electron may be thought of as a free particle with an effective mass m^* moving in a constant potential E_B. The effect of the crystal field is to alter the inertial properties of the electron. Evidently, if we define our scale of energy so that $E_B = 0$, and use the effective mass rather than the actual electronic mass, the free-electron theory discussed in the preceding sections can be used whenever the energy is a quadratic function of the wave-number vector.

For electrons near the top of a band, the energy is also a quadratic function of a wave-number vector, but in this case the wave number is measured relative to the top of the band. That is,

$$\epsilon = E_T - \frac{\hbar^2}{2m^*}(\mathbf{k} - \mathbf{k}_T)^2 \tag{8.4}$$

where E_T is the energy of an electron, and $\mathbf{k}_T$ is its wave-number vector when it is at the top of the band. Equation 8.4 can be rewritten as

$$(E_T - \epsilon) = \frac{\hbar^2}{2m^*}\mathbf{k}'^2 \tag{8.5}$$

where $\mathbf{k}' = \mathbf{k} - \mathbf{k}_T$. This shows that the electron can be treated as a free particle with wave-number vector $\mathbf{k}'$ and energy $(E_T - \epsilon)$.

For the study of the statistical mechanics of electrons in semiconductors, it is convenient to treat the holes left in a valence band when electrons are excited to the conduction band as "particles." It is easy to show that the holes act as positive charge carriers as follows: in a filled band no current can flow, so the sum of all the electron velocities must vanish. That is, if there are N electrons in the band and $\mathbf{v}_i$ is the velocity of the ith electron, then

$$\sum_{i=1}^{N} \mathbf{v}_i = 0 \tag{8.6}$$

Now pick out a particular electron, label it k, and rewrite (8.6) as

$$-\mathbf{v}_k = \sum_{i \neq k}^{N} \mathbf{v}_i \tag{8.7}$$

where the summation is now taken over all values of i except $i = k$. The right-hand side of (8.7) is just the sum of all electron velocities for a band that is missing one electron that is, containing one hole. The hole, therefore, acts as a positive charge carrier.

3.9   IMPURITY LEVELS IN SEMICONDUCTORS

The electronic properties of semiconductors are profoundly affected by impurities, and an understanding of semiconductor statistics requires some knowledge of how the distribution of energy levels depends on these impurities.

Let us replace one of the tetravalent atoms in a semiconducting element (e.g., germanium) by a pentavalent element such as phosphorous. Four of the five valence electrons take part in the tetrahedral bonding in the diamond structure of the semiconductor. The fifth electron, however, is not needed for covalent bonding and, at absolute zero of temperature, just remains in the vicinity of the impurity ion attracted by its positive charge. To a first approximation, the electron-impurity ion system behaves like a hydrogen atom, and the binding energy of the electron to the impurity can be computed from quantum theory in a way analogous to that for computing the ionization energy of hydrogen. The fact that this system is embedded in a crystal can be accounted for by remembering that in a dielectric medium the force of attraction between opposite charges of magnitude e is $e^2/\kappa r^2$ rather than e^2/r^2, where κ is the dielectric constant. Working out the hydrogen-atom problem with the potential $e^2/\kappa r$ gives the ionization energy in electron volts as

$$E_I = \frac{13.6}{\kappa^2} \text{ eV} \qquad (9.1)$$

Using typical values of κ for semiconductors (11.9 for silicon; 16.1 for germanium), we find that the ionization energy is of the order of 0.05 to 0.10 eV. When the impurity ionizes, the electron must go into the conduction band. This means that the energy level of an electron attached to the impurity is about 0.05 to 0.10 eV below the lowest level of the conduction band.

It is important to note that the ionization energy is in the range of thermal energies at room temperature. Thermal agitation is, therefore, sufficient to ionize a significant fraction of impurity atoms, thereby releasing electrons to the conduction band, which can contribute to electrical conductivity. Pentavalent impurities of the type just discussed are called "donors," since they donate electrons to the conduction band. The semiconductor is then called "an extrinsic *n*-type conductor," since conduction is by negative electrons rather than positive holes, and the conduction is not an intrinsic property of the pure crystal.

Trivalent impurities also introduce impurity energy levels into the crystal. In this case, the impurity needs an extra electron to complete the tetrahedral bonding requirements, and it can get this electron from the valence band. The impurity level is now about 0.05 to 0.10 eV above the highest energy of the valence band. When the impurity becomes ionized, it leaves a hole in the valence band, which contributes to conductivity as if it were a positive charge, and we have an "extrinsic *p*-type semiconductor." Energy-band diagrams for extrinsic conductors are shown in Figure 3.4.

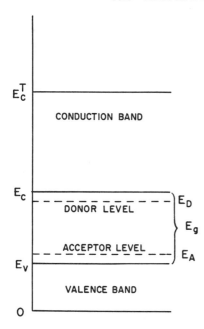

Fig. 3.4. Two-band model for extrinsic semiconductors.

3.10   ELECTRON DISTRIBUTION IN INTRINSIC SEMICONDUCTORS

The main features of the statistics of electrons in pure semiconductors can be understood by studying the simple two-band model shown in Figure 3.5. At absolute zero, all the states in the lower band (I; valence band) are filled with electrons, while the upper band (II; conduction band) is completely empty. The zero of energy is taken to be that of an electron at the bottom of the valence band, E_v is the energy at the top of the valence band, and E_c and E_c^T are the energies at the bottom and at the top of the conduction band, respectively. At nonzero temperatures, a fraction of the electrons will be excited to the conduction band, leaving an equivalent number of holes in the valence band. We will assume that E_v, E_c, E_c^T, and $(E_c^T - E_c)$ are all considerably greater than kT. Furthermore, the energy gap $E_g = E_v - E_c$ will be taken to be large enough so that the conduction band contains only a small fraction of the available electrons.

The purpose of our analysis is to determine the energy distribution functions for the electrons in the conduction band and the holes in the valence band, and to locate the position of the Fermi level on the energy scale. It is clear from the method of developing the particle statistics in Chapter 1 that formulas completely analogous to (3.1) hold for both bands;

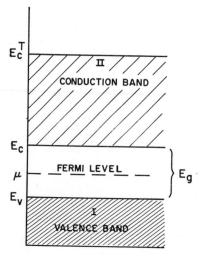

Fig. 3.5. Two-band model of an intrinsic semi-conductor.

that is,

$$N_v(\epsilon) = \frac{\omega_v(\epsilon)}{e^{(\epsilon-\mu)/kT} + 1} \qquad (10.1)$$

$$N_c(\epsilon) = \frac{\omega_c(\epsilon)}{e^{(\epsilon-\mu)/kT} + 1} \qquad (10.2)$$

where $N_v(\epsilon)$ and $N_c(\epsilon)$ are the energy-distribution functions of electrons in the valence band and conduction band, respectively, and $\omega_v(\epsilon)$ and $\omega_c(\epsilon)$ are the corresponding density-of-states-in-energy. As discussed in Section 8, electrons near the bottom of a band can be treated as free particles with kinetic energies $(\epsilon - E_c)$, and electrons near the top of a band can be treated as free particles with kinetic energies $(E_v - \epsilon)$. Since we have assumed that there are only a few electrons in the conduction band, we are interested only in states near the top and bottom of the valence and conduction bands. In these energy ranges, therefore, the density-of-states functions are given by formulas similar to (3.2), except that ϵ is replaced by $(\epsilon - E_c)$ to obtain $\omega_c(\epsilon)$ and by $(E_v - \epsilon)$ to get $\omega_v(\epsilon)$. Therefore

$$\omega_c(\epsilon) = \frac{8\pi\sqrt{2}}{h^3} m_c^{3/2} V \sqrt{\epsilon - E_c} \qquad (10.3)$$

$$\omega_v(\epsilon) = \frac{8\pi\sqrt{2}}{h^3} m_v^{3/2} V \sqrt{E_v - \epsilon} \qquad (10.4)$$

In these equations, the possibility that the effective masses may be different in the two bands has been indicated by the subscripts c and v.

The distribution function for the holes in the valence band can be obtained by remembering that the Fermi function is the probability that a particular state contains an electron. The probability that a state is empty, f_h, is, therefore, just $1 - f$, and we have

$$f_h = 1 - f = 1 - \frac{1}{e^{(\epsilon-\mu)/kT} + 1} = \frac{1}{e^{(\mu-\epsilon)/kT} + 1} \tag{10.5}$$

Multiplying f_h by the density of states, $\omega_v(\epsilon)$, gives the energy distribution function for the holes.

$$N_h(\epsilon) = \frac{\omega_v(\epsilon)}{e^{(\mu-\epsilon)/kT} + 1} \tag{10.6}$$

In order that $N_c(\epsilon)$ and $N_h(\epsilon)$ be small, the denominators in (10.2) and (10.6) must be large compared to unity, and the distribution functions reduce to a Boltzmann form

$$N_c(\epsilon) = \omega_c e^{(\mu-\epsilon)/kT} \tag{10.7}$$

$$N_h(\epsilon) = \omega_v e^{(\epsilon-\mu)/kT} \tag{10.8}$$

The validity of these approximations requires that the Fermi level be in the energy gap, since the exponentials in (10.2) and (10.6) can be greater than unity for all allowed energies only if $E_v < \mu < E_c$.

The actual position of the Fermi level is obtained from the requirement that the total number of electrons in the conduction band is equal to the total number of holes in the valence band. The total number of conduction electrons per unit volume of crystal, n, is obtained by integrating $N_c(\epsilon)$ from E_c to E_c^T and dividing by the volume.

$$n = \frac{8\pi\sqrt{2}}{h^3} m_c^{3/2} \int_{E_c}^{E_c^T} \sqrt{\epsilon - E_c} \, e^{(\mu-\epsilon)/kT} \, d\epsilon \tag{10.9}$$

Changing variables in the integral to $x = (\epsilon - E_c)/kT$ gives

$$n = 4\pi \left(\frac{2m_c kT}{h^2}\right)^{3/2} e^{(\mu-E_c)/kT} \int_0^{(E_c^T - E_c)/kT} \sqrt{x} \, e^{-x} \, dx \tag{10.10}$$

The ratio of the conduction band width to kT is large, and practically no accuracy is lost if the upper limit is replaced by infinity. The value of the integral is then $\sqrt{\pi}/2$ and (10.10) becomes

$$n = 2\left(\frac{2\pi m_c kT}{h^2}\right)^{3/2} e^{(\mu-E_c)/kT} \tag{10.11}$$

A completely analogous calculation can be made for the number of holes per unit volume of crystal with the result

$$p = 2\left(\frac{2\pi m_v kT}{h^2}\right)^{3/2} e^{(E_v - \mu)/kT} \tag{10.12}$$

The Fermi level is now easily obtained by equating n and p and solving for μ to get

$$\mu = \frac{E_v + E_c}{2} + \frac{3kT}{4}\ln\frac{m_v}{m_c} \tag{10.13}$$

If the two effective masses are equal, the Fermi level lies directly in the middle of the energy gap. Actually m_v and m_c are usually not too different, so μ is generally near the center of the gap.

The concentration of conduction electrons, n, and of valence holes, p, can be expressed in terms of the energy gap, $E_g = E_c - E_v$. Multiplying (10.11) and (10.12) eliminates the Fermi level from the equations and we get

$$np = 4\left(\frac{2\pi kT}{h^2}\right)^3 (m_c m_v)^{3/2} e^{-E_g/kT} \tag{10.14}$$

But $n = p$, so taking square roots of both sides of (10.14) gives

$$n = p = 2\left(\frac{2\pi kT}{h^2}\right)^{3/2} (m_c m_v)^{3/4} e^{-E_g/2kT} \tag{10.15}$$

The electron and hole concentrations, therefore, depend only on the magnitude of the energy gap and not on the value of the Fermi level. This is an intuitively reasonable result, since the process of transferring electrons from the valence band to the conduction band depends on thermal excitation, in which individual electrons must acquire a minimum energy of E_g.

Notice that the charge-carrier concentration is an increasing function of temperature. Since the electrical conductivity is proportional to the number of charge carriers, it will, therefore, increase with temperature in a manner controlled by the Boltzmann factor containing E_g. Of course, the increasing amplitude of thermal vibrations at higher temperatures acts to decrease the conductivity, but this effect is overshadowed by the effect of increasing charge-carrier concentration. In metals, the carrier concentration is essentially constant with temperature, and the temperature derivative of the conductivity is controlled largely by the electron-phonon scattering process. The conductivity of metals, therefore, exhibits an approximately linear decrease with temperature.

3.11   ELECTRON STATISTICS IN EXTRINSIC SEMICONDUCTORS

In intrinsic semiconductors, the energy levels in each band form a quasi continuum, and each level can hold two electrons of opposite spin. The Fermi particle statistics can, therefore, be applied directly. For semiconductors containing donor or acceptor impurities, however, some slight modifications are necessary. To show how these modifications come about, it is necessary to go back to the statistical counting procedure used to derive the distribution in energy.

Consider a semiconductor containing both donor and acceptor impurities whose energy-band diagram is shown in Figure 3.4. Now treat the levels in the valence and conduction bands in the same way as for the free-particle levels of Section 13, Chapter 1. The statistical counts in the bands are then given by equations completely analogous to (13.7) of Chapter 1. For the valence band

$$w_v = \prod_i \frac{\omega_v^i!}{N_v^i!\,(\omega_v^i - N_v^i)!} \tag{11.1}$$

where N_v^i is the number of electrons in the ω_v^i levels whose energy is centered on ϵ_i. For the conduction band, the statistical count is

$$w_c = \prod_j \frac{\omega_c^j!}{N_c^j!\,(\omega_c^j - N_c^j)!} \tag{11.2}$$

where N_c^j is the number of electrons in the ω_c^j levels whose energy is centered on ϵ_j.

Now consider the number of ways in which a given number of electrons can be distributed among the acceptor levels. When an electron moves from the valence band to an acceptor level, the acceptor impurity becomes negatively charged. Any attempt to place another electron in that level requires a very high energy because of the electrostatic repulsion. This means that each level can hold only one electron. If N_A is the number of acceptor impurity atoms, and N_A^- is the number of electrons in these levels, then a straightforward application of the Fermi-Dirac counting method would give

$$\frac{N_A!}{N_A^-!\,(N_A - N_A^-)!} \tag{11.3}$$

for the number of ways of putting the N_A^- electrons in the N_A levels. However, because of the existence of electron spin, this result is not quite correct. A neutral trivalent acceptor impurity has an unpaired electron, and this electron has two possible spin states. If the spin of the unpaired electron in a neutral acceptor atom is reversed, a new state is produced. Since the number of neutral acceptors is just the number of empty acceptor levels $(N_A - N_A^-)$, the number of states that can be produced just by reversing spins is $2^{(N_A-N_A^-)}$. To get the statistical count for acceptors, (11.3) must be multiplied by this factor. Therefore, the number of ways of distributing N_A^- electrons among N_A acceptor levels is

$$w_A = \frac{2^{(N_A-N_A^-)}N_A!}{N_A^-!\,(N_A - N_A^-)!} \tag{11.4}$$

Similar considerations hold for the donor levels. When a donor impurity gives up an electron to the conduction band, it becomes ionized, and the energy required to ionize it further is prohibitive. The neutral pentavalent donor has an unpaired electron that can take on two spin values. The number of neutral donors is just the number of electrons in donor levels. Each time the spin of an electron in a donor level is reversed, a new state is produced, and the total number of such possible permutations is $2^{N_D^0}$, where N_D^0 is the number of electrons in donor levels. For donors, therefore, the number of ways of distributing N_D^0 electrons among N_D levels is

$$w_D = \frac{2^{N_D^0}N_D!}{N_D^0!\,(N_D - N_D^0)!} \tag{11.5}$$

The total statistical count is just the product of the separate counts, so that the number of ways of arranging electrons among the available states so that N_v^i are in the ith valence levels, N_c^j are in the jth conduction levels, N_A^- are in acceptor levels and N_D^0 are in donor levels is

$$w = w_v w_c w_A w_D \tag{11.6}$$

Now the distribution-in-energy can be obtained in the usual way by maximizing $\ln w$ subject to the conditions of constant number of electrons and constant total energy. The maximization must be carried out with respect to N_v^i, N_c^i, N_A^-, and N_D^0, so that the variational problem to be solved is

$$\delta \ln w = \sum_i \frac{\partial \ln w}{\partial N_V^i}\,\delta N_V^i + \sum_j \frac{\partial \ln w}{\partial N_C^j}\,\delta N_C^j + \frac{\partial \ln w}{\partial N_A^-}\,\delta N_A^- + \frac{\partial \ln w}{\partial N_D^0}\,\delta N_D^0 = 0 \tag{11.7}$$

$$\delta N = \sum_i \delta N_V^i + \sum_j \delta N_C^j + \delta N_A^- + \delta N_D^0 = 0 \tag{11.8}$$

$$\delta U = \sum_i \epsilon_i \delta N_V^i + \sum_j \epsilon_j \delta N_C^j + E_A \delta N_A^- + E_D \delta N_D^0 = 0 \tag{11.9}$$

where E_A and E_D are the energies of the acceptor and donor levels, respectively. Multiplying (11.8) and (11.9) by the Lagrangian multipliers $-a$ and

$-b$, respectively, adding the results to (11.7), and equating the coefficients of the variations to zero gives

$$\frac{\partial \ln w}{\partial N_V{}^i} = a + b\epsilon_i \tag{11.10}$$

$$\frac{\partial \ln w}{\partial N_C{}^j} = a + b\epsilon_j \tag{11.11}$$

$$\frac{\partial \ln w}{\partial N_A^-} = a + bE_A \tag{11.12}$$

$$\frac{\partial \ln w}{\partial N_D{}^0} = a + bE_D \tag{11.13}$$

The first two of these equations lead directly to the usual Fermi-Dirac distribution for the bands. Performing the differentiations in the last two equations and using Stirling's approximation, however, gives the following results for the distribution in donor and acceptor levels

$$N_D{}^0 = \frac{N_D}{1/2e^{a+bE_D} + 1} \tag{11.14}$$

$$N_A^- = \frac{N_A}{2e^{a+bE_A} + 1} \tag{11.15}$$

The Lagrangian multipliers have the same connection with thermodynamics as derived previously, so these equations may be written as

$$N_D{}^0 = \frac{N_D}{1/2e^{(E_D-\mu)/kT} + 1} \tag{11.16}$$

$$N_A^- = \frac{N_A}{2e^{(E_A-\mu)/kT} + 1} \tag{11.17}$$

The Fermi functions for donor and acceptor levels are defined by $N_D{}^0/N_D$ and N_A^-/N_A, respectively. The probability that an acceptor level contains an electron is therefore

$$f_A = \frac{1}{2e^{(E_A-\mu)/kT} + 1} \tag{11.18}$$

while the probability that a donor level contains an electron is

$$f_D = \frac{1}{1/2e^{(E_D-\mu)/kT} + 1} \tag{11.19}$$

Equation 11.17 gives the number of ionized acceptors and (11.18) gives the probability that an acceptor is ionized. To obtain the number of ionized donors, N_D^+, write

$$N_D^+ = N_D - N_D{}^0 = N_D(1 - f_D) \tag{11.20}$$

$(1 - f_D)$ is just the probability that a donor level is empty and is given by

$$f_D{}^h \equiv (1 - f_D) = \frac{1}{2e^{(\mu - E_D)/kT} + 1} \tag{11.21}$$

and the number of ionized donors is

$$N_D^+ = \frac{N_D}{2e^{(\mu - E_D)/kT} + 1} \tag{11.22}$$

Similarly, the probability that an acceptor level is neutral (empty) is

$$f_A{}^h \equiv (1 - f_A) = \frac{1}{1/2e^{(\mu - E_A)/kT} + 1} \tag{11.23}$$

and the number of empty acceptor levels is

$$N_A{}^0 = N_A - N_A^- = \frac{N_A}{1/2e^{(\mu - E_A)/kT} + 1} \tag{11.24}$$

3.12 MASS ACTION LAWS FOR EXTRINSIC SEMICONDUCTORS

Whenever $(\mu - \epsilon) \gg kT$, the Fermi function reduces to a Boltzmann form. It was shown in Section 10 that this is true for intrinsic semiconductors with a large energy gap, and that this implies that the electron and hole concentrations in the conduction and valence bands, respectively, are small relative to the number of available states. The Fermi level then lies near the center of the energy gap. If small amounts of donor and acceptor impurities are added to such a semiconductor, then the number of electrons and holes in the conduction and valence bands may be increased; but if the impurity concentration is low enough, the electron and hole concentrations will still be small relative to the concentration of available states. The Fermi level will still be somewhere near the center of the gap, and the electron and hole concentrations are given by (10.11) and (10.12). Also, since E_A is near E_V and E_D is near E_C, it follows that $(E_A - \mu)/kT$ and $(\mu - E_D)/kT$ are considerably less than unity. According to (11.17) and (11.22), this means that the impurities are ionized to a considerable extent.

For an extrinsic semiconductor of the type described in the above paragraph, it is convenient to describe the transfer of electrons among the energy states by a set of chemical ionization reactions. Four such reactions are possible: the ionization of a donor in which an electron goes to the conduction band,

$$D^0 \rightleftarrows D^+ + e^-(\text{CB}) \tag{12.1}$$

the ionization of an acceptor, in which an electron goes from the valence band to an acceptor level, leaving behind a hole,

$$A^0 \rightleftarrows A^- + h^+(\text{VB}) \tag{12.2}$$

the transfer of an electron from a neutral donor to a neutral acceptor,

$$A^0 + D^0 \rightleftarrows A^- + D^+ \tag{12.3}$$

and the transfer of an electron from the valence band to the conduction band

$$e^-(\text{VB}) + h^+(\text{CB}) \rightleftarrows e^-(\text{CB}) + h^+(\text{VB}) \tag{12.4}$$

In these equations e^- represents an electron, or a filled state, while h^+ represents a hole, or an empty state. CB and VB stand for conduction band and valence band, respectively.

Now define four equilibrium constants, corresponding to each of the four reactions, as follows:

$$K_D = \frac{n N_D^+}{N_D^{\,0}} \tag{12.5}$$

$$K_A = \frac{p N_A^-}{N_A^{\,0}} \tag{12.6}$$

$$K_{AD} = \frac{N_A^- N_D^+}{N_A^{\,0} N_D^{\,0}} \tag{12.7}$$

$$K_{np} = \frac{np}{\left(\dfrac{2N}{V} - p\right)\left(\dfrac{2N}{V} - n\right)} \tag{12.8}$$

In the last of these equations, N is the number of atoms in the crystal, and $2N/V$ is the number of available states per unit volume in each band, so $(2N/V - p)$ is the concentration of electrons in the valence band, and $(2N/V - n)$ is the number of empty states in the conduction band. But $n \ll 2N/V$ and $p \ll 2N/V$, so n and p can be ignored in the denominator of (12.8) and we can write

$$K = np \tag{12.9}$$

where $K = K_{np} 4N^2/V^2$.

The K's defined by the above equations can be written in terms of the energy parameters of the system by substituting the appropriate expressions for n, N_D^+, $N_D{}^0$, p, N_A^-, and $N_A{}^0$. For the semiconductor we are considering here (large gap, small impurity content) (10.11) and (10.12) for electron and hole concentrations are valid. Using these equations for n and p, and (11.16), (11.17), (11.22), and (11.24) in the right-hand sides of (12.5), (12.6), (12.8), and (12.9), we get

$$\frac{nN_D^+}{N_D{}^0} = \rho_C e^{-E_{CD}/kT} \tag{12.10}$$

$$\frac{pN_A^-}{N_A{}^0} = \rho_V e^{-E_{AV}/kT} \tag{12.11}$$

$$\frac{N_A^- N_D^+}{N_A{}^0 N_D{}^0} = \tfrac{1}{4} e^{E_{DA}/kT} \tag{12.12}$$

$$np = 4\rho_C \rho_V e^{-E_g/kT} \tag{12.13}$$

where the energies in these expressions are defined by

$$\begin{aligned}
E_{CD} &= E_C - E_D \\
E_{AV} &= E_A - E_V \\
E_{DA} &= E_D - E_A \\
E_g &= E_C - E_V
\end{aligned} \tag{12.14}$$

and ρ_C and ρ_V are defined by*

$$\begin{aligned}
\rho_C &= \left(\frac{2\pi m_c kT}{h^2}\right)^{3/2} \\
\rho_V &= \left(\frac{2\pi m_v kT}{h^2}\right)^{3/2}
\end{aligned} \tag{12.15}$$

Equations 12.10 to 12.13 must be satisfied in our semiconducting system. Their value lies in the fact that the right-hand sides are functions of temperature that depend only on the nature of the crystal and its impurities, and not

---

* Examination of (10–11) shows that ρ_C can be given an interesting interpretation. The preexponential factors in this equation give the values n and p would have if $\mu = E_C$. Since, when the energy equals μ, the probability that a state contains an electron is $\tfrac{1}{2}$, the Boltzmann factor is half of the probability that the state E_C contains an electron. ρ_C can then be thought of as an effective density of states for the conduction band. In a similar way, ρ_V can be interpreted as an effective density of states for holes in the valence band.

on concentrations. To illustrate the use of these equations, consider the case in which the energy gap is so large that a negligible number of electrons are excited from the valence band to the conduction band. The number of conduction electrons is then equal to the number of ionized donors

$$N_D^+ = Vn \tag{12.16}$$

and the number of holes is equal to the number of ionized acceptors

$$N_A^- = Vp \tag{12.17}$$

Substituting these equations into (12.10) and (12.11) and remembering that

$$N_D{}^0 = N_D - N_D^+$$
$$N_A{}^0 = N_A - N_A^- \tag{12.18}$$

we get

$$\frac{n^2}{n_D - n} = \rho_C e^{-E_{CD}/kT} \tag{12.19}$$

$$\frac{p^2}{n_A - p} = \rho_V e^{-E_{AV}/kT} \tag{12.20}$$

where $n_D = N_D/V$ and $n_A = N_A/V$ are the concentrations of donors and acceptors, respectively. These equations show how an increase in the impurity concentration increases the concentration of charge carriers. Note that (12.13) is still valid. This means that an increase in concentration of donor impurities increases the electron concentration because of (12.19), but it decreases the hole concentration because of (12.13). Likewise, an increase in acceptor-impurity concentration increases the number of holes, but decreases the number of electrons. It follows that increasing the donor concentration suppresses the ionization of acceptors and vice versa.

3.13   THE RELATION BETWEEN FERMI LEVEL AND IMPURITY CONCENTRATION

In metals, where all electrons treated by particle statistics are in the conduction band, the Fermi level is easily found from the condition that the number of electrons is constant. In semiconductors, we cannot expect the situation to be so simple. The number of electrons in the bands depends on the values of the various energy levels and on the concentration of donor and acceptor impurities. We have already seen in Section 10 that for an intrinsic semiconductor the Fermi level is approximately midway between the valence and conduction levels. It will now be shown how the concentration of impurities affects the Fermi level.

In extrinsic semiconductors the Fermi level is determined by the requirement of electrical neutrality. The number of electrons plus the number of ionized acceptors must equal the number of holes plus the number of ionized donors; that is,

$$n + n_A^- = p + n_D^+ \tag{13.1}$$

where $n_A^- = N_A^-/V$ and $n_D^+ = N_D^+/V$. Substituting from (10.11), (11.12), (11.17), and (11.22) into (13.1) gives

$$2\rho_C e^{(\mu - E_C)/kT} + \frac{n_A}{2e^{(E_A - \mu)/kT} + 1} = 2\rho_V e^{(E_V - \mu)/kT} + \frac{n_D}{2e^{(\mu - E_D)/kT} + 1} \tag{13.2}$$

where ρ_C and ρ_V are defined by (12.15). This equation determines μ. If the impurity concentrations and the energy levels are known, μ can be obtained by graphical or numerical procedures. Simplified solutions for (13.2) can be obtained for several important cases:

Case 1   Weak Ionization. If $(E_A - \mu)$ and $(\mu - E_D)$ are large enough, or if the temperature is low enough, unity in the denominators of the fractions in (13.2) can be neglected relative to the exponentials. Some straightforward algebra then gives

$$\mu = \frac{E_D + E_A}{2} + \frac{kT}{2} \ln \left[\frac{2\rho_V e^{(E_V - E_D)/kT} + n_D/2}{2\rho_C e^{(E_A - E_C)/kT} + n_A/2} \right] \tag{13.3}$$

If n_D and n_A are small enough to be neglected, this equation reduces to

$$\mu = \frac{E_D + E_A}{2} + \frac{kT}{2} \ln \frac{\rho_V}{\rho_C} + \frac{(E_C - E_D) - (E_A - E_V)}{2} \tag{13.4}$$

Now compare this equation to (10.13) for an intrinsic semiconductor. The quantities $(E_C - E_D)$ and $(E_A - E_V)$ are the ionization energies of the donors and acceptors, respectively. They are generally about equal, and small compared to the values of the energy levels themselves. The logarithmic terms in (10.13) and (13.4) are identical. We see, therefore, that for a semiconductor containing a small amount of weakly ionized impurity, the Fermi level is near the middle of the energy gap.

It is of interest to estimate the limits of validity of (13.4). Assuming that the effective masses are equal to the electronic mass, a simple calculation from (12.15) gives $\rho_V = \rho_C = 1.73 \times 10^{20} \, T^{3/2}$ cm^{-3}. Since the ionization energies are of the order of 0.05 to 0.10 eV,

$$2\rho_C e^{(E_A - E_C)/kT} \approx 2\rho_V e^{(E_V - E_D)/kT} \approx 10^{23} \text{ cm}^{-3}$$

at room temperature. The concentration of impurities in semiconductors is always at least three or four orders of magnitude less than this value. In all practical cases, therefore, (13.4) is valid for weakly ionized impurities.

Case 2   Donor Impurities Only. If only donor impurities are present, $n_A^- = 0$ in (13.1). Furthermore, provided the gap width is not too small, there will be many more electrons in the conduction band than in the valence band; that is, because of (12.13), $p \ll n$. The electrical neutrality condition (13.1) then becomes

$$n = n_D^+ \tag{13.5}$$

Utilizing (10.11) and (11.22), this gives

$$2\rho_C e^{(\mu - E_C)/kT} = \frac{n_D}{2e^{(\mu - E_D)/kT} + 1} \tag{13.6}$$

This is readily transformed to a quadratic equation for $e^{\mu/kT}$ whose solution is

$$e^{\mu/kT} = \tfrac{1}{4} e^{E_D/kT} \left\{ -1 + \left[1 + \frac{4n_D}{\rho_C} e^{(E_C - E_D)/kT} \right]^{1/2} \right\} \tag{13.7}$$

From the preceding calculation of ρ_C

$$\frac{4n_D}{\rho_C} e^{(E_C - E_D)/kT} \ll 1$$

for most practical cases, and the square root can be expanded retaining terms to the first order. This gives

$$e^{\mu/kT} = \frac{n_D}{2\rho_C} e^{E_C/kT}$$

or

$$\mu = E_C + kT \ln \frac{n_D}{2\rho_C} \tag{13.8}$$

For n-type semiconductors, the Fermi level is, therefore, close to the conduction level, provided the temperature is not extremely high.

Case 3   Acceptor Impurities Only. If only acceptors are present, $n_D^+ = 0$ and $n \ll p$. Then (13-1) becomes

$$p = n_A^- \tag{13.9}$$

and from (10.12) and (11.15)

$$2\rho_C e^{(E_V - \mu)/kT} = \frac{n_A}{2e^{(E_A - \mu)/kT} + 1} \tag{13.10}$$

Proceeding as we did for the n-type conductor, this leads to

$$\mu = E_V - kT \ln \frac{n_A}{2\rho_V} \tag{13.11}$$

and for p-type conductors the Fermi level is near the valence level.

chapter 4

STATISTICAL-KINETIC THEORY
OF ELECTRON TRANSPORT

4.1 FREE ELECTRONS IN EXTERNAL FIELDS AND TEMPERATURE GRADIENTS

When a system of electrons is in equilibrium, the number of electrons in any volume element is constant in time and there is no net flow. Of course, the electrons are constantly moving about, but on the average, just as many enter a volume element per second as leave it. However, if an electric or magnetic field acts on the system, the electrons will accelerate in a direction determined by the field. A drift motion will thereby be superimposed on the random movements of the electrons, giving rise to a directional net flow.

Now consider an electron with velocity **v** in a state labeled by the wavenumber vector **k**.

Its energy is given by

$$\epsilon_k = \tfrac{1}{2}mv^2 = \frac{\hbar^2 k^2}{2m} \tag{1.1}$$

and its momentum **p** is

$$\mathbf{p} = m\mathbf{v} = \hbar\mathbf{k} \tag{1.2}$$

At a particular time t_0 switch on an electric field $\mathscr{E}$. The field exerts a force $-e\mathscr{E}$ on the electron, where e is the magnitude of the electronic charge, and since the force is the time rate of change of the momentum (by Newton's second law), we have

$$\frac{d\mathbf{p}}{dt} = \hbar\frac{d\mathbf{k}}{dt} = -e\mathscr{E} \tag{1.3}$$

This equation is readily integrated to yield **k** as a function of time. If the initial time of switching is taken to be zero, then (1.3) gives

$$\mathbf{k} = \mathbf{k}_0 - \frac{e\mathscr{E}}{\hbar} t \tag{1.4}$$

where t is the time after switching, and $\mathbf{k}_0$ is the wave-number vector before the field was turned on. Equation 1.4 is valid for all electrons, and it is clear that the effect of the field is to shift the entire Fermi distribution along the field direction by an amount that is the same for all electrons and proportional to the time. According to (1.2), $\mathbf{k} = m\mathbf{v}/\hbar$, so that, in terms of velocities, (1.4) becomes

$$\mathbf{v} = \mathbf{v}_0 - \frac{e\mathscr{E}}{m} t \tag{1.5}$$

which tells us that the field superimposes a constantly increasing velocity on the velocity $\mathbf{v}_0$ that the electron has in the absence of any field.

An analogous situation exists if the electron is acted on by a magnetic field **H**. In this case, the force acting on an electron of velocity **v** is given by electrodynamics as

$$- \frac{e\mathbf{v}}{c} \times \mathbf{H}$$

where c is the velocity of light, and Newton's second law gives

$$\frac{d\mathbf{v}}{dt} = - \frac{e}{mc} \mathbf{v} \times \mathbf{H} \tag{1.6}$$

If the average velocity after turning on the magnetic field is defined by

$$\mathbf{v}_{\text{av}} = \frac{1}{t} \int_0^t \mathbf{v} \, dt \tag{1.7}$$

Then (1.6) can be integrated to give

$$\mathbf{v} = \mathbf{v}_0 - \frac{e}{mc} \mathbf{v}_{\text{av}} \times \mathbf{H}t \tag{1.8}$$

Equation 1.8 tells us that the magnetic field has the effect of superimposing a velocity on the electron that is perpendicular to both the magnetic-field vector and the electron-velocity vector. In terms of wave-number vectors, (1.8) becomes

$$\mathbf{k} = \mathbf{k}_0 - \frac{e}{mc} \mathbf{k}_{\text{av}} \times \mathbf{H}t \tag{1.9}$$

Unlike the electric field, the magnetic field does not affect all electrons equally, irrespective of wave number. This is a result of the fact that magnetic forces are velocity dependent.

Equations 1.5 and 1.8 show that free electrons are constantly accelerated by external fields, and the velocities would reach very high values, unless some mechanism exists that opposes the field effects. We are interested in applying the free-electron model to real metals, and in these systems a variety of mechanisms operate that continually undo the accelerating effects of the external fields.

According to quantum theory, an electron passes through a perfect crystal with no resistance whatever because of the perfect three-dimensional periodicity of the system. A real crystal, however, is not perfect. Anything that disturbs the three-dimensional periodicity may scatter the electron and contributes to the resistance it meets during its flight. Electrons are scattered by atomic vibrations, impurity atoms, vacant lattice sites, interstitials, dislocations, grain boundaries, or anything else that upsets the ideal crystal structure. External fields cannot, therefore, accelerate electrons in metals indefinitely. They eventually interact with an imperfection, transfer energy, and momentum to it, and move out again in some other direction. Each such interaction may be thought of as a collision that erases the electron's memory of its interaction with the field. After each collision, the field has to start afresh to exert its influence on the electron. The final result is a balance between the field, which tries to bring the electrons away from equilibrium, and collisions that try to restore equilibrium.

If the average time between collisions is $\bar{\tau}$, then the average change in k-vectors caused by the field is obtained by replacing t by $\bar{\tau}$ in (1.4) and (1.9). That is, if we define $\delta \mathbf{k} = \mathbf{k} - \mathbf{k}_0$, then

$$\delta \mathbf{k} = -\frac{e\bar{\tau}}{\hbar} \mathscr{E} \qquad \text{(electric field)} \qquad (1.10)$$

$$\delta \mathbf{k} = -\frac{e\bar{\tau}}{mc} \mathbf{k} \times \mathbf{H} \quad \text{(magnetic field)} \qquad (1.11)$$

Of course, there will be a resulting shift in the Fermi distribution. For all practically realizable fields, however, the energy absorbed by the electrons from the field is small compared to the original electronic energy, so that the shift will be small. If $f(\mathbf{k})$ is the probability that an electron is in a state $\mathbf{k}$, we can, therefore, write

$$f(\mathbf{k}) = f_0(\mathbf{k}) + \Delta f(k) \qquad (1.12)$$

where $f_0(\mathbf{k})$ is the equilibrium Fermi function* and $\Delta f \ll f_0$.

---

* In Chapter 3, the subscript 0 was not needed on the Fermi function, since all systems were always in equilibrium.

Now consider a free-electron gas in a temperature gradient. Because temperature is a measure of kinetic energy, the average velocity of electrons in hot regions will be greater than the average velocity in cold regions. As a result, there will be a net flow of electrons from hot to cold regions. These electrons carry kinetic energy with them; therefore, there is a corresponding heat flow.

The temperature gradient also gives rise to a shift in the Fermi distribution. This shift has two origins. First, the distribution function depends on temperature, and if a gradient exists, the distribution will vary with position. Second, the distribution depends on the Fermi level, and to the extent that this varies with temperature, it will vary with position. Of these two effects, the first effect is the more important since (as shown in Chapter 3) the Fermi level varies very slowly with temperature.

There is one more comment to make at this point. Electrons carry both kinetic energy and charge. Therefore, the heat flow caused by a temperature gradient should be accompanied by electrical effects. Conversely, the current caused by electric and magnetic fields should be accompanied by thermal effects. The theory developed in this chapter describes these effects as well as the ordinary electrical and thermal conductivity.

4.2 THE STATISTICAL-KINETIC METHOD

The theory of electron transport in metals can be worked out by treating the conduction electrons as Fermi particles and investigating their kinetics. The particle and energy fluxes are expressed in terms of the particle velocities and the electron-distribution function. The distribution function is then related to the external fields and the temperature gradient. The methods of equilibrium statistical mechanics are not adequate for this task. One reason for this is that the system is no longer at equilibrium so that the distribution function is not given by the equations of Chapter 3. However, for fields and gradients of experimental interest, the perturbed distribution is not very different from the equilibrium distribution. An approximate calculation of gradients of the distribution that is reasonably accurate can, therefore, be made.

A more fundamental difficulty is that equilibrium statistical mechanics has nothing to say about the details of transitions from one state to another. But this is precisely what is of interest in transport theory. When an electron in a particular energy state takes part in a collision, it is scattered into another state, and the rate at which this happens is of prime importance in transport theory. The theory must, therefore, be supplemented by introducing transition probability functions that give the rate at which electrons move from one state to another.

Now let us derive the fundamental flux equations by a kinetic argument. Choose a plane in the system across which the flux is to be calculated, and consider all electrons in a surface area Δs with velocity $\mathbf{v}$. In a small time interval dt, a particle of velocity $\mathbf{v}$ will reach the flux plane if it is anywhere within a distance $\Delta x = \mathbf{v} \cdot \mathbf{i}_n \, dt$, where $\mathbf{i}_n$ is a unit vector normal to the flux plane and Δx is measured in the $\mathbf{i}_n$ direction. Therefore, all of the electrons with velocity $\mathbf{v}$ that are contained in the volume element $\mathbf{v} \cdot \mathbf{i}_n \, dt \, \Delta s$ will cross the area element Δs in time dt. If $N(\mathbf{r}, \mathbf{v}, t) \, d\mathbf{r} \, d\mathbf{v}$ is the number of electrons at time t in a volume element $d\mathbf{r}$ with velocities in the range $\mathbf{v}$ to $\mathbf{v} + d\mathbf{v}$, then the number that cross Δs in time dt with velocity $\mathbf{v}$ is

$$N(\mathbf{r}, \mathbf{v}, t)\mathbf{v} \cdot \mathbf{i}_n \, dt \, \Delta s \, d\mathbf{v} \tag{2.1}$$

The total electric charge crossing Δs is obtained by multiplying (2.1) by $-e$ and integrating over all $\mathbf{v}$:

$$-e\Delta s \, dt \int \mathbf{v} \cdot \mathbf{i}_n N(\mathbf{r}, \mathbf{v}, t) \, d\mathbf{v} \tag{2.2}$$

The flux of electrons in the $\mathbf{i}_n$ direction is the flow per unit area per unit time, which is just (2.2) divided by $\Delta s \, dt$.

$$I_n = -e \int \mathbf{v} \cdot \mathbf{i}_n N(\mathbf{r}, \mathbf{v}, t) \, d\mathbf{v} \tag{2.3}$$

The component of the flux normal to the flux plane is related to the flux vector $\mathbf{I}$ by $I_n = \mathbf{I} \cdot \mathbf{i}_n$ and, therefore,

$$\mathbf{I} = -e \int \mathbf{v} N(\mathbf{r}, \mathbf{v}, t) \, d\mathbf{v} \tag{2.4}$$

The flux of energy is obtained by multiplying (2.1) by the kinetic energy $mv^2/2$ and integrating over $\mathbf{v}$. The result is

$$\mathbf{J} = \frac{m}{2} \int v^2 \mathbf{v} N(\mathbf{r}, \mathbf{v}, t) \, d\mathbf{v} \tag{2.5}$$

where $\mathbf{J}$ is the heat-flux vector.

The flux equations (2.4) and (2.5) constitute the kinetic content of the theory. The statistical content lies in the calculation of the distribution function $N(\mathbf{r}, \mathbf{v}, t)$. For an equilibrium system this function reduces to the equilibrium velocity distribution per unit volume given by (3.39) of Chapter 3. For a nonequilibrium system, $N(\mathbf{r}, \mathbf{v}, t)$ is the distribution resulting from the balance between the effects of the fields and the collisions. To describe the way in which the collisions restore equilibrium, we define a conditional transition probability function $\wedge (\mathbf{v}, \mathbf{v}')$ so that if an electron initially has a velocity $\mathbf{v}$, then the probability that its velocity changes to a value in the range

$\mathbf{v}'$ to $\mathbf{v}' + d\mathbf{v}'$ in a time dt is $\wedge (\mathbf{v}, \mathbf{v}') d\mathbf{v}' \, dt$. The physically reasonable assumption is made that $\wedge (\mathbf{v}, \mathbf{v}')$ does not depend on the electron's position, the external fields or the time.

The rate at which $N(\mathbf{r}, \mathbf{v}, t)$ is changed by the collisions can now be expressed in terms of transition probabilities. The decrease in the number of electrons in a volume element $d\mathbf{r}$ with velocities in a range $d\mathbf{v}$ in a time dt is

$$d\mathbf{r} \, d\mathbf{v} \, dt \int N(\mathbf{r}, \mathbf{v}, t) \wedge (\mathbf{v}, \mathbf{v}') \, d\mathbf{v}' \tag{2.6}$$

The increase in the number of electrons in $d\mathbf{r}$ with velocities in $d\mathbf{v}$ is

$$d\mathbf{r} \, d\mathbf{v} \, dt \int N(\mathbf{r}, \mathbf{v}', t) \wedge (\mathbf{v}', \mathbf{v}) \, d\mathbf{v}' \tag{2.7}$$

Equation 2.6 is the result of counting all particles with velocity $\mathbf{v}$, multiplying by the probability that an electron's velocity will change to $\mathbf{v}'$ and then integrating over all $\mathbf{v}'$. Similarly, counting the electrons with velocity changes from $\mathbf{v}'$ to $\mathbf{v}$, and integrating over all $\mathbf{v}'$, gives (2.7). If (2.6) is subtracted from (2.7) and the result divided by $d\mathbf{r} \, d\mathbf{v} \, dt$, we get the rate at which $N(\mathbf{r}, \mathbf{v}, t)$ changes with time

$$\left(\frac{\partial N(\mathbf{r}, \mathbf{v}, t)}{\partial t} \right)_c = \int N(\mathbf{r}, \mathbf{v}', t') \wedge (\mathbf{v}', \mathbf{v}) \, dv' - N(\mathbf{r}, \mathbf{v}, t) \int \wedge (\mathbf{v}, \mathbf{v}') \, dv' \tag{2.8}$$

The subscript c is put on the derivative to emphasize that this is the time rate of change resulting from collisions. Equation 2.8 will be useful for the development of transport theory. It contains the physical collision mechanisms in the $\wedge (\mathbf{v}, \mathbf{v}')$.

4.3   THE BOLTZMANN TRANSPORT EQUATION

The particle distribution of a system not in equilibrium depends on time, not only because of the collisions, but also because the temperature gradient and the external fields impart a drift velocity to the random-particle motions. The total rate of change of N with time must, therefore, consist of two terms and may be written as

$$\frac{\partial N}{\partial t} (\mathbf{r}, \mathbf{v}, t) = \left(\frac{\partial N}{\partial t} \right)_c + \left(\frac{\partial N}{\partial t} \right)_d \tag{3.1}$$

The first term on the right is the collision derivative discussed in the previous section, while the second term is the drift derivative that originates in the fields and temperature gradient.

The drift derivative can be related to the external influences by following the motion of a given group of electrons. At time t, the number of electrons with positions in $d\mathbf{r}$ and velocities in $d\mathbf{v}$ is

$$N(\mathbf{r}, \mathbf{v}, t) \, d\mathbf{r} \, d\mathbf{v} \tag{3.2}$$

At a later time $t + \Delta t$, the positions and velocities of all the electrons in this group have changed to

$$\mathbf{r} + \Delta\mathbf{r} = \mathbf{r} + \mathbf{v}\,\Delta t$$
$$\mathbf{v} + \Delta\mathbf{v} = \mathbf{v} + \mathbf{a}\,\Delta t \tag{3.3}$$

where $\mathbf{a} = \dot{\mathbf{v}}$ is the particle acceleration. However, the number of electrons in the group has not changed, so that

$$N(\mathbf{r} + \mathbf{v}\,\Delta t, \mathbf{v} + \mathbf{a}\,\Delta t, t + \Delta t) = N(\mathbf{r}, \mathbf{v}, t) \tag{3.4}$$

Expand the left-hand side of this equation in a Taylor series. Since Δt can be arbitrarily small, only terms of the first order in Δt need be retained, and we find that

$$\Delta t \left[\mathbf{v} \cdot \frac{\partial N}{\partial \mathbf{r}} + \mathbf{a} \cdot \frac{\partial N}{\partial \mathbf{v}} + \frac{\partial N}{\partial t} \right] = 0 \tag{3.5}$$

The vector derivatives are defined by

$$\frac{\partial N}{\partial \mathbf{r}} = \frac{\partial N}{\partial x}\mathbf{i} + \frac{\partial N}{\partial y}\mathbf{j} + \frac{\partial N}{\partial z}\mathbf{k}$$
$$\frac{\partial N}{\partial \mathbf{v}} = \frac{\partial N}{\partial v_x}\mathbf{i} + \frac{\partial N}{\partial v_y}\mathbf{j} + \frac{\partial N}{\partial v_z}\mathbf{k} \tag{3.6}$$

The time derivative in (3.5) is just the drift derivative. Therefore

$$\left(\frac{\partial N}{\partial t} \right)_d = - \left[\mathbf{v} \cdot \frac{\partial N}{\partial \mathbf{r}} + \mathbf{a} \cdot \frac{\partial N}{\partial \mathbf{v}} \right] \tag{3.7}$$

This equation connects the transport properties to the external fields through the acceleration $\mathbf{a}$, and to the temperature gradient through the derivative $\partial N/\partial \mathbf{r}$.

For gradients and fields that are constant in time, the system eventually reaches steady state, for which the left-hand side of (3.1) is zero. Actually, the electrons suffer so many collisions per second that steady state is reached very rapidly after the fields are switched on. We can, therefore, substitute (3.7) and (2.8) into (3.1) and set the right-hand side equal to zero to get

$$\mathbf{v} \cdot \frac{\partial N}{\partial \mathbf{r}} + \mathbf{a} \cdot \frac{\partial N}{\partial \mathbf{v}} = \int N(\mathbf{r}, \mathbf{v}', t) \wedge (\mathbf{v}', \mathbf{v}) \, d\mathbf{v}' - N(\mathbf{r}, \mathbf{v}, t) \int \wedge (\mathbf{v}, \mathbf{v}') \, d\mathbf{v}' \tag{3.8}$$

This is the Boltzmann transport equation. It is the equation that must be solved in order to obtain $N(\mathbf{r}, \mathbf{v}, t)$, which can then be substituted into the flux equations (2.4) and (2.5) to get the transport properties. The left-hand side of (3.8) can be reduced to a more useful form without too much trouble. For a system subject to constant electric and magnetic fields with intensities given by $\mathscr{E}$ and $\mathbf{H}$, respectively, the force on an electron is given from electrodynamics by the Lorentz expression

$$m\mathbf{a} = -\left[e\mathscr{E} + \frac{\mathbf{v} \times \mathbf{H}}{c}\right] \tag{3.9}$$

where c is the velocity of light. Furthermore, since $\mathscr{E}$ and $\mathbf{H}$ do not vary with position, the spatial derivative of N must be related to the temperature gradient by

$$\frac{\partial N}{\partial \mathbf{r}} = \frac{\partial N}{\partial T}\frac{\partial T}{\partial \mathbf{r}} = \frac{\partial N}{\partial T}\nabla T \tag{3.10}$$

The use of (3.9) and (3.10) in the left-hand side of (3.8) gives

$$\mathbf{v} \cdot \frac{\partial N}{\partial \mathbf{r}} + \mathbf{a} \cdot \frac{\partial N}{\partial \mathbf{v}} = \nabla T \cdot \mathbf{v}\frac{\partial N}{\partial T} - \frac{e}{m}\mathscr{E} \cdot \frac{\partial N}{\partial \mathbf{v}} - \frac{e}{mc}(\mathbf{v} \times \mathbf{H}) \cdot \frac{\partial N}{\partial \mathbf{v}} \tag{3.11}$$

The right-hand side of (3.8), which is just the collision derivative, is somewhat more difficult to handle. A workable procedure is to introduce a function $\tau(\mathbf{v})$ that is defined by

$$\left(\frac{\partial N}{\partial t}\right)_c = -\frac{N(\mathbf{r}, \mathbf{v}, t) - N^0(\mathbf{v})}{\tau(\mathbf{v})} \tag{3.12}$$

$\tau(\mathbf{v})$ has the dimension of time and is evidently some measure of the rate at which the particle distribution function regresses to its equilibrium value $N^0(\mathbf{v})$.

Integration of (3.12) gives

$$\ln\left[N(t) - N^0\right] = -\frac{t}{\tau(\mathbf{v})} + \ln g(\mathbf{r}, \mathbf{v}) \tag{3.13}$$

where $g(\mathbf{r}, \mathbf{v})$ is an arbitrary function not involving the time. Now consider a system at steady state and at some time, which may be labeled $t = 0$, suddenly turn off all fields and gradients. Let $N(0)$ be the steady state distribution before the fields are switched off. For this situation g is clearly given by

$$g(\mathbf{r}, \mathbf{v}) = N(0) - N^0 \tag{3.14}$$

and (3.13) becomes

$$N(t) - N^0 = \Delta N e^{-t/\tau(\mathbf{v})} \tag{3.15}$$

where we have written ΔN for the initial deviation from equilibrium $[N(0) - N^0]$.

Equation 3.15 provides a straightforward interpretation of $\tau(\mathbf{v})$. It is the characteristic relaxation time for electrons of velocity $\mathbf{v}$ governing the approach to equilibrium of a perturbed distribution. Its value obviously depends on the collision mechanism, and $\tau(\mathbf{v})$ should be different for different mechanisms.

What if several collision mechanisms are operative in the same system? Then each such mechanism contributes to the collision derivative and (3.12) should be written as

$$\left(\frac{\partial N}{\partial t}\right)_c = -\frac{N - N^0}{\tau_1(\mathbf{v})} - \frac{N - N^0}{\tau_2(\mathbf{v})} - \cdots \tag{3.16}$$

where $\tau_j(\mathbf{v})$ refers to the jth mechanism, and there are as many terms on the right-hand side of (3.16) as there are mechanisms. Equation 3.16 can be put in the same form as (3.12) just by defining a combined relaxation time by

$$\frac{1}{\tau(\mathbf{v})} \equiv \frac{1}{\tau_1(\mathbf{v})} + \frac{1}{\tau_2(\mathbf{v})} + \cdots \tag{3.17}$$

In using the relaxation time concept, it is necessary to keep in mind that an overall relaxation time may be the result of several mechanisms whose effects are combined according to (3.17).

With this discussion in mind, let us now equate the right-hand sides of (3.11) and (3.12) and solve for $N(\mathbf{r}, \mathbf{v}, t)$. The result is

$$N(\mathbf{r}, \mathbf{v}, t) = N^0(\mathbf{v}) - \tau(\mathbf{v})\frac{\partial N}{\partial T}\mathbf{v} \cdot \nabla T$$

$$+ \frac{e}{m}\tau(\mathbf{v})\mathscr{E} \cdot \frac{\partial N}{\partial \mathbf{v}} + \frac{e}{mc}\tau(\mathbf{v})[\mathbf{v} \times \mathbf{H}] \cdot \frac{\partial N}{\partial \mathbf{v}} \tag{3.18}$$

We have arrived at a formal solution to the problem of determining $N(\mathbf{r}, \mathbf{v}, t)$. If $\tau(\mathbf{v})$ is computed, the solution becomes explicit. The computation of $\tau(\mathbf{v})$ will be discussed later. For the time being, we will go as far as we can without a detailed theory for the relaxation times.

The last three terms in (3.18) describe the departure of the electron distribution from equilibrium. This description is facilitated by a formal device in which a vector $\boldsymbol{\eta}$ is defined by

$$N = N^0 + \mathbf{v} \cdot \boldsymbol{\eta} \tag{3.19}$$

It is assumed that $\boldsymbol{\eta}$ is a function only of the magnitude of the velocity. Comparing (3.19) and (3.18) shows that

$$\mathbf{v} \cdot \boldsymbol{\eta} = -\tau(\mathbf{v})\frac{\partial N}{\partial T}\mathbf{v} \cdot \nabla T + \frac{e}{m}\tau(\mathbf{v})\mathscr{E} \cdot \frac{\partial N}{\partial \mathbf{v}} + \frac{e}{mc}\tau(\mathbf{v})(\mathbf{v} \times \mathbf{H}) \cdot \frac{\partial N}{\partial \mathbf{v}} \tag{3.20}$$

Expressing the derivatives in (3.20) in terms of $\boldsymbol{\eta}$ by using (3.19) gives

$$\mathbf{v} \cdot \boldsymbol{\eta} = -\tau \mathbf{v} \cdot \nabla T \left[\frac{\partial N^0}{\partial T} + \frac{\partial (\mathbf{v} \cdot \boldsymbol{\eta})}{\partial T} \right]$$

$$+ \frac{e\tau}{m} \boldsymbol{\mathscr{E}} \cdot \left[\frac{\partial N^0}{\partial \mathbf{v}} + \frac{\partial (\mathbf{v} \cdot \boldsymbol{\eta})}{\partial \mathbf{v}} \right]$$

$$+ \frac{e\tau}{mc} [\mathbf{v} \times \mathbf{H}] \cdot \left[\frac{\partial N^0}{\partial \mathbf{v}} + \frac{\partial (\mathbf{v} \cdot \boldsymbol{\eta})}{\partial \mathbf{v}} \right] \qquad (3.21)$$

Because we are dealing with systems in which the deviation from equilibrium is small, we have that

$$\frac{\partial (\mathbf{v} \cdot \boldsymbol{\eta})}{\partial T} \ll \frac{\partial N^0}{\partial T}$$

$$\frac{\partial (\mathbf{v} \cdot \boldsymbol{\eta})}{\partial \mathbf{v}} \ll \frac{\partial N^0}{\partial \mathbf{v}}$$

and to a first approximation, it would seem that the departures from equilibrium can always be neglected in evaluating derivatives of the distribution function. This is true for the first two terms in (3.21). In the third term, however, retention of the derivative $\partial (\mathbf{v} \cdot \boldsymbol{\eta})/\partial \mathbf{v}$ is essential. This can be seen by working out the triple product in the last term:

$$[\mathbf{v} \times \mathbf{H}] \cdot \left[\frac{\partial N^0}{\partial \mathbf{v}} + \frac{\partial (\mathbf{v} \cdot \boldsymbol{\eta})}{\partial \mathbf{v}} \right] = (\mathbf{v} \times \mathbf{H}) \cdot \left(m\mathbf{v} \frac{\partial N^0}{\partial \epsilon} \right) + (\mathbf{v} \times \mathbf{H}) \cdot \frac{\partial (\mathbf{v} \cdot \boldsymbol{\eta})}{\partial \mathbf{v}} \qquad (3.22)$$

where

$$\frac{\partial N^0}{\partial \mathbf{v}} = m\mathbf{v} \frac{\partial N^0}{\partial \epsilon}$$

because the kinetic energy of an electron is $\epsilon = (m/2)(v_x{}^2 + v_y{}^2 + v_z{}^2)$. But $(\mathbf{v} \times \mathbf{H}) \cdot \mathbf{v} = 0$ from vector algebra, so the first term on the right of (3.22) vanishes and (3.21) becomes

$$\mathbf{v} \cdot \boldsymbol{\eta} = -\tau \mathbf{v} \cdot \nabla T \frac{\partial N^0}{\partial T} + \frac{e\tau}{m} \boldsymbol{\mathscr{E}} \cdot \frac{\partial N^0}{\partial \mathbf{v}} + \frac{e\tau}{mc} (\mathbf{v} \times \mathbf{H}) \cdot \frac{\partial (\mathbf{v} \cdot \boldsymbol{\eta})}{\partial \mathbf{v}} \qquad (3.23)$$

Evidently, as long as no magnetic fields are present, derivatives of the distribution function can be treated as if the system were at equilibrium. In the presence of magnetic fields, however, such a procedure would give a zero result for all magnetic effects, and it is necessary to consider the departure from equilibrium in evaluating derivatives.

4.4 FORMAL FLUX EQUATIONS

To get the flux equation for the electric current, substitute (3.18) into (2.4):

$$\mathbf{I} = -e \int \mathbf{v} \left[N^0 - \tau \frac{\partial N}{\partial T} (\mathbf{v} \cdot \nabla T) + \frac{e\tau}{m} \left(\frac{\partial N}{\partial \mathbf{v}} \cdot \mathscr{E} \right) + \frac{e\tau}{mc} (\mathbf{v} \times \mathbf{H}) \cdot \frac{\partial N}{\partial \mathbf{v}} \right] d\mathbf{v} \quad (4.1)$$

The function N^0 is just $N(v_x, v_y, v_z)$, as given in Chapter 3 by (3.39), per unit volume:

$$N^0(\mathbf{v}) = \frac{2m^3}{h^3} [e^{(m/2kT)(v^2 - v_F{}^2)} + 1]^{-1} = \frac{2m^3}{h^3} [e^{(\epsilon - \mu)/kT} + 1]^{-1} \quad (4.2)$$

Since N^0 is symmetric in v, integration of the first term in (4.1) gives zero because of the antisymmetry of vN^0. Therefore

$$\mathbf{I} = e \int \tau \mathbf{v} \frac{\partial N}{\partial T} (\mathbf{v} \cdot \nabla T) \, d\mathbf{v} - \frac{e^2}{m} \int \tau \mathbf{v} \left(\frac{\partial N}{\partial \mathbf{v}} \cdot \mathscr{E} \right) d\mathbf{v}$$

$$- \frac{e^2}{mc} \int \tau \mathbf{v} (\mathbf{v} \times \mathbf{H}) \cdot \frac{\partial N}{\partial \mathbf{v}} \, d\mathbf{v} \quad (4.3)$$

In the same way, the heat-current vector is obtained by combining (3.18) with (2.5).

$$\mathbf{J} = -\frac{m}{2} \int \tau \mathbf{v} v^2 \frac{\partial N}{\partial T} (\mathbf{v} \cdot \nabla T) \, d\mathbf{v}$$

$$+ \frac{e}{2} \int \tau \mathbf{v} v^2 \left(\frac{\partial N}{\partial \mathbf{v}} \cdot \mathscr{E} \right) d\mathbf{v}$$

$$+ \frac{e}{2c} \int \tau \mathbf{v} v^2 (\mathbf{v} \times \mathbf{H}) \cdot \frac{\partial N}{\partial \mathbf{v}} \, d\mathbf{v} \quad (4.4)$$

The entire theory of thermal and electrical conduction is contained in these two flux equations. Note that all of the external influences contribute to the electric and thermal fluxes. Thus, there is an electric current arising not only from the electric field $\mathscr{E}$, but also from the temperature gradient ∇T and the magnetic field H. Similarly, the electric and magnetic fields, as well as the temperature gradient, contribute to the heat flow.

All that is needed to extract the transport coefficients from the flux equations is a computation of the various integrals. In computing these integrals, it will be assumed that the derivatives of the distribution function can be replaced by the values they would have at equilibrium. Since we are dealing with small perturbations of the distribution function from equilibrium, this is a sufficiently accurate approximation.

4.5   THE ELECTRICAL CONDUCTIVITY

To calculate the electrical conductivity, consider a system in which the magnetic field and the temperature gradient are both zero, and the electric field has only a component in the x direction of magnitude $\mathscr{E}_x$. Equation 4.3 then reduces to

$$I = - \frac{e^2 \mathscr{E}_x}{m}\left[\mathbf{i}\int \tau v_x \frac{\partial N}{\partial v_x}\, d\mathbf{v} + \mathbf{j}\int \tau v_y \frac{\partial N}{\partial v_x}\, d\mathbf{v} + \mathbf{k}\int \tau v_z \frac{\partial N}{\partial v_x}\, d\mathbf{v} \right] \quad (5.1)$$

Only the first of the three integrals in this equation is nonzero, since the integrands in the other integrals are odd functions of the velocity, as can be verified by the methods developed below. Then (5.1) becomes

$$I_x = \sigma \mathscr{E}_x$$

where the conductivity σ is

$$\sigma = - \frac{e^2}{m}\int \tau v_x \frac{\partial N}{\partial v_x}\, d\mathbf{v} \quad (5.2)$$

To evaluate this integral, use the relation

$$\epsilon = \tfrac{1}{2}mv^2 = \tfrac{1}{2}m[v_x{}^2 + v_y{}^2 + v_z{}^2] \quad (5.3)$$

to get

$$\frac{\partial N}{\partial v_x} = \frac{\partial N}{\partial \epsilon}\frac{\partial \epsilon}{\partial v_x} = mv_x \frac{\partial N}{\partial \epsilon} \quad (5.4)$$

so that (5.2) becomes

$$\sigma = -e^2 \int \tau v_x{}^2 \frac{\partial N}{\partial \epsilon}\, d\mathbf{v} \quad (5.5)$$

If we assume that τ is a function of the magnitude of the velocity only, then the value of the integral remains unchanged if v_x is replaced by v_y or v_z. Therefore, since $v^2 = v_x{}^2 + v_y{}^2 + v_z{}^2$, (5.5) can be written as

$$\sigma = - \frac{e^2}{3}\int \tau v^2 \frac{\partial N}{\partial \epsilon}\, d\mathbf{v} \quad (5.6)$$

Now express $d\mathbf{v}$ in spherical polar coordinates by

$$d\mathbf{v} = 4\pi v^2\, dv \quad (5.7)$$

or, using (5.3),

$$d\mathbf{v} = \frac{4\pi v}{m}\, d\epsilon \quad (5.8)$$

Equation 5.2, therefore, becomes

$$\sigma = -\frac{8\pi m^2 e^2}{3h^3} \int_0^\infty \tau v^3 \frac{\partial f}{\partial \epsilon}\, d\epsilon \tag{5.9}$$

where N has been replaced by $2m^3f/h^3$ in the derivative according to (4.2) in keeping with our restriction to small deviations from equilibrium. The conductivity is now expressed in terms of a Fermi integral, and if only the first term in the expression for the Fermi integral is retained, we have

$$\int_0^\infty \tau v^3 \frac{\partial f}{\partial \epsilon}\, d\epsilon = -\tau(\mu)[v(\mu)]^3 \tag{5.10}$$

where $v(\mu)$ is the velocity of an electron of energy μ. In terms of the Fermi energy, (5.10) is

$$\int_0^\infty \tau v^3 \frac{\partial f}{\partial \epsilon}\, d\epsilon = -\tau(\mu)\left(\frac{2\mu}{m}\right)^{3/2} \tag{5.11}$$

Substituting (5.11) into (5.9) and neglecting the temperature variation of μ, so that (3.8) of Chapter 3 can be used to relate the Fermi energy to the particle density n, gives

$$\sigma = \frac{ne^2\tau(\mu)}{m} \tag{5.12}$$

The electrical conductivity, therefore, reduces to a very simple form. Note that only the relaxation time of electrons with energy equal to the Fermi energy appears in this equation. Therefore, if we are willing to treat $\tau(\mu)$ as a disposable parameter, no detailed knowledge of the scattering mechanisms is necessary. Of course, a detailed theory is needed in order to compute the numerical value of τ and compare it with experiment.

An approximate quantum mechanical analysis can be developed for the interaction between electrons and lattice vibrations that shows that in a pure metal, $\tau(\mu)$ is inversely proportional to the temperature. Since n is essentially independent of temperature in a metal, this means that the resistivity increases linearly with temperature.

For electron densities of the order of 10^{22} cm^{-3} and specific conductivities of 10^5 ohm^{-1} cm^{-1}, which is characteristic of most metals at room temperature, (5.12) gives a value of about 10^{-14} seconds for τ. The relaxation time is a measure of the average time between scattering collisions, so that multiplying it by the velocity of an electron with energy μ gives a measure of the mean free path between collisions. For a Fermi energy of 4 eV, the room temperature mean free path turns out to be about 150 Å.

4.6   THERMAL CONDUCTIVITY AND THE WIEDEMANN-FRANZ LAW

If the only external influence acting on our system of electrons is a temperature gradient, electrons will move from hot to cold regions. As a result of the spatial rearrangement of electrons caused by the temperature gradient, an internal electric field will be established in the system. This internal field induces an electron flow in the opposite direction to that resulting from the temperature gradient. An analysis of thermal conduction, therefore, requires the use of both the charge flow and heat-flow equations. If the temperature gradient is along the x axis, the heat-flow equation (4.4) becomes

$$J_x = -\frac{m}{2}\int \tau v_x^2 v^2 \frac{\partial N}{\partial T}\, d\mathbf{v}\, \frac{dT}{dx} + \frac{e}{2}\int \tau v_x v^2 \frac{\partial N}{\partial v_x}\, d\mathbf{v}\, \mathcal{E}_x^{\,T} \tag{6.1}$$

where $\mathcal{E}_x^{\,T}$ is the electric field induced by the temperature gradient.

Thermal-conductivity measurements are carried out under open electrical-circuit conditions. The electric-current vector in (4.3), therefore, vanishes. Since, in the present case, $\nabla T \to dT/dx$, $\mathcal{E} \to \mathcal{E}_x^{\,T}$, and $\mathbf{H} = 0$, we can set $\mathbf{I} = 0$ in (4.3) and solve for $\mathcal{E}_x^{\,T}$ with the result that

$$\mathcal{E}_x^{\,T} = \frac{m}{e}\frac{\displaystyle\int \tau v_x^2 \frac{\partial N}{\partial T}\, d\mathbf{v}}{\displaystyle\int \tau v_x \frac{\partial N}{\partial v_x}\, d\mathbf{v}}\,\frac{dT}{dx} \tag{6.2}$$

Now $\mathcal{E}_x^{\,T}$ can be eliminated from (6.1) to give the heat flow in terms of the temperature gradient above. This gives

$$J_x = -\frac{m}{2}\left[\frac{I_2 I_3 - I_1 I_4}{I_2}\right]\frac{dT}{dx} \tag{6.3}$$

where the I's are defined by

$$I_1 = \int \tau v_x^2 \frac{\partial N}{\partial T}\, d\mathbf{v} \tag{6.4}$$

$$I_2 = \int \tau v_x \frac{\partial N}{\partial v_x}\, d\mathbf{v} \tag{6.5}$$

$$I_3 = \int \tau v_x^2 v^2 \frac{\partial N}{\partial T}\, d\mathbf{v} \tag{6.6}$$

$$I_4 = \int \tau v_x v^2 \frac{\partial N}{\partial v_x}\, d\mathbf{v} \tag{6.7}$$

To evaluate these integrals, it is convenient to express them in terms of energy rather than velocity. To do this, use (5.4) and (5.8), replace v_x^2 by $v^2/3$ and remember that $N = 2m^3f/h^3$ and $v = (2\epsilon/m)^{1/2}$. The result is that

$$I_1 = \frac{16\sqrt{2}\,\pi\sqrt{m}}{3h^3} \int \tau\epsilon^{3/2} \frac{\partial f}{\partial T}\, d\epsilon \tag{6.8}$$

$$I_2 = \frac{16\sqrt{2}\,\pi m^{3/2}}{3h^3} \int \tau\epsilon^{3/2} \frac{\partial f}{\partial \epsilon}\, d\epsilon \tag{6.9}$$

$$I_3 = \frac{32\sqrt{2}\,\pi}{3h^3\sqrt{m}} \int \tau\epsilon^{5/2} \frac{\partial f}{\partial T}\, d\epsilon \tag{6.10}$$

$$I_4 = \frac{32\sqrt{2}\,\pi\sqrt{m}}{3h^3} \int \tau\epsilon^{5/2} \frac{\partial f}{\partial \epsilon}\, d\epsilon \tag{6.11}$$

To apply the series-expansion method of evaluating Fermi integrals to I_1 and I_3, $\partial f/\partial T$ must be converted to a derivative with respect to ϵ. It is easy to show that

$$\frac{\partial f}{\partial T} = f^2 e^{(\epsilon-\mu)/kT}\left[\frac{\epsilon-\mu}{kT^2} + \frac{1}{kT}\frac{d\mu}{dT}\right] \tag{6.12}$$

Also, since

$$\frac{\partial f}{\partial \epsilon} = -\frac{f^2}{kT} e^{(\epsilon-\mu)/kT} \tag{6.13}$$

(6.12) can be written as

$$\frac{\partial f}{\partial T} = g(\epsilon)\frac{\partial f}{\partial \epsilon} \tag{6.14}$$

where the function $g(\epsilon)$ is defined by

$$g(\epsilon) = -\frac{d\mu}{dT} - \frac{(\epsilon-\mu)}{T} = \frac{\pi^2 k^2 T}{6\mu_0} - \frac{(\epsilon-\mu)}{T} \tag{6.15}$$

The last equality is obtained by evaluating $d\mu/dT$ from (4.21) of Chapter 3. Substitution of (6.14) into (6.8) and (6.10) gives

$$I_1 = \frac{16\pi\sqrt{2m}}{3h^3} \int \tau g(\epsilon)\epsilon^{3/2} \frac{\partial f}{\partial \epsilon}\, d\epsilon \tag{6.16}$$

$$I_3 = \frac{32\pi\sqrt{2}}{3h^3\sqrt{m}} \int \tau g(\epsilon)\epsilon^{5/2} \frac{\partial f}{\partial \epsilon}\, d\epsilon \tag{6.17}$$

Now all the integrals I_1 to I_4 can be treated by the series expansion formula (4.9) of Chapter 3. In applying this formula to the electrical-conductivity calculation, only the first term in the series was retained since all other terms are negligibly small. The thermal conductivity, however, would vanish if only the first term in the series expansion were retained. To get a nonzero result, it is necessary to retain the first two terms of the expansion in evaluating the integrals I_1 to I_4. A straightforward application of (4.9) of Chapter 3 to (6.9), (6.11), (6.16), and (6.17) gives

$$I_1 = - \frac{16\pi\sqrt{2m}}{3h^3} \left[\tau(\mu)g(\mu)\mu^{3/2} + \frac{(2\pi kT)^2}{24}(\tau g \epsilon^{3/2})'' \right] \qquad (6.18)$$

$$I_2 = - \frac{16\pi\sqrt{2}\,m^{3/2}}{3h^3} \left[\tau(\mu)\mu^{3/2} + \frac{(2\pi kT)^2}{24}(\tau \epsilon^{3/2})'' \right] \qquad (6.19)$$

$$I_3 = - \frac{32\sqrt{2}\,\pi}{3h^3\sqrt{m}} \left[\tau(\mu)g(\mu)\mu^{5/2} + \frac{(2\pi kT)^2}{24}(\tau g \epsilon^{5/2})'' \right] \qquad (6.20)$$

$$I_4 = - \frac{32\pi\sqrt{2}}{3h^3\sqrt{m}} \left[\tau(\mu)\mu^{5/2} + \frac{(2\pi kT)^2}{24}(\tau \epsilon^{5/2})'' \right] \qquad (6.21)$$

In these equations, the double primes indicate that second differentiation of the quantities in the parenthesis have been performed, the second derivatives being evaluated at $\epsilon = \mu$; that is,

$$(\tau g \epsilon^{3/2})'' = \left[\frac{d^2}{d\epsilon^2}(\tau g \epsilon^{3/2}) \right]_{\epsilon = \mu}$$

and so on.

Substituting (6.18) to (6.21) into (6.3) and retaining terms up to the second order in the temperature gives for the heat flux

$$J_x = \frac{8\pi\sqrt{2m}}{9h^3} \{ [g(\mu)(\tau \epsilon^{5/2})'' - (\tau g \epsilon^{5/2})''] - \epsilon[g(\tau \epsilon^{3/2})'' - (\tau g \epsilon^{3/2})''] \} \frac{dT}{dx} \qquad (6.22)$$

But

$$(\tau \epsilon^{5/2})'' = \left[\frac{15}{4}\epsilon^{1/2} + 5\epsilon^{3/2}\frac{d\tau}{d\epsilon} + \epsilon^{5/2}\frac{d^2\tau}{d\epsilon^2} \right]_{\epsilon=\mu}$$

$$(\tau g \epsilon^{5/2})'' = \left[\frac{15}{4}\epsilon^{1/2}\tau g + 5\epsilon^{3/2}\frac{d(\tau g)}{d\epsilon} + \epsilon^{5/2}\frac{d^2(\tau g)}{d\epsilon^2} \right]_{\epsilon=\mu}$$

$$(\tau \epsilon^{3/2})'' = \left[\frac{3}{4}\epsilon^{-1/2}\tau + 3\epsilon^{1/2}\frac{d\tau}{d\epsilon} + \epsilon^{3/2}\frac{d^2\tau}{d\epsilon^2} \right]_{\epsilon=\mu}$$

$$(\tau g \epsilon^{3/2})'' = \left[\frac{3}{4}\epsilon^{-1/2}\tau g + 3\epsilon^{1/2}\frac{d(\tau g)}{d\epsilon} + \epsilon^{3/2}\frac{d^2(\tau g)}{d\epsilon^2} \right]_{\epsilon=\mu}$$

$$(6.23)$$

as can be verified by working out the derivatives. If we use (6.23), (6.22) reduces to

$$J_x = \frac{16\pi^2\sqrt{2}m}{9h^3}(kT)^2\mu^{3/2}\left[g\left(\frac{d\tau}{d\epsilon}\right)_\mu - \left(\frac{d(\tau g)}{d\epsilon}\right)_\mu\right]\frac{dT}{dx}$$

or

$$J_x = \frac{16\pi^3\sqrt{2}m}{9h^3}(kT)^2\mu^{3/2}\tau(\mu)\left(\frac{dg}{d\epsilon}\right)_\mu\frac{dT}{dx} \tag{6.24}$$

From (6.15)

$$\left(\frac{dg}{d\epsilon}\right)_\mu = -\frac{1}{T} \tag{6.25}$$

Putting this into (6.24) and using the first-order expression for μ given by (3.10) of Chapter 3, we obtain

$$J_x = \frac{\pi^2 k^2 T\tau(\mu)n}{3m}\frac{dT}{dx} \tag{6.26}$$

By definition, the coefficient of $-dT/dx$ on the right-hand side of this equation is the thermal conductivity.

$$K = \frac{\pi^2 k^2 T\tau(\mu)n}{3m} \tag{6.27}$$

Table 4.1. Lorenz Number for Metals at $T = 100°C$ in esu

Metal	$L \times 10^{13}$	Metal	$L \times 10^{13}$
Mg	2.57	Cd	2.70
Aℓ	2.47	Sn	2.76
Ni	2.53	W	3.55
Cu	2.60	Ir	2.76
Zn	2.60	Pt	2.88
Mo	3.10	au	2.67
Rh	2.82	Rb	2.85
Pd	2.63	Ri	3.20
Ag	2.63		

From a compilation by A. H. Wilson, *The Theory of Metals*, 2nd ed., Cambridge University Press, Cambridge, England, 1965.

Just as in the case of electrical conductivity, the relaxation time enters into the final result only through its value for $\epsilon = \mu$.

If (6.27) is compared to (5.12) for the conductivity, we see that

$$L \equiv \frac{K}{\sigma T} = \frac{\pi^2 k^2}{3e^2} \tag{6.28}$$

This is the Wiedemann-Franz law and the quantity L is called the Lorenz number. In the free-electron theory it is a universal constant whose value in cgs units is 2.71×10^{-13}. Experimental values of the Lorenz number for several metals are shown in Table 4-1.

4.7 THE ISOTHERMAL HALL EFFECT

The Hall experiment is important because it gives a direct measure of the electron concentration, and of the sign of the charge carrier. It, therefore, can distinguish between electron and hole conduction. It is carried out by imposing an electric field in the x-direction and a magnetic field in the z-direction on a specimen as shown in Figure 4.1. If an electron is moving in

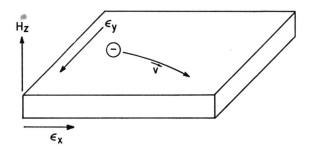

Fig. 4.1. Isothermal Hall effect.

the x-direction, it is deflected by the magnetic field H_z, so that it has a velocity component in the y-direction. Since no current flows in the y-direction, a constant field $\mathscr{E}_y$ is established that is called "the Hall voltage." The Hall coefficient is a measure of the magnitude of the effect and is defined by

$$R_H \equiv \frac{\mathscr{E}_y}{I_x H_z} \tag{7.1}$$

Switching on the magnetic field has the effect of converting the electric field vector from $(\mathscr{E})_x \mathbf{i}$ to $(\mathscr{E}_x \mathbf{i} + \mathscr{E}_y \mathbf{j})$, which is equivalent to rotating the electric field through an angle ϕ given by

$$\tan \phi = \frac{\mathscr{E}_y}{\mathscr{E}_x} \tag{7.2}$$

ϕ is called the Hall angle.

For the isothermal Hall effect, the flux equation for the electric current (4.3) becomes

$$\mathbf{I} = -\frac{e^2}{m} \int \tau \mathbf{v} \left(\frac{\partial N}{\partial \mathbf{v}} \cdot \mathscr{E} \right) d\mathbf{v} - \frac{e^2}{mc} \int \tau \mathbf{v} (\mathbf{v} \times \mathbf{H}) \cdot \frac{\partial N}{\partial \mathbf{v}} d\mathbf{v} \tag{7.3}$$

The first integral is just the ordinary electrical conduction and, as in Section 5,

$$-\frac{e^2}{m} \int \tau \mathbf{v} \left(\frac{\partial N}{\partial \mathbf{v}} \cdot \mathscr{E} \right) d\mathbf{v} = \sigma \mathscr{E} \tag{7.4}$$

The second integral requires a little more attention. As discussed in Section 3, magnetic effects require an explicit calculation of the departure from equilibrium, and the equilibrium part of the derivative $\partial N/\partial \mathbf{v}$ contributes nothing. Using the definition of $\boldsymbol{\eta}$ in (7.3), the second integral, therefore, becomes

$$\int \tau \mathbf{v} (\mathbf{v} \times \mathbf{H}) \cdot \frac{\partial N}{\partial \mathbf{v}} d\mathbf{v} = \int \tau \mathbf{v} (\mathbf{v} \times \mathbf{H}) \cdot \frac{\partial (\mathbf{v} \cdot \boldsymbol{\eta})}{\partial \mathbf{v}} d\mathbf{v} \tag{7.5}$$

An estimate of the derivative in the integral on the right can be obtained by starting with (3.20). Since $\nabla T = 0$ for the present case, this equation is

$$\mathbf{v} \cdot \boldsymbol{\eta} = \frac{e\tau}{m} \mathscr{E} \cdot \frac{\partial N^0}{\partial \mathbf{v}} + \frac{e\tau}{mc} (\mathbf{v} \times \mathbf{H}) \cdot \frac{\partial (\mathbf{v} \cdot \boldsymbol{\eta})}{\partial \mathbf{v}} \tag{7.6}$$

Assume that the magnetic field is small, so that the first term is the major contributor to (7.6). Then, to a first approximation, the second term on the right can be ignored in differentiating $(\mathbf{v} \cdot \boldsymbol{\eta})$. That is,

$$\frac{\partial (\mathbf{v} \cdot \boldsymbol{\eta})}{\partial \mathbf{v}} = \frac{e\tau}{m} \frac{\partial}{\partial \mathbf{v}} \left(\mathscr{E} \cdot \frac{\partial N^0}{\partial \mathbf{v}} \right) \quad \text{(first approximation)} \tag{7.7}$$

But

$$\frac{\partial N^0}{\partial \mathbf{v}} = m\mathbf{v} \frac{\partial N^0}{\partial \epsilon} \tag{7.8}$$

so (7.7) becomes

$$\frac{\partial (\mathbf{v} \cdot \boldsymbol{\eta})}{\partial \mathbf{v}} = e\tau \frac{\partial N^0}{\partial \epsilon} \frac{\partial}{\partial \mathbf{v}} (\mathscr{E} \cdot \mathbf{v}) = e\tau \frac{\partial N^0}{\partial \epsilon} \mathscr{E} \tag{7.9}$$

Substitution of (7.9) into (7.6) results in

$$\mathbf{v} \cdot \boldsymbol{\eta} = e\tau \frac{\partial N^0}{\partial \epsilon} \mathscr{E} \cdot \mathbf{v} + \frac{e^2 \tau^2}{mc} \frac{\partial N^0}{\partial \epsilon} \mathscr{E} \cdot (\mathbf{v} \times \mathbf{H}) \qquad (7.10)$$

A second approximation can be obtained by differentiating (7.10) and treating everything as a constant but $\mathbf{v}$. The result is

$$\frac{\partial (\mathbf{v} \cdot \boldsymbol{\eta})}{\partial \mathbf{v}} = e\tau \frac{\partial N^0}{\partial \epsilon} \mathscr{E} + \frac{e^2 \tau^2}{mc} \frac{\partial N^0}{\partial \epsilon} (\mathscr{E} \times \mathbf{H}) \qquad (7.11)$$

This approximation procedure can be continued indefinitely. For our purposes, (7.11) is adequate, and substituting it into (7.5) gives

$$\int \tau \mathbf{v} (\mathbf{v} \times \mathbf{H}) \cdot \frac{\partial N}{\partial \mathbf{v}} \, d\mathbf{v} = e \int \tau^2 \mathbf{v} \frac{\partial N^0}{\partial \epsilon} (\mathbf{v} \times \mathbf{H}) \cdot \mathscr{E} \, d\mathbf{v} + \mathbf{a}(H^2) \qquad (7.12)$$

where $\mathbf{a}(H^2)$ is the integral arising from the second term in (7.11). Since $\mathbf{H} = H_z \mathbf{k}$ and $\mathscr{E} = \mathscr{E}_x \mathbf{i} + \mathscr{E}_y \mathbf{j}$, working out the product $(\mathbf{v} \times \mathbf{H}) \cdot \mathscr{E}$ reduces (7.12) to

$$\int \tau \mathbf{v} (\mathbf{v} \times \mathbf{H}) \cdot \frac{\partial N}{\partial \mathbf{v}} \, d\mathbf{v} = e \int \tau^2 \mathbf{v} \frac{\partial N^0}{\partial \epsilon} (v_y \mathscr{E}_x - v_x \mathscr{E}_y) \, d\mathbf{v} H_z + \mathbf{a}(H^2) \qquad (7.13)$$

An application of the series expansion formula for the Fermi integral can now be made to the first integral on the right with the result that

$$\int \tau \mathbf{v} (\mathbf{v} \times \mathbf{H}) \cdot \frac{\partial N}{\partial \mathbf{v}} \, d\mathbf{v} = \frac{\tau(\mu)}{e} \sigma H_z (\mathscr{E}_y \mathbf{i} - \mathscr{E}_x \mathbf{j}) + \mathbf{a}(H^2) \qquad (7.14)$$

Finally, we can substitute (7.4) and (7.14) into (7.3) to obtain the current flow

$$\mathbf{I} = \sigma \mathscr{E} - \frac{e H_z \tau(\mu) \sigma}{mc} (\mathscr{E}_y \mathbf{i} - \mathscr{E}_x \mathbf{j}) + \mathbf{a}(H^2) \qquad (7.15)$$

If we write this in component form, remembering that $I_y = I_z = 0$,

$$I_x = \sigma \mathscr{E}_x - \frac{e H_z \tau(\mu) \sigma}{mc} \mathscr{E}_y + a_x(H^2) \qquad (7.16)$$

$$\mathscr{E}_y = -\frac{e H_z \tau(\mu)}{mc} \mathscr{E}_x + \frac{a_y(H^2)}{\sigma} \qquad (7.17)$$

The Hall voltage contains a term linear in H_z. The current, however, does not; the lowest power of the magnetic field it contains is two, as can be seen by substituting (7.17) for $\mathscr{E}_y$ into (7.16). Since we have postulated that the magnetic field is small, quadratic terms in H_z will be neglected. Equations

7.16 and 7.17 then give

$$I_x = \sigma \mathscr{E}_x \tag{7.18}$$

$$\mathscr{E}_y = -\frac{eH_z\tau(\mu)}{mc}\mathscr{E}_x \tag{7.19}$$

The Hall coefficient then becomes

$$R_H = -\frac{e\tau(\mu)}{mc\sigma} \tag{7.20}$$

or, substituting for σ from (5.12),

$$R_H = -\frac{1}{nec} \tag{7.21}$$

The Hall coefficient contains only the electron density, the velocity of light, and the electronic charge. Furthermore, if the current were carried by holes instead of electrons, the only effect would be to convert $-e$ to $+e$. The Hall coefficient is, therefore, negative for electronic conduction and positive for hole conduction. Its measurement, therefore, yields the sign of the charge carrier as well as the density of free electrons.

4.8   ELECTRICAL CONDUCTIVITY IN INTRINSIC SEMICONDUCTORS

The theory developed thus far in this chapter presumes that only one type of charge carrier, namely the electron, is present in our system. However, it is obvious that the theory can be generalized to treat a system containing electrons and holes if the holes are treated as particles. This is done by working out flux equations for holes in the same way as for electrons. The total flux is then the sum of the electron flux and hole flux. Doing this for electrical conduction, the conductivity of a semiconductor containing holes and electrons is given by a formula containing two terms similar to those on the right-hand side of (5.2). The electron and hole conductivity can be treated separately and then combined. For the electron conductivity, (5.2) can be written as

$$\sigma_e = -\frac{e^2}{m}\int \tau v_x \frac{\partial N_c}{\partial v_x}\, d\mathbf{v} \tag{8.1}$$

where N_c is the directional velocity distribution function per unit volume for electrons in the conduction band, and the integration is performed over the entire conduction band. The function N_c is related to $N_c(\epsilon)/V$ of (10.2, Chapter 3) with a density-of-states function given by (10.3). If the energy of the electrons in the conduction band is measured relative to E_c so that an energy scale ϵ' is defined by

$$\epsilon' = \epsilon - E_c \tag{8.2}$$

then (10.2, Chapter 3) becomes

$$N_c(\epsilon') = \frac{\omega_c(\epsilon')}{e^{(\epsilon'-\mu')/kT} + 1} = \omega_c(\epsilon')f(\epsilon') \tag{8.3}$$

where

$$\mu' = \mu - E_c \tag{8.4}$$

and the distribution functions have precisely the same form as in the previous sections of this chapter. This means that the transformations leading from (5.2) to (5.9) are valid and we have

$$\sigma_e = -\frac{8\pi m_c^2 e^2}{3h^3} \int_0^\infty \tau v^3 \frac{\partial f(\epsilon')}{\partial \epsilon'} \, d\epsilon' \tag{8.5}$$

The kinetic energy is $m_c v^2/2 = \epsilon'$, so (8.5) becomes

$$\sigma_e = -\frac{8\pi m_c^2 e^2}{3h^3} \left(\frac{2}{m_c}\right)^{3/2} \int_0^\infty \tau \epsilon'^{3/2} \frac{\partial f(\epsilon')}{\partial \epsilon'} \, d\epsilon \tag{8.6}$$

To evaluate the integral, we make the Boltzmann approximation in which the exponential in (8.3) is much greater than unity (see equation 10.7, Chapter 3). Then

$$f(\epsilon') = e^{(\mu'-\epsilon')/kT} \tag{8.7}$$

$$\frac{\partial f(\epsilon')}{\partial \epsilon'} = -\frac{1}{kT} e^{(\mu'-\epsilon')/kT} \tag{8.8}$$

Furthermore, we assume that τ is a constant that we call τ_c. This is not too serious an approximation since the integrand is largest near the bottom of the band and decreases rapidly as ϵ' increases. The integral now becomes

$$\int_0^\infty \tau \epsilon'^{3/2} \frac{\partial f(\epsilon')}{\partial \epsilon'} \, d\epsilon' = -\frac{\tau_c}{kT} e^{\mu'/kT} \int_0^\infty \epsilon'^{3/2} e^{-\epsilon'/kT} \, d\epsilon'$$

$$= -\tau_c (kT)^{3/2} e^{\mu'/kT} \int_0^\infty x^{3/2} e^{-x} \, dx$$

$$= -\frac{3\sqrt{\pi}}{4} \tau_c (kT)^{3/2} e^{\mu'/kT}$$

Equation 8.6, therefore, reduces to

$$\sigma_e = \frac{4(2\pi^3 m_c)^{1/2} e^2 \tau_c}{h^3} (kT)^{3/2} e^{(\mu-E)/kT} \tag{8.9}$$

Using (10.11) of Chapter 3 for the density of conduction electrons, (8.6) becomes

$$\sigma_e = \frac{ne^2\tau_c}{m_c} \tag{8.10}$$

Note that this has the same form as (5.12).

Now we calculate σ_h, the hole conductivity, in precisely the same way. Instead of (8.3), we have

$$N_h(\epsilon'') = \omega_v(\epsilon'')f(\epsilon'') \tag{8.11}$$

where

$$f(\epsilon'') = \frac{1}{e^{(\epsilon''-\mu'')/kT} + 1}$$

$$\epsilon'' = E_v - \epsilon \tag{8.12}$$

$$\mu'' = E_v - \mu$$

Making the Boltzmann approximation and assuming τ is a constant τ_v, repeating the above development, and using (10.12) of Chapter 3, the hole conductivity becomes

$$\sigma_h = \frac{pe^2\tau_v}{m_v} \tag{8.13}$$

The total conductivity of an intrinsic semiconductor is the sum of (8.7) and (8.10):

$$\sigma = e^2\left[\frac{\tau_c}{m_c}n + \frac{\tau_v}{m_v}p\right] \tag{8.14}$$

Substitution of (10.15) of Chapter 3 into this equation displays the temperature dependence of the conductivity as

$$\sigma = A(T)e^{-E_g/2kT} \tag{8.15}$$

where $A(T)$ is defined as

$$A(T) = 2\left(\frac{2\pi kT}{h^2}\right)^{3/2}(m_c m_v)^{3/4}\left[\frac{\tau_c}{m_c} + \frac{\tau_v}{m_v}\right] \tag{8.16}$$

Equation 8.15 shows that the conductivity of intrinsic semiconductors is an increasing exponential function of temperature, and permits the value of the energy gap to be computed from the measurement of σ as a function of τ. The factor $A(T)$ is a slowly varying function of temperature. The precise temperature variation depends on the evaluation of the relaxation times; however, in all cases the exponential factor dominates and E_g can be computed from a plot of $\ln \sigma$ versus $1/T$ with satisfactory accuracy.

It is important to note an essential difference between the theory developed here and that worked out for metals. In metals, only the value of the relaxation time at the Fermi level enters into the final equations, and it is not necessary to know the functional dependence of τ on energy. In semiconductors, however, a knowledge of τ as a function of ϵ is required if we are to evaluate the integrals involved accurately. We have chosen τ to be constant in arriving at (8.7) and (8.13). In fact, this is not the best choice that could have been made. A quantum theoretic treatment of the scattering of electrons by lattice vibrations in semiconductors shows that τ is proportional to $\epsilon^{-1/2}$; that is, it is inversely proportional to the velocity. Therefore, $v\tau$ is a constant, and we can define a distance λ_e by

$$\lambda_e = v\tau \tag{8.17}$$

that is independent of energy. Since τ measures the mean time between electron-phonon collisions, λ_e has the interpretation of a mean free path. Now let us calculate the electron conductivity using (8.17) instead of the assumption $\tau = $ constant. The integral in (8.6) now becomes

$$\int_0^\infty \tau \epsilon'^{3/2} \frac{\partial f(\epsilon')}{\partial \epsilon'} d\epsilon' = -\frac{\lambda_e}{kT} e^{\mu'/kT} \int_0^\infty \frac{\epsilon'^{3/2}}{v} e^{-\epsilon'/kT} d\epsilon'$$

$$= -\frac{\lambda_e}{kT} \left(\frac{m_c}{2}\right)^{1/2} e^{\mu'/kT} \int_0^\infty \epsilon' e^{-\epsilon'/kT} d\epsilon'$$

$$= -\lambda_e kT \left(\frac{m_c}{2}\right)^{1/2} e^{\mu'/kT} \tag{8.18}$$

where we have replaced τ by λ_e/v. Putting (8.18) into (8.6) gives the electron conductivity;

$$\sigma_e = \frac{4e^2\lambda_e n}{3(2\pi m_c kT)^{1/2}} \tag{8.19}$$

where we have used (10.11) of Chapter 3. This has the same form as (8.7) if we identify τ_c as

$$\tau_c \rightleftarrows \frac{4\lambda_e m_c}{3(2\pi m_c kT)^{1/2}} \tag{8.20}$$

In the same way, if we take a constant mean free path for holes, $\lambda_h = \tau v$, the hole conductivity becomes

$$\sigma_h = \frac{4e^2\lambda_h p}{3(2\pi m_v kT)^{1/2}} \tag{8.21}$$

and the total conductivity becomes, instead of (8.14),

$$\sigma = \frac{4e^2}{3(2\pi kT)^{1/2}} \left[\frac{\lambda_e}{m_c^{1/2}} n + \frac{\lambda_h}{m_v^{1/2}} p\right] \tag{8.22}$$

For intrinsic semiconductors $n = p$, and using (10.15) of Chapter 3 gives

$$\sigma = A'(T)e^{-E_g/2kT} \tag{8.23}$$

where

$$A'(T) = \frac{16\pi kTe^2}{h^3}(m_c m_v)^{3/4}\left[\frac{\lambda_e}{m_c^{1/2}} + \frac{\lambda_h}{m_v^{1/2}}\right] \tag{8.24}$$

Note that in arriving at the conductivity formula (8.23), the only step in which the fact that the semiconductor is intrinsic was used was in setting $n = p$ in (8.22). This means that (8.22) is also valid for extrinsic semiconductors. The electron and hole concentrations would then still be given by (10.11) and (10.12) of Chapter 3, but the Fermi level would be determined by the impurity content.

The Hall coefficient in semiconductors can be computed in a straight-forward manner using the methods of Section 7, the Boltzmann approximation of the distribution function, and the assumption of constant mean free path. We leave it to the reader to show that for electron conduction in semiconductors

$$R_H = -\frac{3\pi}{8}\frac{1}{nec} \tag{8.25}$$

whereas for hole conduction

$$R_H = \frac{3\pi}{8}\frac{1}{pec} \tag{8.26}$$

chapter 5

ORDER–DISORDER ALLOYS

5.1 ORDER-DISORDER STRUCTURES

A binary order-disorder structure is defined as a two-component crystal with the following properties:

1. At absolute zero of temperature the atoms of each component separately occupy the sites of a lattice, so that the structure consists of two interpenetrating sublattices, each sublattice containing only one type of atom. The structure is then said to be completely ordered.

2. At sufficiently high temperatures, both types of atoms are distributed throughout both sublattices at random. The structure is then in the completely disordered state. Two examples of order-disorder structures are shown in Figure 5.1. Figure 5.1A represents the structure of β-brass (CuZn). If we ignore the identity of the two types of atoms, the structure is body-centered cubic (BCC). The BCC lattice consists of two interpenetrating simple cubic (SC) sublattices, one sublattice consisting of the cube corners, and the other sublattice consisting of cube centers, as shown in the figure. In the completely ordered state, the body centers of the unit cubes in the BCC structure are all occupied by atoms of one type (say Cu) while the cube corners are all occupied by atoms of the other type (say Zn). The two sublattices are completely equivalent and it is immaterial which type of atom we assign to a given sublattice. In the completely disordered state, any site may be occupied by an atom of either type, the probability that a given site contains an atom of a given type being one-half.

Figure 5.1B shows the lattice structure of Cu_3Au. The complete lattice, which is face-centered cubic, can be resolved into two sublattices, one formed

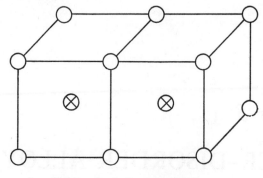

A.- B.C.C. β-CuZn Structure

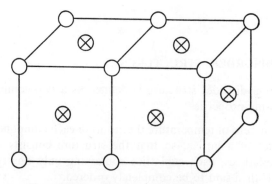

B.- F.C.C. Cu_3Au Structure

Fig. 5.1. Order-disorder lattice structures.

from the cube-corner sites, and the other formed from the cube-face-center sites. The cube corners form a SC lattice, while the cube-face sites form a body-centered-tetragonal (BCT) lattice. The BCT sublattice contains three times as many sites as the SC sublattice. In the completely ordered structure, the SC sublattice sites are all occupied by Au atoms, and the BCT sites are all occupied by Cu atoms. In contrast with the β-brass structure, the two sublattices in this case are not equivalent. In the completely disordered state, all atoms may occupy any site; the probability that a given site contains an Au atom is 1/4 while the probability that it contains a Cu atom is 3/4.

5.2   THE ORDER-DISORDER TRANSITION

Let us consider an AB alloy of the β-brass type that is completely ordered at absolute zero. The energy of the ordered state must then be lower than that of the disordered state. This means that AB bonds are energetically favored over AA or BB bonds. As the temperature is raised above absolute zero, the thermal vibrations exert a disrupting influence on the perfectly ordered crystal so that some AA and BB bonds will form. This is just a result of the fact that the entropy is larger for disordered systems than for ordered systems. At a given temperature, the free energy is a minimum and a balance between the energy and entropy requirements is established so that, in general, the crystal is partially ordered. The degree of order is bound to decrease with increasing temperature, and a temperature must exist above which the degree of order is negligible. However, changes in the degree of order can be observed only if the energy of the ordered state is not too much less than that of the disordered state. If the energy difference between the two states is too large, the crystal will remain ordered right up to the melting point. A system that illustrates this point is sodium chloride, which may be regarded as an order-disorder alloy (Na^+Cl^-) with an ordering energy much greater than kT_m (T_m = melting point). We can expect that alloys exhibiting order-disorder behavior have an energy in the disordered state that exceeds the energy in the ordered state by something less than the thermal energy at the melting point.

The order-disorder transition can be observed directly in a number of alloys by X-ray diffraction techniques. At low temperatures, the ordered arrangement of the alloy constituents gives rise to superlattice lines that are not observed at high temperatures. A number of other changes in physical properties accompany the transition from the ordered to the disordered state, the most important of which is the existence of an anomalous heat capacity. If the heat capacity of an order-disorder alloy is measured as a function of temperature, it is found that at low temperatures, a typical Debye-type heat-capacity–temperature curve is obtained. As the temperature is increased, however, a temperature range is reached where the heat capacity rapidly increases above the Debye curve, and at a reasonably well-defined critical temperature, the heat capacity abruptly dips to a value characteristic of a normal crystal. This behavior is illustrated in Figure 5.2.

The order-disorder transition is a typical cooperative phenomenon, in which the ease with which the transition occurs increases rapidly with the extent to which it has already occurred. A qualitative understanding of the nature of this cooperative action can be based on the fact that an AB bond is favored over an AA or a BB bond. If we start with a perfectly ordered AB alloy, and label the two sublattices α and β, then all A atoms are on α sites and all B atoms are on β sites. Now replace a B atom on a β site by an A atom. This "wrong" A atom is surrounded by α sites. But since AB bonds are energetically easier to form than AA bonds, we see that replacing an A atom by a B atom on one of these sites requires less energy than would be the case if the original β site were occupied by a B atom. The existence of some

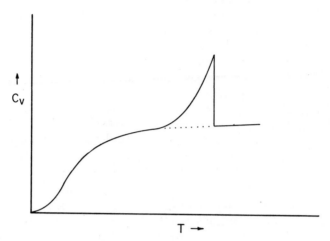

Fig. 5.2. Heat capacity of an order-disorder crystal.

disorder in the crystal thus promotes further disorder. As we raise the temperature, therefore, equal temperature increments produce larger and larger increases in disorder, and the catastrophic behavior shown in Figure 5.2 results.

The statistical-mechanical problem for order-disorder systems consists of establishing a quantitative measure of the degree of order, determining the degree of order as a function of temperature, and computing the effect of order on thermodynamic properties.

5.3 DESCRIPTION OF THE DEGREE OF ORDER

Consider a two-component order-disorder alloy with composition $A_n B_m$. In the completely ordered state, all A atoms are on one sublattice (labeled α) and all B atoms are on the other sublattice (labeled β). In the partially ordered state, however, this is not true, and the number of atoms of each type on each sublattice must be specified. Let

$N_{A\alpha}$ = number of A atoms on α sites
$N_{A\beta}$ = number of A atoms on β sites
$N_{B\beta}$ = number of B atoms on β sites
$N_{B\alpha}$ = number of B atoms on α sites
N_A = total number of A atoms in the crystal
N_B = total number of B atoms in the crystal

N = total number of atoms in the crystal

$\gamma_A = \dfrac{n}{n+m}$ = fraction of A atoms in the crystal

$\gamma_B = \dfrac{m}{n+m}$ = fraction of B atoms in the crystal

The following relations are obvious:

$$N_A + N_B = N \tag{3.1}$$

$$\gamma_A + \gamma_B = 1 \tag{3.2}$$

$$N_{A\alpha} + N_{B\alpha} = N_{A\alpha} + N_{A\beta} = N_A = \gamma_A N \tag{3.3}$$

$$N_{A\beta} + N_{B\beta} = N_{B\alpha} + N_{B\beta} = N_B = \gamma_B N \tag{3.4}$$

also, from (3.3) or (3.4),

$$N_{A\beta} = N_{B\alpha} \tag{3.5}$$

The fractions of sites of a given sublattice occupied by atoms of a given type are defined by

$$f_{A\alpha} = \frac{N_{A\alpha}}{\gamma_A N} \tag{3.6}$$

$$f_{A\beta} = \frac{N_{A\beta}}{\gamma_B N} \tag{3.7}$$

$$f_{B\beta} = \frac{N_{B\beta}}{\gamma_B N} \tag{3.8}$$

$$f_{B\alpha} = \frac{N_{B\alpha}}{\gamma_A N} \tag{3.9}$$

In a perfectly ordered crystal, $f_{A\alpha} = f_{B\beta} = 1$, whereas in a completely disordered crystal, $f_{A\alpha} = \gamma_A$ and $f_{B\beta} = \gamma_B$. We now define a parameter R by the relation

$$R \equiv \frac{f_{A\alpha} - \gamma_A}{1 - \gamma_A} \tag{3.10}$$

Clearly, $R = 1$ for perfect order, and $R = 0$ for complete disorder. Also, from (3.3) to (3.5), it follows that

$$\gamma_A(1 - f_{A\alpha}) = \gamma_B(1 - f_{B\beta}) \tag{3.11}$$

and using this relation it is easy to convert (3.10) to

$$R = \frac{f_{B\beta} - \gamma_B}{1 - \gamma_B} \tag{3.12}$$

The parameter R is therefore invariant with respect to the interchange $(A\alpha) \leftrightarrow (B\beta)$. R is a measure of the degree of order defined by the occupation of atom types on the sublattices and is called the long-range-order parameter, since it describes the extent to which the sublattices are filled with the "right" kinds of atoms and says nothing about local configurations.

An alternative method of describing the degree of order is to count the number of nearest-neighbor AB pairs. The number of unlike pairs will be a maximum in a perfectly ordered crystal, and a minimum in a completely disordered crystal. Such a description of order is essentially local, since it is concerned not with the occupancy of an entire sublattice, but with nearest neighbor configurations. Let Q_{AB} be the number of atom pairs such that an A atom is on an α site and a B atom is on an adjacent β site. Similarly, Q_{BA} is the number of pairs with a B atom on an α site and an A atom on a β site. Then the fraction of unlike pairs in the crystal is

$$q \equiv \frac{Q_{AB} + Q_{BA}}{Q} \tag{3.13}$$

where Q is the total number of pairs in the crystal, and is given by

$$Q = Q_{AB} + Q_{BA} + Q_{AA} + Q_{BB}$$
$$= \tfrac{1}{2}zN \tag{3.14}$$

Q_{AA} and Q_{BB} being the number of AA and BB pairs, respectively, and z being the coordination number ($z = 8$ for β-brass; $z = 12$ for Cu_3Au). In a completely ordered state, q has a maximum value q_m and in the completely disordered state, q has a minimum value q_0. The short-range-order parameter σ is defined in terms of q by

$$\sigma \equiv \frac{q - q_0}{q_m - q_0} \tag{3.15}$$

For an AB alloy (e.g., β-brass) $q_m = 1$ and $q_0 = 1/2$, so (3.15) becomes

$$\sigma = 2q - 1 \quad (AB \text{ alloy}) \tag{3.16}$$

The short-range-order parameter is zero for complete disorder and unity for perfect order, but it is only at these two points that it equals the long-range-order parameter. The configurational state of the crystal is defined by numbering the lattice sites and specifying the type of atom on each site. Clearly, a large number of states are consistent with a given set of sublattice occupation numbers $[N_{A\alpha}, N_{A\beta}, N_{B\alpha}, N_{B\beta}]$, and many different sets $[Q_{AA}, Q_{BB}, Q_{AB}, Q_{BA}]$ are consistent with a given set of N's. The long-range-order parameter must therefore be a function of the average short-range-order parameter, the average being taken over all configurational states consistent with a given set $[N_{A\alpha}, \dots]$. The connection between the long-range-order parameter and the average of the short-range-order parameter is easily studied by recognizing that the unweighted average probability that a given pair consists of unlike atoms is just the product of the probabilities that adjacent sites are occupied by unlike atoms. Thus, if $N_{A\alpha}$, $N_{B\beta}$, and so on are given, the probability that a particular α site contains an A atom is just

$$f_{A\alpha} = \frac{N_{A\alpha}}{\gamma_A N}$$

since $\gamma_A N$ is the number of α sites. The probability that an adjacent β site contains a B atom is

$$f_{B\beta} = \frac{N_{B\beta}}{\gamma_B N}$$

The product of these two probabilities is just the average probability that A is on α and B is on β; that is,

$$\frac{\bar{Q}_{AB}}{Q} = f_{A\alpha} f_{B\beta} \tag{3.17}$$

where $\bar{Q}_{AB}$ is the average of Q_{AB} over all states consistent with the given N's. Similarly

$$\frac{\bar{Q}_{BA}}{Q} = f_{B\alpha} f_{A\beta} \tag{3.18}$$

so that

$$\bar{q} = \frac{\bar{Q}_{AB} + \bar{Q}_{BA}}{q} = f_{A\alpha} f_{B\beta} + f_{B\alpha} f_{A\beta} \tag{3.19}$$

The f's can be expressed in terms of the long-range-order parameter R as follows: from (3.10) and (3.12) we get $f_{A\alpha}$ and $f_{B\beta}$;

$$f_{A\alpha} = \gamma_A + \gamma_B R \tag{3.20}$$

$$f_{B\beta} = \gamma_B + \gamma_A R \tag{3.21}$$

and from (3.3) and (3.4), $f_{B\alpha}$ and $f_{A\beta}$ can be written as

$$f_{B\alpha} = \frac{\gamma_B}{\gamma_A} [1 - f_{B\beta}] \tag{3.22}$$

$$f_{A\beta} = \frac{\gamma_A}{\gamma_B} [1 - f_{A\alpha}] \tag{3.23}$$

which, when combined with (3.20) and (3.21), become

$$f_{B\alpha} = \gamma_B(1 - R) \tag{3.24}$$

$$f_{A\beta} = \gamma_A(1 - R) \tag{3.25}$$

Now substituting (3.20), (3.21), (3.24), and (3.25) into (3.19) gives

$$\bar{q} = 2\gamma_A\gamma_B + (\gamma_A - \gamma_B)^2 R + 2\gamma_A\gamma_B R^2 \tag{3.26}$$

From (3.15), the average short-range-order parameter is

$$\bar{\sigma} = \frac{\bar{q} - q_0}{q_m - q_0} \tag{3.27}$$

and substitution of (3.26) into (3.27) gives the relation between R and $\bar{\sigma}$. In the special case of an AB alloy, $\gamma_A = \gamma_B = 1/2$, and (3.26) becomes

$$\bar{q} = \tfrac{1}{2}(1 + R^2) \quad (AB \text{ alloy}) \tag{3.28}$$

also,

$$\bar{\sigma} = 2\bar{q} - 1 \quad (AB \text{ alloy}) \tag{3.29}$$

so that

$$\bar{\sigma} = R^2 \quad (AB \text{ alloy}) \tag{3.30}$$

Equations 3.26 and 3.27 can be used to calculate q_0 and q_m, since when $R = 1$, $\bar{\sigma}$ can be unity only if $\bar{q} = q_m$, and when $R = 0$, $\bar{\sigma}$ can be zero only if $\bar{q} = q_0$. Applying these conditions to (3.26) and (3.27) gives

$$q_0 = 2\gamma_A\gamma_B \tag{3.31}$$

$$q_m = (\gamma_A + \gamma_B)^2 \tag{3.32}$$

It must be emphasized that the averages considered above are not statistical-mechanical averages over states. They are simple arithmetic averages over all configurations consistent with a given value of R. (Such averages will be denoted by a bar in this chapter.) The relation between short- and long-range order given by (3.30) is therefore not true in general for the statistical-thermodynamic values of these quantities, except within the framework of the Bragg-Williams approximation, which will be developed subsequently.

The long- and short-range-order parameters provide a reasonably satisfactory basis for the description of order-disorder systems. Our next task is to incorporate this description into a statistical-mechanical framework from which we can obtain thermodynamic results. In doing this, we will treat only the 50-50 AB alloy in detail; the extension to more complex systems is, in principle, quite straightforward.

5.4   THE ORDER-DISORDER PARTITION FUNCTION

The central problem of the statistical mechanics of order-disorder systems is the evaluation of the partition function. In the theory of monatomic crystals, there was no need to consider the occupancy of lattice sites, since the interchange of two identical atoms did not alter the quantum state of the crystal. For order-disorder alloys, however, it is necessary to take the distribution of atom types into account. This is done by constructing a canonical ensemble in which R is identical for every member of the ensemble. The partition function for this ensemble can then be evaluated, provided certain approximations are made. From the partition function we then obtain the free energy for a particular degree of long-range order. To get the equilibrium degree of order as a function of temperature, the free energy is then minimized with respect to R.

If $Z(R)$ is the partition function for a given degree of long-range order, then

$$Z(R) = \sum_{k,v} e^{-[W_k + E_v{}^k]/kT} \tag{4.1}$$

where W_k is the potential energy of the kth configurational state (for given R) when all atoms are at rest, and $E_v{}^k$ is the vth vibrational state of the crystal in the kth configurational state. The vibrational energy is given by

$$E_v{}^k(R) = \sum_{j=1}^{\infty} (n_j + \tfrac{1}{2}) h v_j{}^k \tag{4.2}$$

where n_j are the integers that specify the quantum levels of the normal modes, and $v_j{}^k$ is the frequency of the jth normal mode of the crystal with configuration k. The vibrational partition function for a given configurational state is readily evaluated (see Chapter 2, Section 3) and is given by

$$Z_v{}^k(R) \equiv \sum_{v} e^{-E_v{}^k/kT} = e^{-\epsilon_0{}^k/kT} \prod_{j=1}^{3N} (1 - e^{-hv_j{}^k/kT})^{-1} \tag{4.3}$$

where $\epsilon_0{}^k$ is the vibrational zero point energy for the kth configuration. If we write (4.1) as

$$Z(R) = \sum_{k} Z_v{}^k(R) e^{-W_k/kT} \tag{4.4}$$

it is quite clear that a complete evaluation of $Z(R)$ requires an analysis of the dependence of the normal mode frequencies on the configurational state. Such an analysis would be very difficult to carry out and, in fact, no general theory in which the vibrations are treated along with the configurational states has been constructed. Instead, it is generally assumed that the vibrational and configurational parts of the problem can be separated. This is done by working with the configurational partition function, defined by

$$Z_c = \sum_k e^{-W_k/kT} \tag{4.5}$$

It is then assumed that the thermodynamic functions computed from Z_c can be added to those obtained from vibrational considerations. In practice, this means that when experimental data are examined, some estimate of the contribution of the vibrations to the data must be made. This estimate is then subtracted from the total, and the residue is compared with theory based on Z_c. An approach of this kind can be successful only if the vibrational partition function is much less sensitive to the degree of order than the configurational partition function. That this is approximately correct can be seen from the fact that, for example, in β-brass, the effect of order on the heat capacity is most pronounced at temperatures from 550 to 750°K. In this temperature range, the heat capacity can be represented by the high-temperature approximation (4.16, Chapter 2), the leading term of which is independent of the frequency spectrum. Furthermore, it was shown in Chapter 2 that the thermodynamic functions, particularly the heat capacity, are remarkably insensitive to the precise form of the distribution function. Therefore, we can expect that an analysis based on the separation of configurational and vibrational partition functions will give a description of order-disorder phenomena that is at least qualitatively correct. It turns out that this expectation is justified.

To get the degree of order of an equilibrium system, we just minimize the Helmholtz free energy obtained from Z_c with respect to R. That is, we solve the equation

$$\frac{\partial A_c(R)}{\partial R} = 0 \tag{4.6}$$

where

$$A_c(R) = -kT \ln Z_c \tag{4.7}$$

The equilibrium value of R obtained from (4.6) can then be inserted into the free-energy function to give the equilibrium Helmholtz free energy.

The partition function, Z_c, cannot be evaluated unless the energy levels W_k are known. The basic assumption made in order-disorder theory is that the energy consists of nearest-neighbor pairwise contributions. If $-v_{AA}$, $-v_{BB}$, and $-v_{AB}$ are the nearest-neighbor interaction energies for AA, BB, and AB pairs, respectively, then W_k is given by

$$W_k = -v_{AA}Q_{AA} - v_{BB}Q_{BB} - v_{AB}(Q_{AB} + Q_{BA}) \tag{4.8}$$

The interaction energies are chosen so that v_{AA}, v_{BB}, and v_{AB} are positive constants, since the state of zero energy is taken to be that in which all atoms are very far apart.

Equation 4.8 can be written in several alternate ways that are often useful. From (3.13)

$$Q_{AB} + Q_{BA} = qQ \qquad (4.9)$$

and from (3.14) and (4.9)

$$Q_{AA} + Q_{BB} = Q(1 - q) \qquad (4.10)$$

In a 50-50 AB alloy, $Q_{AA} = Q_{BB}$, so that (4.10) gives

$$Q_{AA} = Q_{BB} = \frac{Q}{2}(1 - q) \qquad (4.11)$$

Substituting (4.9) and (4.11) into (4.8) gives

$$W_k = -\frac{Q}{2}(v_{AA} + v_{BB}) - qQ(v_{AB} - \tfrac{1}{2}v_{AA} - \tfrac{1}{2}v_{BB}) \qquad (4.12)$$

The ordering energy v is defined by

$$v \equiv v_{AB} - \tfrac{1}{2}(v_{AA} + v_{BB}) \qquad (4.13)$$

It is the energy change accompanying the formation of an AB pair in the crystal. With this definition, (4.12) becomes

$$W_k = w_0 - Qvq \qquad (4.14)$$

where

$$w_0 \equiv -\frac{Q}{2}(v_{AA} + v_{BB}) \qquad (4.15)$$

is the average energy of pure A and pure B. From (3.16) and (4.14) the energy can also be written in terms of the short-range-order parameter:

$$W_k = w_0 - \frac{Qv}{2}(1 + \sigma) \qquad (4.16)$$

Another expression for the energy that is often used is obtained by assigning a parameter S_i to each lattice site so that $S_i = +1$ if the site contains an A atom and $S_i = -1$ if it contains a B atom. Then, since the product S_iS_j is always $+1$ if two adjacent sites i and j contain like atoms and always -1 if they contain unlike atoms,

$$\sum_{\langle i,j \rangle} S_iS_j = Q_{AA} + Q_{BB} - (Q_{AB} + Q_{BA})$$

$$= Q - 2(Q_{AB} + Q_{BA}) \qquad (4.17)$$

$$= Q(1 - 2q)$$

the summation being carried out over all nearest neighbor pairs. Then

$$qQ = \tfrac{1}{2}Q - \tfrac{1}{2}\sum_{\langle i,j \rangle} S_iS_j \qquad (4.18)$$

and putting this into (4.14) gives

$$W_k = w_0 - \frac{Qv}{2} + \frac{v}{2} \sum_{\langle i,j \rangle} S_i S_j \tag{4.19}$$

or

$$W_k = -\frac{Q}{2} v_{AB} + \frac{v}{2} \sum_{\langle i,j \rangle} S_i S_j \tag{4.20}$$

Finally, the energy can be written in another way that is sometimes useful, by defining two parameters α_i and β_j such that $\alpha_i = 1$ if the ith site of the α sublattice contains an A atom and is zero otherwise. Similarly, $\beta_j = 1$ if the jth atom of the β sublattice contains an A atom and is zero otherwise. Also, let $\lambda_{ij} = 1$ if i and j are nearest neighbor sites and $\lambda_{ij} = 0$ otherwise. Then we have

$$\sum_{i,j} \lambda_{ij} \alpha_i \beta_j = Q_{AA} = \frac{Q}{2}(1 - q) \tag{4.21}$$

since there is a term $+1$ for every nearest neighbor AA pair all other terms being zero. The sum in (4.21) is taken over all sites of both sublattices. Combining (4.21) with (4.14) gives

$$W_k = w_0 - Qv + 2v \sum_{i,j} \lambda_{ij} \alpha_i \beta_j \tag{4.22}$$

Accurate evaluation of the partition function given by (4.5) is very difficult. A knowledge is required of the number of configurational states that have a given energy, subject to the restriction that the long-range order is fixed. The combinatorial problem that must be solved to obtain this information is extremely complex, and approximation methods must be used. Two general approximation methods will be treated in this chapter.

5.5 THE KIRKWOOD METHOD

In 1938 Kirkwood developed a method by which the configurational free energy, A_c, can be developed as a series in the long-range-order parameter. Let $\overline{W}$ be the unweighted average of the energy over all microstates. That is,

$$\overline{W} = \frac{1}{g(R)} \sum_k W_k \tag{5.1}$$

where $g(R)$ is the total number of configurational states consistent with a given R. $g(R)$ is easily evaluated since it is just the number of ways of arranging $N_{A\alpha}$ A atoms and $N_{B\alpha}$ B atoms on the α sublattice, and $N_{A\beta}$ A atoms and $N_{B\beta}$ B atoms on the β sublattice. Since we are restricting ourselves to AB

alloys, the number of α sites is equal to the number of β sites, and $g(R)$ is given by

$$g(R) = \frac{(N/2)! \, (N/2)!}{N_{A\alpha}! \, N_{B\alpha}! \, N_{A\beta}! \, N_{B\beta}!} \tag{5.2}$$

By using (3.6) to (3.9), and (3.20), (3.21), (3.24), and (3.25), and remembering that $\gamma_A = \gamma_B = 1/2$ for an AB alloy, we see that

$$N_{A\alpha} = N_{B\beta} = \frac{N}{4}(1 + R) \tag{5.3}$$

$$N_{A\beta} = N_{B\alpha} = \frac{N}{4}(1 - R) \tag{5.4}$$

Using (5.3) and (5.4) allows us to write $g(R)$ as a function of R:

$$g(R) = \frac{\left(\frac{N}{2}\right)! \, \left(\frac{N}{2}\right)!}{\left[\frac{N}{4}(1 + R)\right]! \, \left[\frac{N}{4}(1 - R)\right]! \, \left[\frac{N}{4}(1 + R)\right]! \, \left[\frac{N}{4}(1 - R)\right]!} \tag{5.5}$$

Now we expand the exponential in the partition function in a Taylor series about $\overline{W}$. The result is

$$
\begin{aligned}
e^{-W_k/kT} &= e^{-\overline{W}/kT}\left[1 - \frac{1}{kT}(W_k - \overline{W}) + \frac{1}{2}\frac{(W_k - \overline{W})^2}{(kT)^2} - \cdots\right] \\
&= e^{-\overline{W}/kT} \sum_{j=0}^{\infty}\left(\frac{-1}{kT}\right)^j \frac{(W_k - \overline{W})^j}{j!}
\end{aligned} \tag{5.6}
$$

If (5.6) is substituted into (4.5), the result is

$$Z_c = e^{-A_c/kT} = e^{-\overline{W}/kT} \sum_{k} \sum_{j=0}^{\infty}\left(\frac{-1}{kT}\right)^j \frac{(W_k - \overline{W})^j}{j!} \tag{5.7}$$

The jth moment of the distribution of energy about the average $\overline{W}$ is defined by

$$M_j \equiv \frac{1}{g(R)} \sum_{k} (W_k - \overline{W})^j \tag{5.8}$$

and therefore (5.7) can be written as

$$e^{-A_c/kT} = g(R)e^{-\overline{W}/kT} \sum_{j=0}^{\infty}\left(\frac{-1}{kT}\right)^j \frac{M_j}{j!} \tag{5.9}$$

Now define a power series in $x \equiv (-1/kT)$ by the relation

$$\sum_{n=1}^{\infty} \frac{B_n}{n!} x^n = \ln \left(\sum_{j=0}^{\infty} \frac{M_j}{j!} x^j \right) \qquad (5.10)$$

where the B_n are to be determined. To obtain the B_n in terms of the M_j, differentiate both sides of (5.10) with respect to x. From this we get

$$\sum_{n=1, j=0}^{\infty} \frac{n}{n! \, j!} B_n M_j x^{n+j-1} = \sum_{j=0}^{\infty} \frac{j}{j!} M_j x^{j-1} \qquad (5.11)$$

The B_n are readily evaluated by equating coefficients of equal powers of x in both series, from which we get

$$B_1 M_0 = M_1$$
$$B_1 M_1 + B_2 M_0 = M_2$$
$$B_3 M_0 + 2B_2 M_1 + B_1 M_2 = M_3$$
$$B_4 M_0 + 3B_3 M_1 + 3B_2 M_2 + B_1 M_3 = M_4 \qquad (5.12)$$
$$\vdots \qquad \qquad \vdots$$

But from the definition of the moments, we see that

$$M_0 = 1$$
$$M_1 = 0 \qquad (5.13)$$

so that solving (5.12) for the B's gives

$$B_1 = 0$$
$$B_2 = M_2$$
$$B_3 = M_3$$
$$B_4 = M_4 - 3M_2^{\,2} \qquad (5.14)$$
$$\vdots \qquad \qquad \vdots$$

Substitution of (5.14) into (5.10), replacing x by $-1/kT$, and taking exponentials shows that the series in (5.9) can be expressed as

$$\sum_{j=0}^{\infty} \left(\frac{-1}{kT} \right)^{j} \frac{M_j}{j!} \exp \left[\frac{M_2}{2(kT)^2} - \frac{M_3}{3! \, (kT)^3} + \frac{M_4 - 3M_2^{\,2}}{4! \, (kT)^4} - \cdots \right] \qquad (5.15)$$

Putting this into (5.9) and taking logarithms, the free energy becomes

$$A_c = -kT \ln g(R) + \overline{W} - \frac{M_2}{2kT} + \frac{M_3}{6(kT)^2} - \cdots \qquad (5.16)$$

All that is necessary now is to write the various terms on the right-hand side of this equation as functions of R. The moments will be considered later, but the first two terms can be treated immediately. If Stirling's approximation is applied to the logarithm of the right-hand side of (5.5), it is found that

$$\ln g(R) = \frac{N}{2} [2 \ln 2 - (1 + R) \ln (1 + R) - (1 - R) \ln (1 - R)] \qquad (5.17)$$

To obtain $\overline{W}$, it is only necessary to take averages in (4.14);

$$\overline{W} = w_0 - Qv\bar{q} \qquad (5.18)$$

But $\bar{q}$ is given by (3.28), so that

$$\overline{W} = w_0 - \frac{Qv}{2}(1 + R^2) \qquad (5.19)$$

If we let $W(0)$ be the energy for zero order, then

$$W(0) = -\frac{Q}{4}[v_{AA} + v_{BB} + 2v_{AB}] \qquad (5.20)$$

and (5.19) becomes

$$\overline{W} = W(0) - \frac{Qv}{2} R^2 \qquad (5.21)$$

Substitution of (5.17) and (5.21) in (5.16) gives

$$A_c = -\frac{NkT}{2} [2 \ln 2 - (1 + R) \ln (1 + R) - (1 - R) \ln (1 - R)]$$

$$+ W(0) - \frac{Nzv}{4} R^2 - \frac{M_2}{2kT} + \frac{M_3}{6(kT)^2} - \cdots \qquad (5.22)$$

where Q has been replaced by $Nz/2$. This gives the configurational free energy for a given R, and if the value of R corresponding to equilibrium is inserted into (5.22), all of the equilibrium thermodynamic properties can be computed.

To determine the equilibrium degree of long-range order, just set the derivative of A_c with respect to R equal to zero. This gives

$$\ln \frac{(1 + R)}{(1 - R)} = \frac{zv}{kT} R + \frac{1}{N(kT)^2} \frac{\partial M_2}{\partial R} - \frac{1}{3N(kT)^3} \frac{\partial M_3}{\partial R} + \cdots \qquad (5.23)$$

which determines the equilibrium value of R as a function of temperature.

The energy is readily obtained by the application of (4.4) in Chapter 2, which for our problem is conveniently written as

$$U_c = \frac{\partial \left(\dfrac{A_c}{kT} \right)}{\partial \left(\dfrac{1}{kT} \right)} \tag{5.24}$$

From the derivation of this equation (see Chapter 2) it is clear that the differentiation is purely formal, in that all quantities but $1/kT$ are treated as constants during the differentiation. Then, dividing (5.22) by kT and applying (5.24) gives

$$U_c = W(0) - \frac{Nzv}{4} R^2 - \frac{M_2}{kT} + \frac{M_3}{2(kT)^2} - \cdots \tag{5.25}$$

for the configurational energy, and

$$S_c = \frac{U_c - A_c}{T}$$

$$= \frac{Nk}{2} [2 \ln 2 - (1 + R) \ln (1 + R) - (1 - R) \ln (1 - R)]$$

$$- \frac{1}{T} \left[\frac{M_2}{2kT} - \frac{1}{3} \frac{M_3}{(kT)^2} + \cdots \right] \tag{5.26}$$

for the configurational entropy. The heat capacity is obtained by differentiating (5.25).

$$C_V{}^c \equiv \left(\frac{\partial U_c}{\partial T} \right)_V = -\frac{nzv}{2} R \frac{dR}{dT} - \frac{d}{dT} \left[\frac{M_2}{kT} - \frac{M_3}{2(kT)^2} + \cdots \right] \tag{5.27}$$

The infinite series in these formulae are convergent, and although the convergence is not very rapid, useful equations are obtained by truncating the series at various points. The accuracy of the formulae, of course, depends on the number of terms retained.

5.6   THE BRAGG-WILLIAMS APPROXIMATION

The Bragg-Williams approximation is obtained by neglecting all terms containing second- and higher-order moments in the equations of the previous section. A comparison of (5.25) and (5.21) shows that this is equivalent to assuming that the configurational energy of the crystal is equal to the average energy $\overline{W}$. At first sight, this may seem like a crude assumption, but the Bragg-Williams approximation succeeds in displaying the main qualitative features of the order-disorder transition. Also, the mathematical development is quite straightforward. The Bragg-Williams method will therefore be treated in some detail in this section. Methods of solving the order-disorder problem to a higher degree of accuracy will be discussed in later sections.

Neglecting all terms containing M_2, M_3, and so on in (5.23), (5.24), (5.26), (5.27), and (5.28), we get the thermodynamic equations for the Bragg-Williams approximation.

$$A_c = W(0) - \frac{Nzv}{4} R^2$$

$$+ \frac{NkT}{2} [(1 + R) \ln (1 + R) + (1 - R) \ln (1 - R) - 2 \ln 2] \quad (6.1)$$

$$\ln \frac{(1 + R)}{(1 - R)} = \frac{zv}{kT} R \quad (6.2)$$

$$U_c = W(0) - \frac{Nzv}{4} R^2 \quad (6.3)$$

$$S_c = \frac{Nk}{2} [2 \ln 2 - (1 + R) \ln (1 + R) - (1 - R) \ln (1 - R)] \quad (6.4)$$

$$C_V{}^c = -\frac{Nzv}{2} R \frac{dR}{dT} \quad (6.5)$$

To show the relationship between the thermodynamic quantities and the degree of order, it is desirable to remove the constants that clutter up the above equations by defining the following quantities:

$$a \equiv \frac{4[A_c - W(0)]}{Nzv} \quad (6.6)$$

$$s \equiv \frac{2S_c}{Nk} \quad (6.7)$$

$$u \equiv \frac{4[U_c - W(0)]}{Nzv} \quad (6.8)$$

$$\varphi \equiv \frac{2kT}{zv} \quad (6.9)$$

Equations 6.1 to 6.5 then become

$$a = u - \varphi s \tag{6.10}$$

$$\ln \frac{(1 + R)}{(1 - R)} = 2\frac{R}{\varphi} \tag{6.11}$$

$$u = -R^2 \tag{6.12}$$

$$s = 2\ln 2 - (1 + R)\ln(1 + R) - (1 - R)\ln(1 - R) \tag{6.13}$$

$$C_V{}^c = -NkR\frac{dR}{d\varphi} \tag{6.14}$$

The quantities a and u are dimensionless free energy and energy, respectively, measured from the energy of the completely disordered structure in units of $Nzv/4$. A dimensionless entropy s measured in units of $Nk/2$, and φ is a dimensionless temperature.

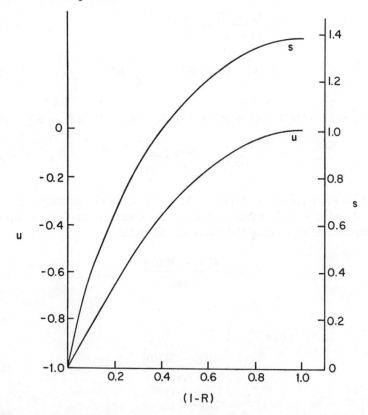

Fig. 5.3. Energy and entropy as a function of disorder for an AB alloy in the BW approximation.

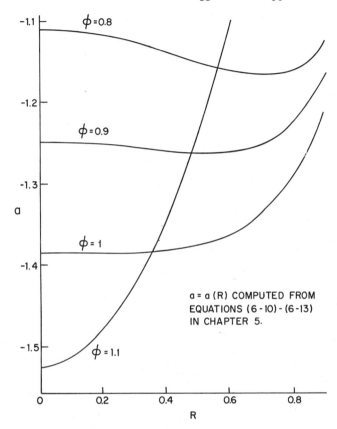

Fig. 5.4. Free energy as a function of order in the BW approximation.

Plots of u and s are shown as functions of $(1 - R)$ in Figure 5.3. The entropy increases from zero at perfect order to the entropy of mixing for a random solution ($2 \ln 2$) at complete disorder. Similarly, u increases from zero to the energy of mixing for a random solution.

In Figure 5.4, the dimensionless free energy a is shown as a function of R for several values of the dimensionless temperature φ. The curves show that for temperatures where $\varphi < 1$, there is a minimum in the free energy. As the temperature is raised, the minimum shifts to lower values of R, until at $\varphi = 1$, the minimum coincides with $R = 0$. Also, for $\varphi \leq 1$, the free energy approaches $R = 0$ with a zero slope. This means that above $\varphi = 1$, there is no nonzero value of R for which the slope is zero, and $\varphi = 1$ defines a critical temperature above which there is no long-range order. From (6.9) we see that this temperature is given by

$$1 = \frac{2kT_c}{zv}$$

or

$$T_c = \frac{zv}{2k} \tag{6.15}$$

From the free-energy curves in Figure 5.4, the critical temperature is clearly that temperature at which the following conditions are simultaneously satisfied

$$\left. \begin{array}{l} R = 0 \\[2mm] \dfrac{\partial A_c}{\partial R} = 0 \\[2mm] \dfrac{\partial^2 A_c}{\partial R^2} = 0 \end{array} \right\} \quad \text{at} \quad T = T_c \tag{6.16}$$

Application of these conditions to the free-energy expression (6.1) or its equivalent (6.10) shows that the critical temperature is given by $\varphi = 1$. This is in agreement with (6.15), which was arrived at from a graphical analysis of the free-energy function.

Equation 6.11 determines the equilibrium value of R as a function of temperature. At high temperatures R is small, so the logarithms may be approximated by the first three terms of their series expansions. That is,

$$\ln (1 + R) \simeq R - \tfrac{1}{2}R^2 + \tfrac{1}{3}R^3$$

$$\ln (1 - R) \simeq -R - \tfrac{1}{2}R^2 - \tfrac{1}{3}R^3$$

so that for high enough temperatures, (6.11) becomes

$$2R + \tfrac{2}{3}R^3 = \frac{2}{\varphi} R$$

or

$$R^2 = 3\left(\frac{1 - \varphi}{\varphi}\right) \quad \text{(high } T) \tag{6.17}$$

When $R = 0$, $\varphi = 1$, showing once again that the critical temperature is given by (6.15).

A more accurate high-temperature approximation can be obtained by including more terms in the logarithmic expansions. Thus if we take

$$\ln (1 + R) \simeq R - \tfrac{1}{2}R^2 + \tfrac{1}{3}R^3 - \tfrac{1}{4}R^4 + \tfrac{1}{5}R^5$$

$$\ln (1 - R) \simeq -R - \tfrac{1}{2}R^2 - \tfrac{1}{3}R^3 - \tfrac{1}{4}R^4 - \tfrac{1}{5}R^5 \tag{6.18}$$

then (6.11) becomes

$$R^2 = \frac{5}{6}\left\{\left[1 + \frac{36}{5}\frac{(1 - \varphi)}{\varphi}\right]^{1/2} - 1\right\} \quad (R < 0.5) \tag{6.19}$$

This equation is an accurate approximation to (6.9) up to $R = 1/2$, where it is in error by about 6%.

Equation 6.11 can be written in another form that is convenient for obtaining numerical solutions of R versus φ. From the definition of the hyperbolic tangent

$$\tanh \frac{R}{\varphi} = \frac{e^{2R/\varphi} - 1}{e^{2R/\varphi} + 1} \tag{6.20}$$

it is easy to see that (6.11) can be converted to

$$R = \tanh \frac{R}{\varphi} \tag{6.21}$$

Equation 6.20 can be solved for φ to give

$$\varphi = \frac{R}{\tanh^{-1} R} \tag{6.22}$$

so that numerical values of φ versus R can be calculated from standard tables of the hyperbolic tangent. A set of R versus φ values computed in this way are given in Table 5.1 and plotted in Figure 5.5. The figure shows that at low temperatures R decreases slowly with temperature, but as the temperature increases the decrease in R becomes more and more rapid. Most of the change in order with temperature takes place in a relatively narrow temperature range; R changes from about 0.8 to zero in a range of φ corresponding to only 27% of the range from absolute zero to the critical temperature. Experimental values of R in β-brass are given in Table 5.2.

With the help of the tabulated values of R as a function of φ in Table 5.1, the energy, entropy and free energy are readily determined as functions of temperature. These are shown graphically in Figure 5.6 in the dimensionless forms defined by equations (6.10) to (6.13). The energy and entropy at first rise slowly with increasing temperature, and then, for $\varphi > 0.4$ increase very rapidly. The free energy is almost constant until $\varphi \simeq 0.5$ and then steadily drops to the value characteristic of a random solution at $\varphi = 1$. This behavior is characteristic of cooperative phenomena.

At low temperatures R/φ is large, so that a low-temperature approximation to (6.21) can be obtained by using the following series expansion for the hyperbolic tangent

$$\tanh \frac{R}{\varphi} = 1 - 2e^{-2R/\varphi} + 2e^{-4R/\varphi} - \cdots \tag{6.23}$$

For values of φ as high as 0.9 (corresponding to $R \simeq 0.53$) the third term in this series is only 6% of the second term, so that for $R > 0.5$ it is sufficiently

Table 5.1. Long-Range-Order Parameter as a Function of Temperature in the Bragg-Williams Approximation

R	$\tan h^{-1} R$	$\varphi = T/T_c$
0.0000	0 0	1.000
0.02999	0.03	0.9997
0.04996	0.05	0.9992
0.09967	0.10	0.9967
0.14889	0.15	0.9926
0.19738	0.20	0.9869
0.24492	0.25	0.9797
0.29131	0.30	0.9710
0.33638	0.35	0.9611
0.37995	0.40	0.9499
0.42190	0.45	0.9376
0.46212	0.50	0.9242
0.53705	0.60	0.8951
0.60437	0.70	0.8634
0.66404	0.80	0.8301
0.71630	0.90	0.7959
0.76159	1.00	0.7616
0.80050	1.10	0.7277
0.83365	1.20	0.6947
0.86172	1.30	0.6629

Table 5.1   (*Continued*)

R	tan h $^{-1}$R	$\varphi = {}^T/T_c$
0.88535	1.40	0.6324
0.90515	1.50	0.6034
0.92167	1.60	0.5760
0.94681	1.80	0.5260
0.96403	2.00	0.4820
0.98661	2.50	0.3946
0.99505	3.00	0.3317
0.99933	4.00	0.2498
0.99999	6.00	0.1667
1.0000	0 0	0.000

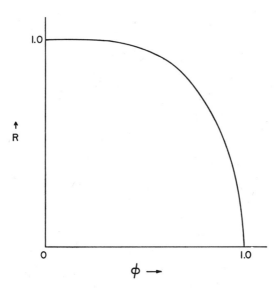

Fig. 5.5. Long-range order parameter versus reduced temperature in the BW approximation.

Table 5.2. Experimental Values of Long-Range Order in β-Brass ($T_c = 736°K$)

		$T_c = 736°K$		
$\phi = T/T_c$	R	$\phi = T/T_c$	R
0.3909	1.001	0.9336	0.675
0.4729	0.990	0.9418	0.673
0.5075	0.983	0.9419	0.656
0.5498	0.987	0.9541	0.613
0.5839	0.960	0.9562	0.614
0.6521	0.954	0.9665	0.555
0.6726	0.944	0.9673	0.556
0.7086	0.924	0.9741	0.520
0.7460	0.915	9.9741	0.514
0.7673	0.896	0.9795	0.480
0.8030	0.870	0.9803	0.460
0.8241	0.851	0.9837	0.435
0.8452	0.850	0.9870	0.416
0.8241	0.826	0.9899	0.395
0.8723	0.795	0.9901	0.392
0.8790	0.788	0.9934	0.334
0.9112	0.722	0.9949	0.310
0.9005	0.742	0.9965	0.273
0.9132	0.721	0.9965	0.259
0.9270	0.689		

From J. C. Norwell and J. Als-Nielsen, *Phys. Rev.*, **B277** (1970).

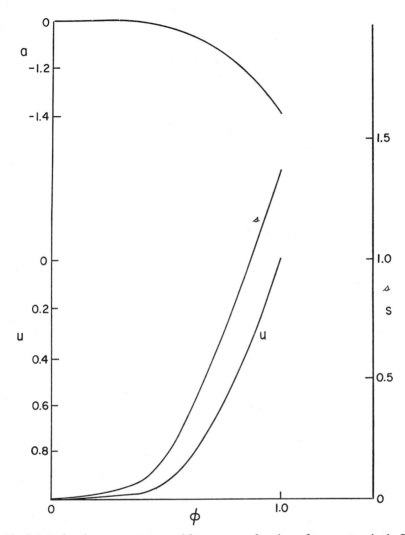

Fig. 5.6. Reduced energy, entropy, and free energy as functions of temperature in the BW approximation.

accurate to retain only the first two terms in (6.23). Then (6.22) becomes

$$R = 1 - 2e^{-2\,R/\varphi} \tag{6.24}$$

or

$$\varphi = \frac{2R}{\ln 2 - \ln(1 - R)} \qquad (R > 0.5) \tag{6.25}$$

We will now compute the heat capacity as a function of temperature from the calculated values of R versus φ. First we must evaluate the derivative $dR/d\varphi$ from (6.11):

$$\ln(1 + R) - \ln(1 - R) = \frac{2R}{\varphi}$$

$$\frac{dR}{1 + R} + \frac{dR}{1 - R} = \frac{2dR}{\varphi} - \frac{2R\,d\varphi}{\varphi^2}$$

or

$$\frac{dR}{d\varphi} = \frac{R(1 - R^2)}{\varphi(1 - \varphi - R^2)} \tag{6.26}$$

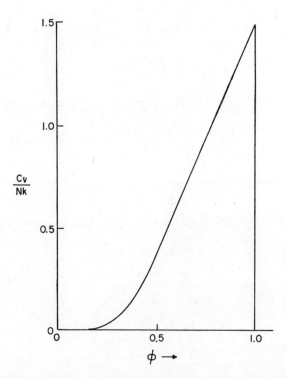

Fig. 5.7. Configurational heat capacity of an AB alloy in the BW approximation.

Putting this into (6.14) gives

$$C_V{}^c = -Nk \frac{R^2(1 - R^2)}{\varphi(1 - \varphi - R^2)} \tag{6.27}$$

a form that permits $C_V{}^c$ to be computed from the table of R versus φ values. A plot of $C_V{}^c$ versus φ in units of Nk is shown in Figure 5.7. Note that the maximum value of $C_V{}^c$ is $1.5Nk$, as can be seen by differentiating the high-temperature expression for the R versus φ relation and substituting it into (6.14). Thus, using (6.19),

$$2R \frac{dR}{d\varphi} = -\frac{3}{\varphi^2}\left[1 + \frac{36}{5}\left(\frac{1 - \varphi}{\varphi}\right)\right]^{-1/2}$$

so that as $\varphi \to 1$, $R\, dR/d\varphi \to 3/2$ and $C_V{}^c \to 3Nk/2$.

5.7 THE SECOND-MOMENT APPROXIMATION

Although the Bragg-Williams approximation reproduces the general features of order-disorder systems, it provides only a semiquantitative description of such systems. Its limitations are a result of neglecting the distribution of possible energies of the system and replacing all energies by the average energy. A more accurate theory is obtained if we retain the second moment in the Kirkwood equations. To do this, an explicit calculation of M_2 is required. The calculation is a straightforward lattice-counting exercise whose result is

$$M_2 = \frac{Nzv^2}{8}(1 - R^2)^2 \tag{7.1}$$

from which

$$\frac{\partial M_2}{\partial R} = -\frac{Nzv^2}{2}(1 - R^2)R \tag{7.2}$$

and

$$\frac{d}{dT}\left(\frac{M_2}{kT}\right) = -\frac{Nzv^2}{8kT}\left[\frac{(1 - R^2)^2}{T} + 4R(1 - R^2)\frac{dR}{dT}\right] \tag{7.3}$$

Substituting (7.1) into (5.24) and neglecting terms containing the higher moments $M_3, \ldots$, and so on, we obtain

$$\ln \frac{(1 + R)}{(1 - R)} = \frac{zv}{kT}R\left[1 - \frac{v}{2kT}(1 - R^2)\right] \tag{7.4}$$

which determines R as a function of T in the Kirkwood second-moment approximation. The heat capacity is obtained by combining (7.3) and (5.27), and again neglecting moments higher than the second;

$$C_V{}^c = -\frac{Nzv}{2} R \frac{dR}{dT}\left[1 - \frac{v}{kT}(1 - R^2)\right] + \frac{Nzv^2}{8kT^2}(1 - R^2)^2 \qquad (7.5)$$

To get a high-temperature approximation to (7.4), we proceed just as in the Bragg-Williams case. If we expand the logarithms to the third order in R, (7.4) becomes

$$2 + \tfrac{2}{3}R^2 = \frac{zv}{kT} - \frac{zv^2}{2(kT)^2}(1 - R^2) \qquad (7.6)$$

At $R = 0$, this equation determines a critical temperature T_c given by

$$\frac{2kT_c}{zv} = 1 - \frac{v}{2kT_c} \qquad (7.7)$$

from which

$$\frac{2kT_c}{zv} = \frac{1}{2}\left[1 \pm \left(1 - \frac{4}{z}\right)^{1/2}\right] \qquad (7.8)$$

The appropriate root to be chosen is determined by requiring (7.8) to reduce to the Bragg-Williams result when the second term on the right of (7.7) is neglected. Without this second term, the term $4/z$ would not appear under the radical in (7.8), and the right-hand side would reduce to $(1 \pm 1)/2$. Choosing the positive sign would then give the Bragg-Williams result, so we must take (7.8) to be

$$\frac{2kT_c}{zv} = \frac{1}{2}\left[1 + \left(1 - \frac{4}{z}\right)^{1/2}\right] \qquad (7.9)$$

From the definition of the critical temperature in the Bragg-Williams approximation

$$\frac{2k}{zv} = \frac{1}{T_c(\text{BW})} \qquad (7.10)$$

Combining this with (7.9)

$$\frac{T_c}{T_c(\text{BW})} = \frac{1}{2}\left[1 + \left(1 - \frac{4}{z}\right)^{1/2}\right] \qquad (7.11)$$

For $z = 8$, this becomes

$$\frac{T_c}{T_c(\text{BW})} = 0.854 \qquad (7.12)$$

which shows that for a given ordering energy v, the second-moment approximation gives a lower critical temperature than the Bragg-Williams approximation.

The variation of the long-range-order parameter with temperature is determined by (7.4). If we treat R as the independent variable, we can take advantage of the numerical work done for the Bragg-Williams case by defining a function $f(R)$ by

$$f(R) = 2R \ln \frac{(1-R)}{(1+R)} = \frac{R}{\tanh^{-1} R} \qquad (7.13)$$

In the Bragg-Williams approximation, $f(R) = T/T_c$ (BW), so $f(R) = \varphi$(BW) which is given as the last entry in Table 5.1.

Combining (7.13) with (7.4) results in

$$\frac{zv}{2kT}\left[1 - \frac{v}{2kT}(1-R^2)\right] = \frac{1}{f(R)} \qquad (7.14)$$

But from the definition of the Bragg-Williams critical temperature,

$$\frac{zv}{2k} = T_c(\text{BW}); \qquad \frac{v}{2k} = \frac{T_c(\text{BW})}{z} \qquad (7.15)$$

Also, the critical temperature in the present approximation is related to that in the Bragg-Williams approximation by

$$T_c(\text{BW}) = \kappa T_c \qquad (7.16)$$

where κ is a constant determined by (7.11) (equal to 1.172 for $z = 8$). Using (7.15) and (7.16), we rewrite (7.14) as

$$\kappa \frac{T_c}{T}\left[1 - \frac{\kappa}{z}\frac{T_c}{T}(1-R^2)\right] = \frac{1}{f(R)} \qquad (7.17)$$

Solving (7.17) for T/T_c gives

$$\frac{T}{T_c} = \frac{\kappa}{2}f(R)\left[1 + \left(1 - \frac{4(1-R^2)}{3f(R)}\right)^{1/2}\right] \qquad (7.18)$$

where the positive square root was taken to ensure that $T/T_c \rightarrow 1$ as $R \rightarrow 0$.

To compute the heat capacity, it is necessary to evaluate the derivative in the right-hand side of (7.5). From (7.17) it is easy to show that the derivative is

$$\frac{dR}{dT} = \frac{f(R)}{T_c}\frac{2\left(\dfrac{T}{T_c}\right) - f(R)}{\left(\dfrac{T}{T_c}\right)^2 + 2R\dfrac{\kappa}{z}[f(R)]^2} \qquad (7.19)$$

An important difference between the Kirkwood and Bragg-Williams approximations for the heat capacity can be seen immediately by examining

(7.5). Above the critical temperature, $R = 0$ and $dR/dT = 0$, so (7.5) reduces to

$$C_V{}^c = \frac{Nzv^2}{8kT^2} \quad (T > T_c) \tag{7.20}$$

That is, the ordering phenomenon contributes to the heat capacity above T_c in the Kirkwood approximation, whereas it does not in the Bragg-Williams approximation.

Now compute the heat capacity at the critical temperature. From (7.6), we have at high temperatures

$$2R\frac{dR}{dT} = -3\kappa\frac{T_c}{T^2} + 2R\frac{dR}{dT}\frac{3\kappa^2 T_c{}^2}{zT^3} + \frac{6\kappa^2 T_c{}^2}{zT^3}(1 - R^2) \tag{7.21}$$

where we have used the relation

$$\frac{zv}{2kT_c} = \kappa \tag{7.22}$$

At $T = T_c$, (7.21) reduces to

$$R\frac{dR}{dT}\bigg|_{T=T_c} = \frac{\dfrac{3\kappa^2}{2T_c} - \dfrac{3\kappa}{2T_c}}{1 - \dfrac{3\kappa^2}{z}} \tag{7.23}$$

Putting this in (7.5), setting $R = 0$ and again using (7.22), we obtain the maximum heat capacity

$$C_V{}^c(T \to T_c) = Nk\kappa \left[\frac{\dfrac{3\kappa}{2} - \dfrac{3\kappa^2}{z}}{1 - \dfrac{3\kappa^2}{z}}\right]\left(1 - \frac{2\kappa}{z}\right) + Nk\frac{\kappa^2}{2z} \tag{7.24}$$

For $z = 8$, $\kappa = 1.172$, and (7.24) gives

$$C_V{}^c(T \to T_c) = 2.03Nk \tag{7.25}$$

In the second-moment approximation, therefore, the peak in the heat-capacity curve is higher than in the Bragg-Williams approximation.

5.8 THE QUASI-CHEMICAL APPROXIMATION

In a 50-50 AB alloy, the energy of the kth microstate can be written as

$$W_k = 2vQ_{AA} - v_{AB}Q \tag{8.1}$$

as can be seen by combining (4.13), the definition of the ordering energy, and the expression for the conservation of the number of pairs, which is

$$Q = Q_{AA} + Q_{BB} + (Q_{AB} + Q_{BA})$$
$$Q_{AA} = Q_{BB}$$

For a crystal with a given long-range order R, the configurational energy is just the statistical average of (8.1).

$$U_c(R) = \langle W_k \rangle = 2v\langle Q_{AA} \rangle - v_{AB}Q \tag{8.2}$$

Triangular brackets are used to indicate averages, rather than bars, to emphasize that the average is statistical mechanical and not arithmetic. The average is taken over all of the $g(R)$ microstates consistent with a given R.

If $U_c(R)$ is known, the free energy can be computed by integrating (5.24) to get

or

$$\left. \frac{A_c}{kT} \right|_{1/kT=0}^{1/kT} = \int_0^{1/kT} U_c \, d\left(\frac{1}{kT}\right) \tag{8.3}$$

$$A_c = A_c(T \to \infty) + kT \int_0^{1/kT} U_c \, d\left(\frac{1}{kT}\right) \tag{8.4}$$

From the relation between the free energy and the partition function

$$A_c = -kT \ln \sum_{k=1}^{g(R)} e^{-W_k/kT}$$

$$A_c(T \to \infty) = -kT \ln g(R) \tag{8.5}$$

so (8.4) becomes

$$A_c = kT \int_0^{1/kT} U_c \, d\left(\frac{1}{kT}\right) - kT \ln g(R) \tag{8.6}$$

In the quasi-chemical method, U_c is computed by treating the formation of pairs as a chemical reaction of the type

$$(AA) + (BB) \rightleftarrows (AB) + (BA) \tag{8.7}$$

where (AB) represents a pair of atoms such that an A atom is on an α site and a B atom is on an adjacent β site. The other symbols in the chemical equation have analogous meanings.

The energy change accompanying the conversion of two like pairs, (AA) and (BB), to two unlike pairs, (AB) and (BA), is

$$-v_{AB} - v_{BA} + v_{AA} + v_{BB} = -2v_{AB} + v_{AA} + v_{BB}$$
$$= -2v \tag{8.8}$$

Now let us regard (8.7) as representing a gaseous chemical reaction in which the various pairs are analogous to diatomic molecules. Since we have agreed to ignore vibrational effects in the order-disorder theory, we can take the free energy of reaction for (8.7) to be just the transformation energy $-2v$. Then, from the theory of the equilibrium constant for gas reactions, we have at equilibrium

$$\frac{\langle Q_{AA}\rangle\langle Q_{BB}\rangle}{\langle Q_{AB}\rangle\langle Q_{BA}\rangle} = e^{-2v/kT} \tag{8.9}$$

The limitation on this formula is that it describes a model in which the atom pairs are treated as independent entities. The pairs are clearly not independent since they fill a lattice. However, we can expect that the quasi-chemical model will lead to more accurate results than those obtained from the Bragg-Williams approximation. The Bragg-Williams approximation neglects completely the statistical-mechanical weighting of pair formation. Actually, we will see that the quasi-chemical method gives results that are similar to those obtained from the Kirkwood method.

The utility of (8.9) lies in the fact that it can be converted into a form that gives $\langle Q_{AA}\rangle$ as a function of R; therefore, the energy (8.2) and finally the free energy (8.6) can be obtained as a function of the long-range-order parameter. The equilibrium relation between R and T can then be obtained in the usual way by minimizing the free energy.

The number of pairs that have an A atom on an α-site is $zN_{A\alpha}$. Therefore

$$\langle Q_{AB}\rangle + \langle Q_{AA}\rangle = zN_{A\alpha}$$

Similarly

$$\langle Q_{BA}\rangle + \langle Q_{BB}\rangle = zN_{B\alpha} \tag{8.10}$$

Now let x be defined by

$$x \equiv e^{2v/kT} \tag{8.11}$$

Then solving (8.10) for $\langle Q_{AB}\rangle$ and $\langle Q_{BA}\rangle$, substituting the result in (8.9), and remembering that $\langle Q_{AA}\rangle = \langle Q_{BB}\rangle$, we obtain

$$\frac{\langle Q_{AA}\rangle^2}{[zN_{A\alpha} - \langle Q_{AA}\rangle][zN_{B\alpha} - \langle Q_{AA}\rangle]} = \frac{1}{x} \tag{8.12}$$

This is a quadratic equation for $\langle Q_{AA}\rangle$ whose solution is

$$\langle Q_{AA}\rangle = \frac{zN}{4(x-1)}\left\{\left[1 + \frac{16N_{A\alpha}N_{B\alpha}}{N^2}(x-1)\right]^{1/2} - 1\right\} \tag{8.13}$$

or, recalling (5.3) and (5.4) for the relations between $N_{A\alpha}$ and $N_{B\alpha}$ and R,

$$\langle Q_{AA}\rangle = \frac{zN}{4(x-1)}\left\{[1 + (1-R^2)(x-1)]^{1/2} - 1\right\} \tag{8.14}$$

Combining (8.14) with (8.2) gives the energy in terms of R

$$U_c(R) = \frac{vzN}{2(x-1)}\{[1 + (1 - R^2)(x - 1)]^{1/2} - 1\} - v_{AB}Q \quad (8.15)$$

so that the free energy (8.6) becomes

$$A_c = \frac{vzNkT}{2}\int_0^{1/kT} \frac{\{[1 + (1 - R^2)(x - 1)]^{1/2} - 1\}}{(x - 1)}$$

$$\times\, d\left(\frac{1}{kT}\right) - v_{AB}Q - kT \ln g(R) \quad (8.16)$$

The integration can be performed by making a change of variable defined by

$$\alpha^2 \equiv 1 + (1 - R^2)(x - 1) = 1 + (1 - R^2)(e^{2v/kT} - 1) \quad (8.17)$$

which transforms the integral in (8.16) into

$$\frac{1}{v}\int_1^\alpha \frac{(1 - R^2)\alpha\, d\alpha}{(\alpha + 1)(\alpha^2 - R^2)} = \frac{(1 - R^2)}{2(1 + R)v}\int_1^\alpha \frac{d\alpha}{(\alpha - R)}$$

$$+ \frac{(1 - R^2)}{2(1 - R)v}\int_1^\alpha \frac{d\alpha}{(\alpha + R)} - \frac{(1 - R^2)}{(1 - R^2)v}\int_1^\alpha \frac{d\alpha}{(\alpha + 1)} \quad (8.18)$$

The right-hand side of (8.18) comes from a reduction of the integrand on the left-hand side to partial fractions. Performing the integrations and substituting into (8.16) gives

$$A_c = \frac{zNkT}{4}\left[(1 - R)\ln\frac{\alpha - R}{1 - R} + (1 + R)\ln\frac{\alpha + R}{1 + R} - 2\ln\frac{\alpha + 1}{2}\right]$$

$$- Qv_{AB} - kT \ln g(R) \quad (8.19)$$

To obtain the equilibrium R versus T relation, set the derivative of (8.19) equal to zero:

$$\frac{\partial A_c}{\partial R} = \frac{zNkT}{4}\left[\ln\frac{\alpha + R}{\alpha - R} - \ln\frac{1 + R}{1 - R} + \frac{\partial\alpha}{\partial R}\left(\frac{1 - R}{\alpha + R} + \frac{1 + R}{\alpha + R} - \frac{2}{\alpha + 1}\right)\right.$$

$$\left. + \frac{1 + R}{\alpha + R} - \frac{1 - R}{\alpha - R}\right] - kT\frac{\partial \ln g}{\partial R} = 0 \quad (8.20)$$

From (8.17)

$$\frac{\partial\alpha}{\partial R} = \frac{R(1 - \alpha^2)}{\alpha(1 - R^2)} \quad (8.21)$$

and from (5.17)

$$\frac{\partial \ln g}{\partial R} = -\frac{N}{2}\ln\frac{(1 + R)}{(1 - R)} \quad (8.22)$$

Substitution of (8.21) and (8.22) into (8.20) leads to

$$\ln \frac{(1 + R)}{(1 - R)} = \frac{z}{z - 2} \ln \frac{(\alpha + R)}{(\alpha - R)} \tag{8.23}$$

This is the desired functional relation between the equilibrium value of the long-range-order parameter and the temperature.

The critical temperature is determined by expanding the logarithms in series and letting $R \to 0$ as follows

$$2R + \tfrac{2}{3}R^3 + \cdots = \frac{z}{z - 2}\left[2\frac{R}{\alpha} + \frac{2}{3}\frac{R^3}{\alpha^3} + \cdots\right]$$

or

$$1 + \frac{R^2}{3} + \cdots = \frac{z}{z - 2}\left[\frac{1}{\alpha} + \frac{1}{3}\frac{R^2}{\alpha^2} + \cdots\right] \tag{8.24}$$

Now let $R \to 0$. Then $\alpha \to e^{2v/kT_c}$, and (8.24) gives

$$\frac{2v}{kT_c} = \ln \frac{z}{z - 2} \tag{8.25}$$

It is of interest to compare the values of the critical temperature given by the various approximations. This is done in Table 5.3 where the critical

Table 5.3. Critical Temperatures According to Various Approximations

Approximation	kT_c/v
Bragg-Williams	4
Kirkwood (2nd moment)	3.366
Quasi-chemical	3.483

temperature is given in units of v/k for a crystal in which $z = 8$. The agreement between the Kirkwood and the quasi-chemical methods is fairly good. It can be considerably improved by including third moments in the development of the Kirkwood method. In fact, if the free energy given by (8.19) is expressed in series form by expanding the logarithms in $(\alpha - 1)$ and then expanding α in v/kT, it can be shown that the result is identical to the Kirkwood series up to terms cubic in (v/kT).* Thus, although the quasi-chemical

---

* See, for example, Fowler R. and Guggenheim E. A. *Statistical Thermodynamics* Cambridge University Press, Cambridge, 1952, p. 580.

method is based on a simplified physical model, its accuracy is comparable to that of the Kirkwood method for practical purposes, since including moments higher than the third would enormously increase the tedious complexity of the mathematical operations.

Equation 8.23 can be used to compute the long-range-order parameter as a function of temperature as follows: solving for α, (8.23) gives

$$\alpha = R \frac{\left(\dfrac{1 + R}{1 - R}\right)^{z-2/z} + 1}{\left(\dfrac{1 + R}{1 - R}\right)^{z-2/z} - 1} \tag{8.26}$$

so that a table of α versus R can be constructed. From (8.17)

$$\frac{kT}{v} = 2 \ln \frac{1 - R^2}{\alpha^2 - R^2} \tag{8.27}$$

and from the α versus R, the temperature can be computed in units of v/k. Combining (8.27) with (8.25), we can compute the reduced temperature to be

$$\frac{T}{T_c} = 2\left(\ln \frac{z}{z - 2}\right)\left(\ln \frac{1 - R^2}{\alpha^2 - R^2}\right) \tag{8.28}$$

Once R is computed as a function of T, the free energy and the other thermodynamic functions can be obtained from (8.19). Further detail will not be pursued here, since the results are similar to those obtained from the Kirkwood method.

chapter 6

VACANCIES AND INTERSTITIALS

IN MONATOMIC CRYSTALS

6.1 INTRODUCTION AND CHOICE OF ENSEMBLE

In simple metals and rare gas solids, it is generally agreed that the lattice vacancy is the predominant type of point defect. Direct and indirect measurements of the vacancy concentration as a function of temperature and pressure exist, and these measurements will undoubtedly be increased and extended. In more open structures, interstitials and more complex point defects may also be important.

Point defects have a number of important effects on crystal properties. They are involved in the processes of diffusion, they contribute to electrical and thermal resistivity, and they play a role in void growth during plastic deformation. Through their interaction with dislocations they have an effect on the mechanical properties of metals, and on microstructure. The importance of point defects rests on their critical presence in many metallurgical and solid state phenomena. Thus, the questions of their concentration and its relation to thermodynamic factors is of great interest.

In this chapter, the statistical-thermodynamic theory of a pure monatomic crystal containing vacant lattice sites and interstitial atoms is developed. The next three sections are concerned with crystals in which the monovacancy is the only defect. This allows us to present the theory in its most transparent form. Divacancies and interstitials are introduced in Section 6, and Section 7 displays some numerical results for the contribution of defects to crystal properties.

It is important to point out that the theory of defect concentration and defect thermodynamic functions is not a classical thermodynamic theory in any sense. As required by the Phase Rule of Gibbs, the pressure and temperature are sufficient to determine *all* classical thermodynamic quantities. Point-defect theory requires, therefore, statistical mechanics and crystal structure. As a result, the so-called defect functions are statistical-mechanical rather than thermodynamic quantities, and the extent to which they satisfy the classical thermodynamic differential relations must be investigated.

In working out the statistical mechanics of defect crystals, it turns out to be most convenient to use the pressure ensemble. The reason for this is that defect-concentration formulae are always expressed in terms of Gibbs free energies, and the thermodynamic independent variables used are usually temperature and pressure. Also, in an ensemble in which vacancies enter into the definition of state, the volume varies over members of the ensemble. All this makes the pressure ensemble our most logical starting point. The equations can be developed in a simple and elegant fashion from the pressure ensemble, and no unique calculational difficulties arise even if we wish to compute properties from partition functions. This latter point is true because once the formulae are derived, the canonical partition function can be used to compute defect properties.

For convenience, we rewrite the basic pressure-ensemble equations in the following form (see Chapter 1, Section 10).

$$f_j(V_i) = \frac{1}{Z_p} \exp\left\{ -\frac{1}{kT} [E_j + PV_i] \right\} \tag{1.1}$$

$$f_j(V_i) = \frac{N_j(V_i)}{X} \tag{1.2}$$

$$\sum_{j,i} f_j(V_i) = 1 \tag{1.3}$$

$$Z_p = e^{-G/kT} \tag{1.4}$$

In these equations, $f_j(V_i)$ is the probability that in an ensemble representing a system at temperature T and pressure P, $N_j(V_i)$ out of X ensemble members will be in a state whose energy is E_j and whose volume is V_i. The pressure-ensemble partition function, Z_p, determines the Gibbs free energy of the system by (1.4).

6.2   THE VACANCY CONCENTRATION

The problem before us is to compute the concentration of vacancies in a monatomic crystal in equilibrium. To do this, consider an ensemble of X crystals, all of which have N identical atoms. Each crystal can have any volume V_i, any number of vacancies N_v, and can exist in any of a set of quantum states with energy E_j. The probability that a crystal exists in a state characterized by particular values of V_i, N_v, and E_j is given by the pressure ensemble as

$$f_{j,i,N_v} = \frac{1}{Z_p} \exp\left\{-\frac{1}{kT}[E_j + PV_i]\right\} \tag{2.1}$$

If we multiply this by the number of distinct wave functions for a crystal with N_v vacancies in a state with energy E_j and volume V_i, we get the probability that the crystal has the values (N_v, E_j, V_i). This degeneracy factor will be written as

$$w(N_v)\Omega(N_v, E_j, V_i)$$

where $\Omega(N_v, E_j, V_i)$ is the degeneracy for a crystal with N_v vacancies, and $w(N_v)$ is the number of ways of distributing N_v vacancies and N atoms on the crystal lattice. Multiplying (2.1) by this degeneracy factor we get

$$f(N_v, E_j, V_i) = w(N_v)\Omega(N_v, E_j, V_i)f_{j,i,N_v}$$

or

$$f(N_v, E_j, V_i) = \frac{1}{Z_p} w(N_v)\Omega(N_v, E_j, V_i) \exp\left\{-\frac{1}{kT}[E_j + PV_i]\right\} \tag{2.2}$$

Now if (2.2) is summed over all values of energy and volume for a given N_v, we obtain the probability that the crystal has N_v vacancies as

$$f(N_v) = \frac{1}{Z_p} w(N_v) \sum_{E_j, V_i} \Omega(N_v, E_j, V_i) \exp\left\{-\frac{1}{kT}[E_j + PV_i]\right\} \tag{2.3}$$

In performing this sum, it is assumed that $\Omega(N_v, E_j, V_i)$ depends only on the number of vacancies, and not on the configuration of their distribution over the crystal lattice. We also assume that all lattice sites are equivalent in the sense that the energy and volume do not depend on the distance between vacancies. This ignores the formation of divacancies that will be considered in a later section.

It is convenient to define a free energy $G\{N_v\}$ by

$$e^{-G\{N_v\}/kT} \equiv \sum_{E_j, V_i} \Omega(N_v, E_j, V_i) \exp\left\{-\frac{1}{kT}[E_j + PV_i]\right\} \tag{2.4}$$

and to write (2.3) as

$$f(N_v) = \frac{1}{Z_p} w(N_v)e^{-G\{N_v\}/kT} \tag{2.5}$$

Since $\sum_{N_v} f(N_v) = 1$, and since the total free energy is related to Z_p by (1.4), summation of (2.5) leads to

$$e^{-G/kT} = \sum_{N_v} w(N_v)e^{-G\{N_v\}/kT} \tag{2.6}$$

G is the total Gibbs free energy of the crystal, and $G\{N_v\}$ is the Gibbs free energy of a crystal containing N_v vacancies except for the configurational contribution of the vacancies.

Equations 2.5 and 2.6 are the two basic relations of the statistical thermodynamics of vacancies, and enable us to obtain the vacancy-concentration formulas and the thermodynamic functions of a crystal containing vacancies.

As usual, we identify the most probable value with the equilibrium value, and therefore maximize (2.5) to obtain the equilibrium number of vacancies. Thus, requiring that

$$\left[\frac{\partial \ln f(N_v)}{\partial N_v}\right]_{N_v=\bar{N}_v} = 0 \tag{2.7}$$

where $\bar{N}_v$ is the equilibrium number of vacancies, (2.5) gives

$$\left[\frac{\partial \ln w(N_v)}{\partial N_v}\right]_{N_v=\bar{N}_v} = \frac{1}{kT}\left[\frac{\partial G\{N_v\}}{\partial N_v}\right]_{N_v=\bar{N}_v} \tag{2.8}$$

In performing these differentiations, the number of atoms is held constant so that vacancies are formed by transferring atoms to the crystal surface.

The derivative on the right-hand side is similar to a partial atomic free energy. It is called the free energy of vacancy formation and given the symbol $G_v{}^f$

$$G_v{}^f \equiv \left[\frac{\partial G\{N_v\}}{\partial N_v}\right]_{N_v=\bar{N}_v} \tag{2'9}$$

since it is the increase in Gibbs free energy upon adding one vacancy to a crystal containing $\bar{N}_v$ vacancies.

To get the left-hand side of (2.8), we write

$$w(N_v) = \frac{(N + N_v)!}{N!\,N_v!} \tag{2.10}$$

because $w(N_v)$ is the number of ways of arranging N atoms and N_v vacancies on the lattice. Using Stirling's approximation, (2.10) gives

$$\ln w(N_v) = (N + N_v)\ln (N + N_v) - N \ln N - N_v \ln N_v \tag{2.11}$$

from which

$$\frac{\partial \ln w(N_v)}{\partial N_v} = \ln \frac{(N + N_v)}{N_v} \tag{2.12}$$

Putting (2.9) and (2.12) into (2.8) gives the vacancy-concentration formula:

$$\frac{\bar{N}_v}{N + \bar{N}_v} = e^{-G_v{}^f/kT} \tag{2.13}$$

which shows that the vacancy concentration increases with temperature according to a Boltzmann factor of the Gibbs free energy of formation, as expected.

Equation 2.13 is often derived by more elementary methods. However, the method used here clearly shows the nature of the quantities involved, and provides a sound basis for the investigation of the thermodynamic functions of the crystal.

It is clear that the method of finding the equilibrium number of vacancies by maximizing $f(N_v)$ of (2.5) is completely equivalent to finding the maximum term in the sum (2.6) that gives the crystal free energy. In fact, if we define a free energy $G(N_v)$ for a crystal containing N_v vacancies by

$$G(N_v) = G\{N_v\} - kT \ln w(N_v) \tag{2.14}$$

our method is equivalent to minimizing this expression. That is, the equilibrium concentration of vacancies is given by

$$\frac{\partial G(N_v)}{\partial N_v} = 0 \tag{2.15}$$

where the derivative refers to a change in the number of vacancies in the crystal interior by transferring atoms to or from the surface. Using (2.14) and (2.15) directly saves us from the cumbersome notation that results when the pressure ensemble probability functions are written for complex systems. Therefore, these equations will be used to determine equilibrium-defect distributions from now on.

6.3 THE CRYSTAL FREE ENERGY

The analysis of the thermodynamics of crystals containing vacancies requires an expression for the crystal free energy that contains the vacancy concentration. Such an expression can be derived from (2.6). The procedure consists of showing that the sum in (2.6) can be written as a Gaussian distribution about the equilibrium vacancy concentration, and that the spread of the distribution contributes a negligible amount to the free energy. It is then only necessary to take the nonconfigurational part of the free energy as linear in the vacancy concentration to arrive at the desired result.

We rewrite (2.6) in the form

$$e^{-G/kT} = \sum_{N_v} h(N_v) \tag{3.1}$$

where

$$h(N_v) \equiv w(N_v)e^{-G\{N_v\}/kT} \tag{3.2}$$

It will now be shown that $h(N_v)$ forms a very narrow Gaussian distribution about the equilibrium value $\bar{N}_v$. Using this distribution, we can sum (3.1) to find the crystal free energy.

To arrive at the Gaussian form of (3.2), write the ratio

$$\frac{h(N_v)}{h(\bar{N}_v)} = \frac{w(N_v)}{w(\bar{N}_v)} e^{[G\{\bar{N}_v\}-G\{N_v\}]/kT} \tag{3.3}$$

and define the deviation from the mean number of vacancies by

$$\Delta N_v \equiv N_v - \bar{N}_v \tag{3.4}$$

Also, $G\{N_v\}$ will be taken as linear in N_v so that

$$G\{N_v\} = G\{\bar{N}_v\} + \Delta N_v G_v{}' \tag{3.5}$$

From (2.10) the ratio of the statistical counts is

$$\frac{w(N_v)}{w(\bar{N}_v)} = \frac{(N + N_v)! \; \bar{N}_v!}{(N + \bar{N}_v)! \; N_v!} \tag{3.6}$$

so that, using (3.5) and (3.6), (3.3) becomes

$$\frac{h(N_v)}{h(\bar{N}_v)} = \frac{(N + N_v)! \; \bar{N}_v!}{(N + \bar{N}_v)! \; N_v!} [e^{-G_v{}'/kT}]^{\Delta N_v} \tag{3.7}$$

or, since the exponential is given by (2.13),

$$\frac{h(N_v)}{h(\bar{N}_v)} = \frac{(N + N_v)! \; \bar{N}_v!}{(N + \bar{N}_v)! \; N_v!} \left(\frac{\bar{N}_v}{N + \bar{N}_v}\right)^{\Delta N_v} \tag{3.8}$$

The desired Gaussian distribution is readily obtained if ΔN_v is restricted to a range of values that are all small relative to $\bar{N}_v$. We will adopt this restriction, which leads to a result that shows $h(N_v)$ to be quite small even for small values of ΔN_v. Then it will be shown that $h(N_v)$ decreases with increasing deviation of N_v from $\bar{N}_v$ for all values of ΔN_v, and that the restriction of $\Delta N_v/\bar{N}_v \ll 1$ is sufficient to obtain all significant contributions to the sum in (3.1) defining the free energy.

Taking logarithms of (3.8), using Stirling's approximation, and performing a little algebra gives

$$\ln \frac{h(N_v)}{h(\bar{N}_v)} = (N + \bar{N}_v) \ln \left(1 + \frac{\Delta N_v}{N + \bar{N}_v}\right) - \bar{N}_v \ln \left(1 + \frac{\Delta N_v}{\bar{N}_v}\right)$$

$$+ \Delta N_v \left[\ln \left(1 + \frac{\Delta N_v}{N + \bar{N}_v}\right) - \ln \left(1 + \frac{\Delta N_v}{\bar{N}_v}\right)\right] \tag{3.9}$$

Expansion of the logarithms with retention of only the first term $(\ln (1 + x) \simeq x)$ converts (3.9) to

$$\ln \frac{h(N_v)}{h(\bar{N}_v)} = -\frac{\Delta N_v{}^2}{\bar{N}_v} \frac{N}{N + \bar{N}_v}$$

or, since $\bar{N}_v \ll N$, to a sufficient approximation,

$$\ln \frac{h(N_v)}{h(\bar{N}_v)} = -\frac{\Delta N_v{}^2}{\bar{N}_v}$$

so that

$$h(N_v) = h(\bar{N}_v)e^{-(\Delta N_v{}^2)/(\bar{N}_v)} \tag{3.10}$$

which is the required Gaussian distribution.

Now let us compute the exponential in (3.10) for $\Delta N_v/\bar{N}_v = 10^{-4}$, a value for which the above derivation is certainly valid. Then, since $\bar{N}_v$ in crystals at temperatures for which the vacancy concentration is detectable is at least of the order of 10^{16}, $\Delta n_v \approx 10^{12}$. These values give

$$e^{-10^8}$$

for the exponential which is certainly negligible. This shows that (3.10) is indeed a very sharply peaked distribution.

It is now only necessary to show that $h(N_v)$ decreases with increasing deviation of N_v from the equilibrium value $\bar{N}_v$, regardless of the magnitude of the deviation. To do this, first consider positive values of ΔN_v and rewrite (3.8) as

$$\frac{h(N_v)}{h(\bar{N}_v)} = \frac{(N + \bar{N}_v + \Delta N_v)! \, \bar{N}_v!}{(N + \bar{N}_v)! \, (\bar{N}_v + \Delta N_v)!} \left(\frac{\bar{N}_v}{N + \bar{N}_v}\right)^{\Delta N_v} \tag{3.11}$$

or

$$\frac{h(N_v)}{h(\bar{N}_v)} = \prod_{j=1}^{\Delta N_v} \frac{(N + \bar{N}_v + j)}{(\bar{N}_v + j)} \left(\frac{\bar{N}_v}{N + \bar{N}_v}\right)^{\Delta N_v} \tag{3.12}$$

Since there are ΔN_v factors in the product, (3.12) is a product of the following factors

$$\frac{N + \bar{N}_v + j}{\bar{N}_v + j} \cdot \frac{\bar{N}_v}{N + \bar{N}_v} = \frac{N\bar{N}_v + \bar{N}_v{}^2 + N_v j}{N\bar{N}_v + \bar{N}_v{}^2 + \bar{N}_v j + Nj}$$

Each of these factors is less than unity, and the more of them there are, the smaller is the value of (3.12). We have therefore shown that $h(N_v)$ decreases with ΔN_v for all positive values of ΔN_v.

If ΔN_v is negative, we write (3.8) in the form

$$\frac{h(N_v)}{h(\bar{N}_v)} = \frac{(N + \bar{N}_v - |\Delta N_v|)! \, \bar{N}_v!}{(N + \bar{N}_v)! \, (\bar{N}_v - |\Delta N_v|)!} \left(\frac{N + \bar{N}_v}{\bar{N}_v}\right)^{|\Delta N_v|} \tag{3.13}$$

or

$$\frac{h(N_v)}{h(\bar{N}_v)} = \prod_{j=1}^{|\Delta N_v|} \frac{(\bar{N}_v - |\Delta N_v| + j)}{(N + \bar{N}_v - |\Delta N_v| + j)} \left(\frac{N + \bar{N}_v}{\bar{N}_v}\right)^{|\Delta N_v|} \tag{3.14}$$

But each of the factors in the product is less than unity, and the product clearly decreases with increasing $|\Delta N_v|$ more rapidly than the factor $(N + \bar{N}_v/\bar{N}_v)^{|\Delta N_v|}$ increases.

We have therefore shown that $h(N_v)$ decreases with the deviation of N_v from the equilibrium value for all values of the deviation. Since $h(N_v)$ is very small relative to $h(\bar{N}_v)$ even for small deviations, a negligible error is made if the Gaussian distribution of (3.10) is adopted for all values of N_v.

Now substitute (3.10) into (3.1) to obtain

$$e^{-G/kT} = h(\bar{N}_v) \sum_{\Delta N_v} e^{-(\Delta N_v^2/\bar{N}_v)} \tag{3.15}$$

To evaluate the sum, we replace it by an integral as follows:

$$\sum_{\Delta N_v} e^{-\Delta N_v^2/\bar{N}_v} = \int_{-\infty}^{\infty} e^{-x^2/\bar{N}_v}\, dx$$
$$= \sqrt{\pi \bar{N}_v}$$

and therefore (3.15) becomes

$$e^{-G/kT} = h(\bar{N}_v)(\pi \bar{N}_v)^{1/2} \tag{3.16}$$

From the definition of $h(\bar{N}_v)$ given by (3.2), taking logarithms of (3.16) results in

$$\frac{G}{kT} = -\ln w(\bar{N}_v) + \frac{G\{\bar{N}_v\}}{kT} + \tfrac{1}{2}\ln(\pi \bar{N}_v) \tag{3.17}$$

All that remains is to evaluate the first log term. This is easily done from (2.11), which gives

$$\ln w(\bar{N}_v) = N \ln\left(1 + \frac{\bar{N}_v}{N}\right) + \bar{N}_v \ln\left(\frac{N + \bar{N}_v}{\bar{N}_v}\right) \tag{3.18}$$

The log in the first term can be approximated by the first term in its series expansion, and the second term can be expressed in terms of the free energy of vacancy formation by using (2.13). Doing this, (3.18) gives

$$\ln w = \bar{N}_v + \bar{N}_v \frac{G_v^f}{kT} \tag{3.19}$$

Thus, combining (3.19) and (3.17),

$$G = G\{\bar{N}_v\} - \bar{N}_v G_v^f - \bar{N}_v kT \tag{3.20}$$

where we have neglected $(1/2)\ln(\pi \bar{N}_v)$ relative to $\bar{N}_v$. Taking $G\{\bar{N}_v\}$ as linear in the number of vacancies, that is,

$$G\{\bar{N}_v\} = G^0 + \bar{N}_v G_v^f \tag{3.21}$$

we finally get

$$G = G^0 - \bar{N}_v kT \tag{3.22}$$

Equation 3.21 states that each vacancy contributes an amount to the configurationless free energy $G\{\bar{N}_v\}$ that is equal to the free energy of vacancy formation, G^0 being the free energy of a crystal without any vacancies. This means that we assume the vacancy concentration to be low enough that the vacancies do not influence each other. Vacancy concentrations are of the order of 10^{-4} atomic percent in simple crystals, thereby justifying the form of (3.21), and the approximations based on $\bar{N}_v/N \ll 1$ and $\ln \bar{N}_v \ll \bar{N}_v$.

Equation 3-22 enables us to obtain all the thermodynamic functions of a crystal as functions of the vacancy concentration, and to derive the differential relations among the vacancy-formation quantities. In this connection, it is of interest to note from (3.20) that the last two terms represent the configurational contribution of the vacancies to the crystal free energy. Thus the vacancy-configurational entropy is

$$S_v^{\ c} = k\bar{N}_v\left[1 + \frac{G_v^{\ f}}{kT}\right] \tag{3.23}$$

As will be seen below, the existence of this term means that the crystal entropy is not just the sum of the entropy of a vacancy free crystal and the entropy of formation of the vacancies.

6.4 VACANCIES AND THERMODYNAMIC FUNCTIONS

Equation 3.22 shows that each vacancy contributes $-kT$ to the Gibbs free energy of a monatomic crystal. This equation enables us to find the vacancy contributions to all other thermodynamic functions through the usual thermodynamic formulae. For the entropy, volume, energy, enthalpy, specific heat at constant pressure, specific heat at constant volume, thermal expansion and compressibility, these formulae are:

$$S = -\left(\frac{\partial G}{\partial T}\right)_P \tag{4.1}$$

$$V = \left(\frac{\partial G}{\partial P}\right)_T \tag{4.2}$$

$$U = G + TS - PV \tag{4.3}$$

$$H = U + PV = G + TS \tag{4.4}$$

$$C_P = \left(\frac{\partial H}{\partial T}\right)_P \tag{4.5}$$

$$C_V = \left(\frac{\partial U}{\partial T}\right)_V = C_P - TV\frac{\alpha^2}{\beta} \tag{4.6}$$

$$\alpha = \frac{1}{V}\left(\frac{\partial V}{\partial T}\right)_P \tag{4.7}$$

$$\beta = -\frac{1}{V}\left(\frac{\partial V}{\partial P}\right)_T \tag{4.8}$$

It is convenient to define defect-formation quantities in terms of the free energy of vacancy formation, just as the above crystal quantities are written in terms of the crystal free energy. Thus we define:

The entropy of vacancy formation:

$$S_v^f \equiv -\left(\frac{\partial G_v^f}{\partial T}\right)_P \tag{4.9}$$

The volume of vacancy formation:

$$V_v^f \equiv \left(\frac{\partial G_v^f}{\partial P}\right)_T \tag{4.10}$$

The energy of vacancy formation:

$$U_v^f \equiv G_v^f + TS_v^f - PV_v^f \tag{4.11}$$

The enthalpy of vacancy formation:

$$\begin{aligned} H_v^f &\equiv U_v^f + PV_v^f \\ &= G_v^f + TS_v^f \end{aligned} \tag{4.12}$$

The specific heat at constant pressure of vacancy formation:

$$(C_v^f)_P \equiv \left(\frac{\partial H_v^f}{\partial T}\right)_P \tag{4.13}$$

The specific heat at constant volume of vacancy formation:

$$(C_v^f)_V \equiv \left(\frac{\partial U_v^f}{\partial T}\right)_V \tag{4.14}$$

The thermal expansion of vacancy formation:

$$\alpha_v^f \equiv \frac{1}{V_v^f}\left(\frac{\partial V_v^f}{\partial T}\right)_P \tag{4.15}$$

The compressibility of vacancy formation:

$$\beta_v^f = -\frac{1}{V_v^f}\left(\frac{\partial V_v^f}{\partial P}\right)_T \tag{4.16}$$

These definitions ensure that the defect-formation quantities will follow the usual rules of thermodynamics. However, it is not generally true that the defect quantities defined above give the contribution per vacancy to the corresponding crystal quantities. Although this is true for the energy and enthalpy, for example, it is not true for the free energy and entropy.

To use (4.1) through (4.8) in conjunction with (3.22), we need the derivatives of the vacancy concentration with respect to pressure and temperature.

From (2.13), and using the definitions of the defect-formation quantities, we have

$$\left(\frac{\partial N_v}{\partial T}\right)_P = \frac{N_v H_v^f}{kT^2} \tag{4.17}$$

$$\left(\frac{\partial N_v}{\partial P}\right)_T = -\frac{N_v V_v^f}{kT} \tag{4.18}$$

From this point on, we have dropped the bar over N_v for the sake of convenience, with the understanding that all vacancy concentrations in the thermodynamic formulae refer to equilibrium.

Using (4.1) through (4.18), it is only a matter of some algebra to derive the crystal thermodynamic functions from (3.22) in the following forms:

The entropy:

$$S = S^0 + \frac{N_v}{T}[H_v^f + kT] \tag{4.19}$$

or

$$S = S^0 + S_v^c + N_v S_v^f \tag{4.20}$$

where S_v^c is the vacancy-configurational entropy given by (3.23), and the superscript 0 refers, as usual, to a hypothetical defect-free crystal.

The volume:

$$V = V^0 + N_v V_v^f \tag{4.21}$$

The energy:

$$U = U^0 + N_v U_v^f \tag{4.22}$$

The enthalpy:

$$H = H^0 + N_v H_v^f \tag{4.23}$$

The heat capacity at constant pressure:

$$C_P = C_P^0 + N_v\left[(C_v^f)_P + \frac{(H_v^f)^2}{kT^2}\right] \tag{4.24}$$

The thermal expansion

$$\alpha = \alpha^0 + \frac{N_v V_v^f}{V_0}\left[\alpha_v^f - \alpha^0 + \frac{H_v^f}{kT^2}\right] \tag{4.25}$$

The compressibility:

$$\beta = \beta^0 + \frac{N_v V_v^f}{V_0}\left[\beta_v^f - \beta^0 + \frac{V_v^f}{kT}\right] \tag{4.26}$$

In deriving (4.25) and (4.26), terms quadratic in N_v were neglected, and the following approximation was used:

$$\left(1 + \frac{N_v V_v^f}{V_0}\right)^{-1} = 1 - \frac{N_v V_v^f}{V_0} \tag{4.27}$$

The heat capacity at constant volume can now be obtained from (4.6), using (4.24), (4.25), and (4.27). Treating all vacancy contributions as small compared to the corresponding crystal quantities, and retaining terms only to the first order in the vacancy concentration, the result is

$$
C_V = C_V{}^0 + N_v \left[(C_v{}^f)_P + \frac{(H_v{}^f)^2}{kT^2} \right]
$$

$$
+ \frac{N_v V_v{}^f}{V_0} (C_V{}^0 - C_p{}^0) \left[\frac{2\alpha_v{}^f}{\alpha^0} + \frac{2H_v{}^f}{\alpha^0 kT^2} - \frac{\beta_v{}^f}{\beta^0} - \frac{V_v{}^f}{\beta^0 kT} \right] \quad (4.28)
$$

With the above formulae, the effect of vacancies on the thermodynamic functions can be investigated. Note that only for the volume, the energy, and the enthalpy can the vacancy effect be written as an incremental addition of vacancy-formation quantities. For the other thermodynamic functions, the formulae are more complex.

6.5   THE VACANCY-FORMATION FUNCTIONS

The vacancy-concentration formula (2.13) is often written, with the aid of (4.11), as

$$
\frac{N_v}{N} = e^{S_v{}^f/k} e^{-U_v{}^f/kT} e^{-PV_v{}^f/kT} \quad (5.1)
$$

where we have neglected N_v relative to N. It is this form that is used in the experimental determination of vacancy-formation energies and volumes. The formation energy is obtained by measuring a quantity that is proportional to the vacancy concentration as a function of temperature at constant (usually atmospheric) pressure. The formation energy is then obtained from an Arrhenius $\ln N_v - (1/T)$ plot. The formation volume is obtained from the slope of a plot of $\ln N_v$ versus P at constant temperature.

The applicability of these methods is simplest when the formation energy, entropy, and volume are independent of the temperature and pressure, at least within the accuracy of the experiments. The question of the variation of the Gibbs free energy of vacancy formation with temperature and pressure is therefore of considerable importance. If an Arrhenius plot, or a $\ln n_v$ versus P plot, exhibits a curvature, two causes may be operative. The first is that the temperature or pressure dependences of the vacancy-formation quantities are being reflected in the curvature. The second is that a process other than vacancy formation may be affecting the measurement. Such processes might include, for example, impurity-vacancy binding or divacancy formation. Without some theoretical guide, it is difficult to distinguish between these two possibilities.

It is important to note that a linear Arrhenius plot does not in itself guarantee that $U_v{}^f$ is independent of temperature. If the vacancy-formation energy is linear in the temperature, a linear Arrhenius plot will result. Likewise, if the formation volume is linear in the reciprocal of the pressure, the $\ln n_v$ versus P plot will be linear.

In this and the following section, we will develop the theory of the vacancy-formation quantities in such a way as to give some physical insight into the factors that determine them. It will also be shown that for most monatomic crystals, the entropy, energy, and volume of vacancy formation are constants to a good degree of approximation.

To get the Gibbs free energy of formation, and therefore all the other formation quantities, we start with (2.9).

$$G_v{}^f = \left[\frac{\partial G\{N_v\}}{\partial N_v} \right]_{N_v = \bar{N}_v} \tag{5.2}$$

Since we are dealing with crystals in which the vacancy concentration is small, we will take the derivative in this definition to be independent of the vacancy concentration. It then becomes equal to the difference in free energy between a crystal having one vacancy and a crystal having no vacancies. Thus

$$G_v{}^f = G' - G^0 \tag{5.3}$$

where G' is the Gibbs free energy of a crystal containing a single vacancy and G^0 is the Gibbs free energy of a perfect crystal.

The reason that it is legitimate to define G' and G^0 in this way is that $G\{N_v\}$ does not include the configurational contribution to the crystal free energy. Thus, we can use canonical ensemble theory to calculate the Helmholtz free energy, and add to it PV to get

$$G' = A' + PV' \tag{5.4}$$

$$G^0 = A^0 + PV^0 \tag{5.5}$$

where the primes again refer to a crystal containing one vacancy, and 0 refers to a perfect crystal. The Helmholtz free energies are given by (3.13) of Chapter 2 as

$$A' = E_0' + kT \sum_{j=1}^{3N} \ln\left(1 - e^{-h\nu_j'/kT}\right) \tag{5.6}$$

$$A^0 = E_0{}^0 + kT \sum_{j=1}^{3N} \ln\left(1 - e^{-h\nu_j{}^0/kT}\right) \tag{5.7}$$

in an obvious notation.

It is clear from the above that all of the vacancy-formation quantities defined in (4.9) through (4.14) refer to differences between a fictitious perfect crystal, and a crystal containing a single vacancy; that is,

$$S_v^f = S' - S^0 \tag{5.8}$$

$$V_v^f = V' - V^0 \tag{5.9}$$

$$U_v^f = U' - U^0 \tag{5.10}$$

$$H_v^f = H' - H^0 \tag{5.11}$$

$$(C_v^f)_P = C_P' - C_P{}^0 \tag{5.12}$$

$$(C_v^f)_V = C_V' - C_v{}^0 \tag{5.13}$$

Where again the prime refers to a crystal containing one vacancy and the 0 to a perfect crystal.

Because of the presence of V_v^f in front of the derivatives in (4.15) and (4.16), the thermal expansion and compressibility of vacancy formation are not given by simple difference formulae, but rather by

$$V_v^f \alpha_v^f = \alpha' V' - \alpha^0 V^0 \tag{5.14}$$

$$V_v^f \beta_v^f = \beta' V' - \beta^0 V^0 \tag{5.15}$$

The entire apparatus of canonical ensemble theory and the quasi-harmonic theory of crystals is now available for the computation of the vacancy-formation quantities and, therefore, for the vacancy concentration as a function of temperature and pressure.

The Debye temperature represents the maximum normal mode vibration frequency, so that, for temperatures above θ_D, the $h\nu/kT$ are all small. In the high-temperature limit, the formulae (5.6) and (5.7) can be considerably simplified. Since vacancies exist in appreciable concentrations only at high temperatures, we will adopt the approximation that

$$e^{-h\nu_j'/kT} \simeq 1 - \frac{h\nu_j'}{kT} \tag{5.16}$$

and

$$e^{-h\nu_j^0/kT} \simeq 1 - \frac{h\nu_j{}^0}{kT} \tag{5.17}$$

With this approximation, (5.6) and (5.7) become

$$A' = E_0' + kT \sum_{j=1}^{3N} \ln \frac{h\nu_j'}{kT} \tag{5.18}$$

$$A^0 = E_0{}^0 + kT \sum_{j=1}^{3N} \ln \frac{h\nu_j{}^0}{kT} \tag{5.19}$$

We can now write the Gibbs free energy of vacancy formation from the above, and (5.3) to (5.5) as

$$G_v^f = (E_0' - E_0{}^0) + kT \sum_{j=1}^{3N} \ln \frac{\nu_j'}{\nu_j{}^0} + P(V' - V^0) \tag{5.20}$$

From the derivation of (5.20), the first two terms constitute the Helmholtz free energy of vacancy formation

$$A_v{}^f = (E_0' - E_0{}^0) + kT \sum_{j=1}^{3N} \ln \frac{v_j'}{v_j{}^0} \qquad (5.21)$$

so that using the thermodynamic relations

$$S_v{}^f = -\left(\frac{\partial A_v{}^f}{\partial T}\right)_V \qquad (5.22)$$

$$U_v{}^f = A_v{}^f + TS_v{}^f \qquad (5.23)$$

gives

$$S_v{}^f = -k \sum_{j=1}^{3N} \ln \frac{v_j'}{v_j{}^0} \qquad (5.24)$$

$$U_v{}^f = (E_0' - E_0{}^0) \qquad (5.25)$$

and therefore, comparing (4.11) to (5.20) requires that

$$V_v{}^f = (V' - V^0) \qquad (5.26)$$

In deriving (5.24) through (5.26), the physically reasonable assumption is made that the vibration frequencies and zero point energies are explicit functions of only the crystal volume, and depend on the temperature and pressure only implicitly through the volume.

To investigate the temperature dependence of the free energy of formation, we will expand it in a Taylor series, retaining terms to the second order in the temperature and pressure. The origin of the expansion will be taken at zero pressure and the temperature at the melting point to give

$$G_v{}^f(P, T) = G_v{}^f(0, T_m) + P\left(\frac{\partial G_v{}^f}{\partial P}\right)_{0,T_m}$$

$$+ (T - T_m)\left(\frac{\partial G_v{}^f}{\partial T}\right)_{0,T_m} + \frac{P^2}{2}\left(\frac{\partial^2 G_v{}^f}{\partial P^2}\right)_{0,T_m}$$

$$+ \frac{(T - T_m)^2}{2}\left(\frac{\partial^2 G_v{}^f}{\partial T^2}\right)_{0,T_m} + P(T - T_m)\left(\frac{\partial^2 G_v{}^f}{\partial T\,\partial P}\right)_{0,T_m} \qquad (5.27)$$

The appearance of this equation can be simplified by using the definitions of the defect-formation quantities in the previous section to give

$$G_v{}^f(P, T) = \tilde{G}_v{}^f - (T - T_m)\tilde{S}_v{}^f + P\tilde{V}_v{}^f - \frac{(T - T_m)^2}{2T_m}$$

$$\times (\tilde{C}_v{}^f)_P - \tfrac{1}{2}P^2\tilde{\beta}_v{}^f\tilde{V}_v{}^f + P(T - T_m)\tilde{\alpha}_v{}^f\tilde{V}_v{}^f \qquad (5.28)$$

where the tilde signifies values at zero pressure and the melting temperature. The heat capacity at constant pressure in the fourth term was introduced by means of the thermodynamic relation

$$\left(\frac{\partial S_v^f}{\partial T}\right)_P = \frac{(C_v^f)_P}{T} \tag{5.29}$$

Experimental measurements of the vacancy concentration as functions of temperature and pressure in face-centered cubic metals show that $U_v^f \approx$ 1 eV, $S^f \approx k$ and $V_v^f \approx v/2$, where v is the atomic volume. The accuracy of the measurements is not sufficient to detect any temperature or pressure variations arising from the quadratic terms in (5.28). Indeed, if it is assumed that $\tilde{\beta}_v^f$ and $\tilde{\alpha}_v^f$ are roughly of the same order as the measured compressibility and thermal expansion, a simple calculation shows that the last two terms are negligible relative to $P\tilde{V}_v^f$ up to pressures of the order of 50,000 atmospheres. Welch has shown that for copper, α_v^f and β_v^f are actually quite close in value to α^0 and β^0.* His calculations were based on atomistic considerations in which the ion-core interaction energies, electron distributions and lattice vibrations were analyzed for perfect and defect crystals.

An estimate of the fourth term can be made from an analysis of the heat capacity of defect formation. The heat capacities at constant pressure and constant volume are related to each other by

$$(C_v^f)_P = (C_v^f)_V + T\left[\frac{V'\alpha'^2}{\beta'} - \frac{V^0\alpha^{0^2}}{\beta^0}\right] \tag{5.30}$$

Using (5.14) and (5.15), and taking $\alpha_v^f \approx \alpha^0$ and $\beta_v^f \approx \beta^0$, it is easy to show that to an excellent approximation,

$$(C_v^f)_P = (C_v^f)_V \tag{5.31}$$

The vacancy-formation heat capacity at constant volume, from (8.8) of Chapter 2 is, for a crystal containing a single vacancy, in the high temperature Debye approximation

$$C_V' = 3Nk\left[1 - \frac{1}{20}\left(\frac{\theta_D'}{T}\right)^2\right] \tag{5.32}$$

and for the perfect crystal

$$C_V^0 = 3Nk\left[1 - \frac{1}{20}\left(\frac{\theta_D^0}{T}\right)^2\right] \tag{5.33}$$

---

* Girifalco L. A. and Welch D. O., *Point Defects and Diffusion in Strained Metals*, Gordon and Breach Science Publishers, New York, 1967. From Welch's equation 3.22 on page 93, it is readily found that $\beta_v^f \simeq 1.34\beta^0$ and $\alpha_v^f \approx 0.95\alpha^0$.

θ'_D differs from $\theta_D{}^0$ because the Debye temperature varies with dilatation. Placing an interior atom on the surface increases the volume of the crystal by one atomic volume. The vacancy-formation volume is less than this, however, because the atoms around the vacant site relax inward toward the vacancy center to seek a new equilibrium configuration that accommodates the missing repulsive forces of the missing atom. This gives rise to a relaxation volume, V^R, which is negative; thus, if v is the atomic volume,

$$V_v{}^f = v + V^R \tag{5.34}$$

The relaxation volume consists of two parts. As the crystal relaxes around the vacancy, any spherical surface centered on the vacancy and anchored in the atoms sweeps out a volume ΔV^∞. This is equal to V^R only if the crystal is infinite. The distortions around the vacancy induce an elastic stress field throughout the crystal. The crystal surface, however, must be stress free, so that image forces must be set up to cancel the surface effects of the vacancy elastic field. These image forces give rise to an additional contribution to the relaxation volume, ΔV^I, which is called the image volume. Thus

$$V_v{}^f = v + \Delta V^\infty + \Delta V^I \tag{5.35}$$

ΔV^∞ and ΔV^I can be computed from elasticity theory by methods developed by Eshelby. Both ΔV^∞ and ΔV^I are negative for vacancies in noble metals. For the copper vacancy $\Delta V^\infty \simeq -0.326v$ and $\Delta V^I \simeq -0.16v$.* Of the three terms in (5.35), only ΔV^I produces a dilatation and, therefore, it is the only volume change that affects the Debye temperature. Using the Gruneisen assumption (see Chapter 2, Sections 11 and 12), we therefore have

$$\frac{\theta'_D - \theta_D{}^0}{\theta_D{}^0} = -\gamma \frac{\Delta V^I}{V^0} \tag{5.36}$$

Combining (5.32), (5.33), and (5.36), we get

$$(C_v{}^f)_V = -\frac{3Nk}{20T^2}\left[\theta_D{}^{02}\left(1 - \frac{\gamma \Delta V^I}{V^0}\right)^2 - \theta_D{}^{02}\right] \tag{5.37}$$

Since $\Delta V^I/V^0 \ll 1$, this can be written as

$$(C_v{}^f)_V = \frac{3}{10}\frac{\Delta V^I}{v}\frac{k\theta_D{}^{02}}{T^2}\gamma \tag{5.38}$$

where $v = V^0/N$ is the atomic volume. For copper, $\theta_D/T_m = 0.23$ and $\gamma \approx 2$, so according to (5.38), the vacancy-formation heat capacity is about $-0.05k$ at the melting point.

* *Ibid.*, Chapter 2.

This calculation is, of course, very rough since it ascribes the entire heat capacity of vacancy formation to the elastic image field volume, which is based on linear elasticity theory. The relaxation of atoms right around the vacancy is, however, too large to be accurately described by elasticity theory. A more rigorous calculation would start with the atomic interaction force constants for the atoms around the vacancy and proceed to a calculation of the altered vibration frequencies from these force constants. Such calculations have been done in attempts to compute the entropy of vacancy formation. They are fraught with problems, however, and the precise results are quite sensitive to the details of the models used.

For our purposes, we can take $(C_v^f)_V$ as negligible, since no improvement in our model will increase the value of $-0.05k$ to the point where the fourth term need be retained in (5.28), at least for close-packed structures.

We will therefore write, for close-packed crystals,

$$G_v^f \simeq \tilde{G}_v^f - (T - T_m)\tilde{S}_v^f + P\tilde{V}_v^f \tag{5.39}$$

or

$$G_v^f = \tilde{U}_v^f - T\tilde{S}_v^f + P\tilde{V}_v^f \tag{5.40}$$

so that Arrhenius plots can indeed be expected to yield the energy of vacancy formation and $\ln N_v$ versus P plots can be expected to give the vacancy-formation volume.

The validity of the approximations in (5.39) depends on the image volume being small, and on the thermal expansion and compressibility of the region around a vacancy not being too greatly different from that of the perfect crystal. Although this seems to be the case in close-packed crystals, it is not true in sodium, which has the more open body-centered cubic structure.

6.6   VACANCIES, DIVACANCIES, AND INTERSTITIALS

Up to this point, it has been assumed that the vacancy is the only defect in our monatomic crystal. For the noble metals, both experimental and theoretical considerations give the result that the free energy of vacancy formation is considerably less than the formation free energy of other defects. However, other point defects certainly exist, and their importance relative to monovacancies can be expected to vary from one material to another. In aluminum, for example, it appears that nearly 40% of the vacant lattice sites are tied up in divacancies at the melting point, whereas in copper at the melting point, only about 0.2% of the vacancies are in divacancies.

Most of our quantitative information about point defects in metals is based on work in face-centered cubic systems. In these systems, only vacancies, and sometimes divacancies, need to be taken into account at equilibrium because the interstitial-formation free energy is so high. There is no reason to expect these results to carry over to more open structures. In sodium, for example, the ion-core radius is small relative to the nearest-neighbor separation distance, so that the interstitial-formation energy can be expected to be much smaller than in close-packed metals. Unfortunately, there has not been enough research done in such open systems to identify the nature and number of all point defects present, but there are strong indications that the alkali metals have a variety of point-defect structures that do not exist in close-packed crystals.

In this section, the coexistence of vacancies, divacancies, and interstitials in a pure, monatomic crystal will be considered. The defect concentrations will be determined by a straightforward generalization of (2.15):

$$\frac{\partial G(V, 2V, I)}{\partial N_v} = 0$$

$$\frac{\partial G(V, 2V, I)}{\partial N_{2v}} = 0 \tag{6.1}$$

$$\frac{\partial G(V, 2V, I)}{\partial N_I} = 0$$

where N_v, N_{2v}, and N_I are the numbers of vacancies, divacancies, and interstitial atoms, respectively, and $G(V, 2V, I)$ is the Gibbs free energy of a crystal containing N atoms, N_v vacancies, N_{2v} divacancies and N_I interstitials. This free energy is given by

$$G(V, 2V, I) = G\{V, 2V, I\} - kT \ln w(V, 2V, I) \tag{6.2}$$

where $G\{V, 2V, I\}$ is the Gibbs free energy except for the configurational contribution and $w(V, 2V, I)$ is the number of ways of distributing N atoms, N_v vacancies, N_{2v} divacancies, and N_I interstitials on a lattice of L normal sites and L interstitial sites, where

$$L = N + N_v + 2N_{2v} \tag{6.3}$$

The configuration-free Gibbs free energy is given in terms of a partition function completely analogous to (2.4).

In (6.1), the differentiations are performed while holding the number of atoms constant, so that the vacancy- and divacancy formation processes consist of removing atoms from the interior of the crystal and placing them on the surface, while the interstitial formation process consists of taking a surface atom and placing it in an interior interstitial position.

Equations 6.1 and 6.2 give the following conditions of defect equilibrium:

$$\frac{\partial \ln w(V, 2V, I)}{\partial N_v} = \frac{G_v{}^f}{kT} \tag{6.4}$$

$$\frac{\partial \ln w(V, 2V, I)}{\partial N_{2v}} = \frac{G_{2v}{}^f}{kT} \tag{6.5}$$

$$\frac{\partial \ln w(V, 2V, I)}{\partial N_I} = \frac{G_I{}^f}{kT} \tag{6.6}$$

where $G_v{}^f$, $G_{2v}{}^f$, and $G_I{}^f$ are the free energies of formation of a vacancy, a divacancy, and an interstitial, respectively, defined by

$$G_v{}^f \equiv \frac{\partial G\{V, 2V, I\}}{\partial N_v} \tag{6.7}$$

$$G_{2v}{}^f \equiv \frac{\partial G\{V, 2V, I\}}{\partial N_{2v}} \tag{6.8}$$

$$G_I{}^f \equiv \frac{\partial G\{V, 2V, I\}}{\partial N_I} \tag{6.9}$$

Since the number of atoms in the crystal is held constant during the differentiation, $G_v{}^f$ is the free-energy change on bringing an atom from an interior lattice site to the crystal surface, $G_{2v}{}^f$ is the free-energy change on bringing two adjacent atoms from interior lattice sites to the surface, and $G_I{}^f$ is the free-energy change on taking a surface atom and placing it in an interior interstitial position.

As usual, we will assume that the concentration of defects is low enough so that interactions among them can be ignored and the formation free energies can be taken as independent of concentration.

The statistical count $w(V, 2V, I)$ is obtained by first counting the number of ways N_{2v} divacancies can be placed in a lattice of L sites, then counting the number of ways N_v vacancies can be placed on the remaining available sites, and finally counting the number of ways N_I interstitials can be placed on available interstitial sites. In performing this count, the restriction is imposed that the different types of defects cannot be nearest neighbors.* The result is that $w(V, 2V, I)$ is the product of three terms

$$w(V, 2V, I) = w_{2v}w_vw_I \tag{6.10}$$

* See Chapter 7 for a detailed analysis of the statistical count.

where

$$w_{2v} = \frac{(z/2)^{N_{2v}}}{N_{2v}!} \prod_{m=0}^{N_{2v}-1} (L - 2m) \tag{6.11}$$

$$w_v = \frac{[L - (z' + 2)N_{2v}]!}{[L - (z' + 2)N_{2v} - N_v]!\, N_v!} \tag{6.12}$$

$$w_I = \frac{[L - z_I N_v - (z_I' + 2)N_{2v}]!}{[L - z_I N_v - (z_I' + 2)N_{2v} - N_I]!\, N_I!} \tag{6.13}$$

In these equations, z is the number of nearest neighbors to a lattice site, z' is the number of nearest-neighbor lattice sites to a vacancy pair, z_I is the number of nearest-neighbor interstitial sites surrounding a vacancy, and z_I' is the number of interstitial sites that are nearest neighbors to a vacancy pair, defined such that $(z_I' + 2)$ are the number of interstitial sites around a divacancy that are not available to interstitials.

The process of differentiating the statistical count is a bit tedious but straightforward. It is only necessary to use approximations of the form $\ln(1 + x) = x$ for small x, and to take the number of defects as small relative to the number of lattice sites. The product in (6.11) is treated as follows

$$\ln \prod_{m=0}^{N_{2v}-1} (L - 2m) = \sum_{m=0}^{N_{2v}-1} \ln (L - 2m) = \sum_{m=0}^{N_{2v}-1} \ln L \left(1 - \frac{2m}{L}\right)$$

$$\simeq \sum_{m=0}^{N_{2v}-1} \left[\ln L - \frac{2m}{L}\right] \simeq N_{2v} \ln L$$

Using these approximations, we get

$$\ln w(V, 2V, I) = N_v + N_v \ln \frac{L}{N_v} + N_{2v} + N_{2v} \ln \frac{L}{N_{2v}} + N_I + N_I \ln \frac{L}{N_I} \tag{6.14}$$

and

$$\frac{\partial \ln w(V, 2V, I)}{\partial N_v} = \ln \frac{L}{N_v} \tag{6.15}$$

$$\frac{\partial \ln w(V, 2V, I)}{\partial N_{2v}} = \ln \frac{zL}{2N_{2v}} \tag{6.16}$$

$$\frac{\partial \ln w(V, 2V, I)}{\partial N_I} = \ln \frac{L}{N_I} \tag{6.17}$$

Combining these with (6.4) to (6.6) we get the defect concentrations as

$$\frac{N_v}{L} = e^{-G_v{}^f/kT} \tag{6.18}$$

$$\frac{N_{2v}}{L} = \frac{z}{2} e^{-G_{2v}{}^f/kT} \tag{6.19}$$

$$\frac{N_I}{L} = e^{-G_I{}^f/kT} \tag{6.20}$$

There is a relation between the vacancy and divacancy concentrations because two monovacancies can combine to form a divacancy and, conversely, a divacancy can dissociate into two monovacancies. This process can be represented in the notation of chemical reactions as

$$V + V \rightleftarrows V_2$$

The equilibrium constant governing the equilibrium concentrations for this reaction is defined by

$$K(2V \rightleftarrows V_2) \equiv \frac{(N_{2v}/L)}{(N_v/L)^2} \tag{6.21}$$

From (6.18) and (6.19), the right-hand side of this is given by

$$\frac{(N_{2v}/L)}{(N_v/L)^2} = \frac{z}{2} e^{-[G_{2v}{}^f - 2G_v{}^f]/kT} \tag{6.22}$$

The quantity G^{vv}, defined by

$$G^{vv} = 2G_v{}^f - G_{2v}{}^f \tag{6.23}$$

is the binding free energy of the vacancy, since it is the free-energy change upon separating a divacancy into the two monovacancies. Equation 6.22 is usually written as

$$\frac{N_{2v}}{L} = \frac{z}{2} \left(\frac{N_v}{L}\right)^2 e^{G^{vv}/kT} \tag{6.24}$$

The total concentration of vacant sites, $N_v{}^T$ is given by

$$N_v{}^T = N_v + 2N_{2v} \tag{6.25}$$

or, using (6.18) and (6.19),

$$\frac{N_v{}^T}{L} = e^{-G_v{}^f/kT}[1 + ze^{-(G_v{}^f + G^{vv})/kT}] \tag{6.26}$$

Having obtained the equilibrium concentration of defects, we can now derive the crystal free energy in a manner completely analogous to that used in arriving at (3.22). If we write the configurationless free energy $G\{V, 2V, I\}$ as linear in the defect concentrations,

$$G\{V, 2V, I\} = G^0 + N_v G_v{}^f + N_{2v} G_{2v}{}^f + N_I G_I{}^f \qquad (6.27)$$

and substitute this, along with (6.14) into (6.2), and use (6.18) to (6.20) the result is

$$G(V, 2V, I) = G^0 - kT[N_v + N_{2v} + N_I] \qquad (6.28)$$

which is a straightforward generalization of (3.22). Just as in Section 4, we can now obtain all the thermodynamic crystal functions in a form that displays the defect concentrations. These are listed below.

$$S = S^0 + \frac{N_v}{T}[H_v{}^f + kT] + \frac{N_{2v}}{T}[H_{2v}{}^f + kT] + \frac{N_I}{T}[H_I{}^f + kT] \quad (6.29)$$

$$V = V^0 + N_v V_v{}^f + N_{2v} V_{2v}{}^f + N_I V_I{}^f \qquad (6.30)$$

$$U = U^0 + N_v U_v{}^f + N_{2v} U_{2v}{}^f + N_I U_I{}^f \qquad (6.31)$$

$$H = H^0 + N_v H_v{}^f + N_{2v} H_{2v}{}^f + N_I H_I{}^f \qquad (6.32)$$

$$C_P = C_P{}^0 + N_v \left[(C_v{}^f)_P + \frac{(H_v{}^f)^2}{kT^2} \right]$$

$$+ N_{2v} \left[(C_{2v}{}^f)_P + \frac{(H_{2v}{}^f)^2}{kT^2} \right] + N_I \left[(C_I{}^f)_P + \frac{(H_I{}^f)^2}{kT^2} \right] \quad (6.33)$$

$$\alpha = \alpha^0 + \frac{N_v V_v{}^f}{V_0} \left[\alpha_v{}^f - \alpha^0 + \frac{H_v{}^f}{kT^2} \right] + \frac{N_{2v} V_{2v}{}^f}{V_0} \left[\alpha_{2v}{}^f - \alpha^0 + \frac{H_{2v}{}^f}{kT^2} \right]$$

$$+ \frac{N_I V_I{}^f}{V_0} \left[\alpha_I{}^f - \alpha^0 + \frac{H_I{}^f}{kT^2} \right] \quad (6.34)$$

$$\beta = \beta^0 + \frac{N_v V_v{}^f}{V_0} \left[\beta_v{}^f - \beta^0 + \frac{V_v{}^f}{kT} \right] + \frac{N_{2v} V_{2v}{}^f}{V_0} \left[\beta_{2v}{}^f - \beta^0 + \frac{V_{2v}{}^f}{kT} \right]$$

$$+ \frac{N_I V_I{}^f}{V_0} \left[\beta_I{}^f - \beta^0 + \frac{V_I{}^f}{kT} \right] \quad (6.35)$$

The defect quantities are defined just as in Section 4, and they are identified by appropriate subscripts. In deriving (6.34) and (6.35), quantities to the second order in the defect concentrations were neglected.

6.7   SOME NUMERICAL RESULTS

A variety of experimental methods have been used to determine the values of energies and entropies of defect formation. These include the measurement of electrical resistivity of quenched and irradiated metals, high-temperature measurements of heat capacity and thermal expansion, comparison of dilatometric and X-ray measurements of thermal expansion, determination of stored energy release of irradiated and cold-worked specimens, diffusion measurements, and internal friction studies.

The various methods agree in that they give roughly comparable results. The accuracy of any of the reported results is, however, open to question, in spite of the fact that they represent a considerable amount of careful and highly competent research. There are two reasons for this: the first is that defects are present in low concentrations and very sensitive experimental methods must be used to see their effects. The second is that in most materials a rich variety of point defects, as well as more extended defects such as dislocations and grain boundaries, exists. The question of sorting out the effect of all the defects and their interactions on the measurements is a very difficult one, so that there is an uncertainty inherent in the interpretation of the experiments.

In spite of these difficulties, some information on the defect parameters has been obtained. Table 6.1 shows these parameters for some metals. Most of the values have been reduced to one significant figure. For those values that seem to be better established, two significant figures were retained.

Table 6.1.   Point-Defect Parameters of Metals

Metal	U_v^f eV	S_v^f	U^{vv} eV	S_{2v}^f
Ag	1			
Cu	1.05	0.4k	0.1	
Au	0.87	0.5k		
Ni	1.4	1.5k	0.3	2k
Aℓ	0.65	0.8k	0.3	1k

From an analysis by A. Seeger and H. Mehrer, *Vacancies and Interstitials in Metals*, Proceedings of the Julich Conference, September 1968, A. Seeger, D. Schumacher, W. Schilling, and J. Diehl, Eds., North-Holland Publishing Company, 1970.

Figure 6.1 shows the vacancy concentration in copper as a function of temperature according to (6.17), using $E_v^f = 1.05$ eV and $S_v^f = 0.4k$. The atomic fraction of vacancies near the melting point is of the order of 10^{-4}, and it decreases rapidly with temperature. This curve shows why the effects of an equilibrium concentration of vacancies are only observable at high temperatures. Because of the low concentration of monovacancies and the low-divacancy binding energy, the divacancy concentration is quite small in copper. Also, because of the high interstitial-formation energy, the concentration of interstitials is negligible. Thus, in copper at equilibrium, the monovacancy is the predominant defect.

Since the volume of vacancy formation is positive, the application of pressure decreases the vacancy concentration. This effect is shown in Figure 6.2 in which the ratio of the vacancy concentration at pressure P and at zero pressure is plotted against pressure (at the melting point) according to

$$\frac{N_v(P, T_m)}{N_v(0, T_m)} = e^{-PV_v^f/kT_m} \qquad (7.1)$$

which is readily obtained from (5.1). V_v^f was taken to be $0.5v$, where v is the atomic volume, and T_m is the melting point.

The vacancy contribution to the heat capacity of copper is displayed in Figure 6.3, which was computed from

$$\frac{C_P - C_P^0}{N} = \frac{N_v}{N} \frac{(U_v^f)^2}{kT^2} \qquad (7.2)$$

This equation is readily obtained from (4.24) since $H_v^f = U_v^f$ at zero pressure, and $(C_v^f)_P$ can be shown to be negligible from the calculations in Section 5.

The contributions of vacancies to the thermal expansion of copper at zero pressure is shown in Figure 6.4, which was obtained from (4.25) $(\alpha_v^f - \alpha^0)$ was neglected. Since the total thermal expansion of copper is about 70×10^{-6}, we see that the vacancies contribute about 0.7% at the melting point. The contribution of vacancies to the compressibility is even smaller.

Although the figures show that the contributions of point defects to the thermodynamic properties of pure copper are small, and generally not detectable with any accuracy using available experimental methods, this must not be taken as true for all metals. The effects are proportional to the defect concentrations, and in many metals they can be considerably higher than in copper. As an example, the concentration of vacancies and divacancies in aluminum is shown in Figure 6.5 to be much higher than in copper. The total concentration of vacant sites at the melting point is about 10^{-3}, which is a factor of 10 greater than for copper. It is interesting to note

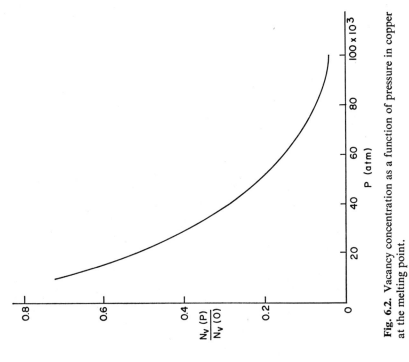

Fig. 6.2. Vacancy concentration as a function of pressure in copper at the melting point.

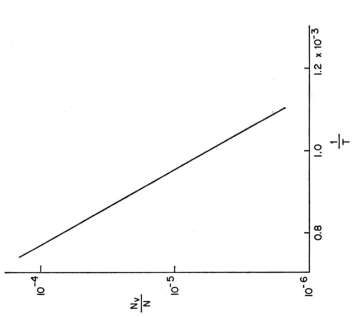

Fig. 6.1. Vacancy concentration in copper.

219

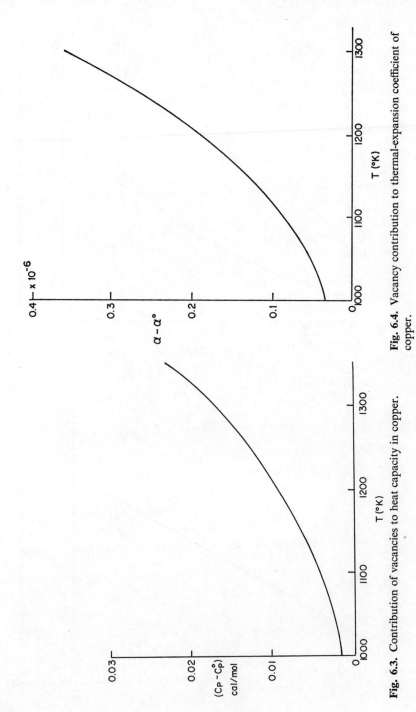

Fig. 6.4. Vacancy contribution to thermal-expansion coefficient of copper.

Fig. 6.3. Contribution of vacancies to heat capacity in copper.

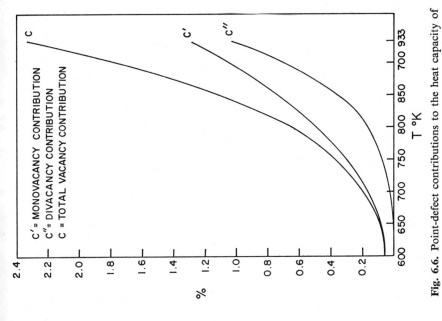

Fig. 6.6. Point-defect contributions to the heat capacity of aluminum.

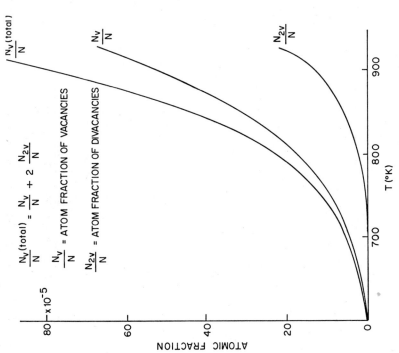

Fig. 6.5. Vacancy and divacancy concentrations in aluminum.

221

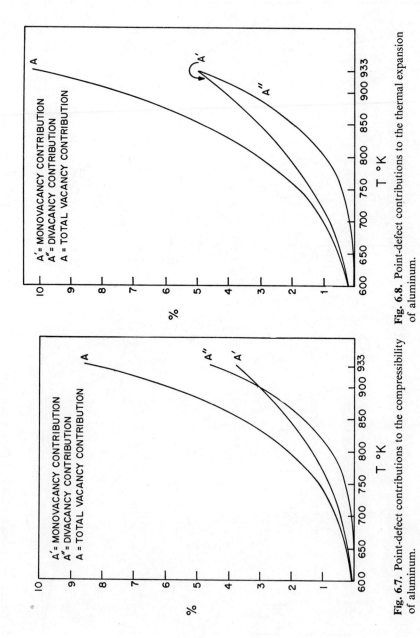

Fig. 6.7. Point-defect contributions to the compressibility of aluminum.

Fig. 6.8. Point-defect contributions to the thermal expansion of aluminum.

that nearly 40% of the vacant sites are tied up in divacancies so that divacancies can by no means be neglected in aluminum. Furthermore, it will be shown in the next chapter that impurities can increase the number of vacant sites. Thus, we can expect that point defects can make significant and measurable contributions to the thermodynamic properties of some metals.

In Figures 6.6 to 6.8, we see that the defect contributions to the heat capacity, compressibility, and thermal expansion in aluminum are much greater than in copper. In fact, the defect contributions are larger than the error in a decent experiment and, therefore, contribute measurable amounts.

chapter 7

POINT DEFECTS IN

DILUTE ALLOYS

7.1 GENERAL COMMENTS AND INTRODUCTION

This chapter is devoted to an analysis of point-defect equilibria in binary alloys for which the concentration of one of the constituents is small relative to the concentration of the other. In such alloys, the minority constituent can itself be regarded as a defect or an impurity.

Our attention will be restricted to one-center and two-center defects; that is, to alterations in the local composition from that of the pure perfect crystal that involve either one atom or a pair of adjacent atoms. The perfect atomic lattice is taken to be one that is fully occupied by atoms of the type of the major constituent, and the perfect interstitial lattice is completely empty. It will be assumed that all atomic sites and all interstitial sites in the perfect crystal are equivalent. The results can easily be generalized to crystals with atoms or interstitial sites that are arranged on two or more non-equivalent sublattices, and to crystals containing more than one type of impurity.

To simplify our discussion, the normal lattice on which the atoms in the pure perfect crystal are distributed will be called the L lattice and consists of L sites, while the interstitial lattice is called the L' lattice and consists of L' sites. The possible one- and two-center defects will then be denoted as follows:

V—a vacant L site (the monovacancy)
V_2—two adjacent vacant L sites (the divacancy)
B—an impurity atom on an L site (the lattice impurity)

B_2—two impurity atoms on adjacent L sites (the lattice diimpurity)

BV—an impurity atom on an L site adjacent to a vacant L site (the vacancy-lattice impurity complex)

A'—a major constituent atom at an L' site (the interstitial)

A'_2—two major constituent atoms on adjacent L' sites (the diinterstitial)

B'—an impurity atom on an L' site (the impurity interstitial)

B'_2—two impurity atoms on L' sites (the impurity diinterstitial)

$A'B'$—a major constituent atom and an impurity atom on adjacent L' sites (the host-impurity interstitial complex)

$A'V$—a major constituent atom on an L' site adjacent to a vacant L site (the interstitial-vacancy complex)

$B'V$—an impurity atom on an L' site adjacent to a vacant L site (the vacancy-interstitial impurity complex)

$A'B$—a major constituent atom on an L' site adjacent to an impurity on an L site (the impurity-interstitial complex)

BB'—an impurity atom on an L site adjacent to an impurity on an L' site (the lattice-interstitial diimpurity)

The first five of these defects occur only on the atomic lattice and will be called substitutional defects; the second five occur on the interstitial lattice and will be called interstitial defects; the last four have one center on the atomic lattice and one center on the interstitial lattice and will be called mixed-lattice defects.

Illustrations of these defects for the face-centered square lattice are shown in Figures 7.1.

It is implicit in the definition of these defects that they are surrounded by atoms of the major constituent (A atoms) and, in fact, the immediate environment of a defect should enter into its definition. This can be seen from the fact that if two vacancies are brought close enough together, they form a divacancy. Thus, there is a limit to how close defects can approach each other without losing their identity and being transformed into something else, so there is a minimum number of A atoms surrounding a defect that must be included in its definition. In general, the stronger the lattice distortion the larger this minimum number. Thus, in a close-packed crystal, the region defining an interstitial will be larger than that defining a lattice vacancy because the interstitial distorts the lattice considerably, while the distortions around the vacancy are smaller.

Just as in the previous chapter, our analysis will be based on the pressure ensemble. However, we now consider a system consisting of the impurity alloy in equilibrium with a gas phase containing A atoms and B atoms. This will permit us to investigate not only the internal defect equilibria, but also the solid-vapor equilibrium and its coupling to the defect equilibria. Thus

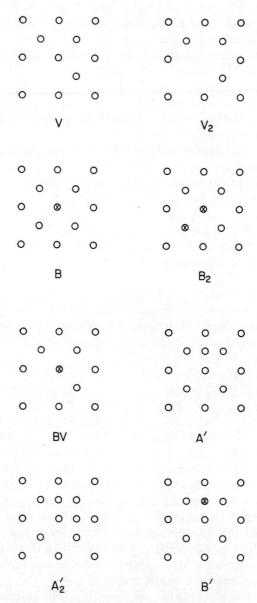

Fig. 7.1. Defects and defect pairs in dilute alloy.

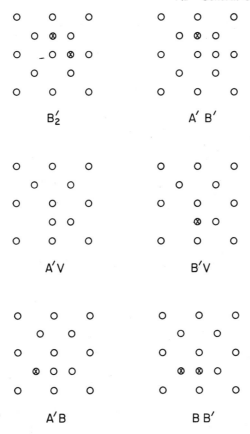

Fig. 7.1. (Continued).

we write the Gibbs free energy G of a crystal containing N_A atoms of type A, N_B atoms of type B and $N_{(i)}$ defects of type (i), in equilibrium with a gas containing $N_A{}^g$ atoms of type A and $N_B{}^g$ atoms of type B as

$$e^{-G/kT} = \sum_{N_A, N_B, N_{(i)}} w(N_A, N_B, N_{(i)})e^{-G\{N_A, N_B, N_{(i)}\}/kT}e^{-G^g(N_A{}^g, N_B{}^g)/kT} \quad (1.1)$$

In this equation, the total numbers of A atoms and of B atoms are constant. $G\{N_A, N_B, N_{(i)}\}$ is the free energy of a crystal containing N_A A atoms, N_B B atoms, and $N_{(i)}$ defects of type (i), except for the configurational contribution arising from the defect distribution. It is defined by a form completely analogous to (2.4) of Chapter 6. $G^g(N_A{}^g, N_B{}^g)$ is the free energy of the gas phase.

The equilibrium-defect concentrations are again obtained by finding the most probable distribution, which is equivalent to determining the maximum term in the sum of (1.1). We therefore define the Gibbs free energy of a system in which a crystal with N_A A atoms, N_B B atoms, and $N_{(i)}$ defects of type (i), is in contact with a gas containing $N_A{}^g$ A atoms and $N_B{}^g$ B atoms as

$$G_{cg}(N_A, N_B, N_{(i)}) = G\{N_A, N_B, N_{(i)}\}$$
$$+ G^g(N_A{}^g, N_B{}^g) - kT \ln w(N_A, N_B, N_{(i)}) \quad (1.2)$$

in complete analogy with (2.14) of Chapter 6, and we identify the thermodynamic Gibbs free energy of the system with (1.2) when N_A, N_B, $N_{(i)}$, $N_A{}^g$, and $N_B{}^g$ have their most probable values. These most probable values are determined by

$$\delta G_{cg}(N_A, N_B, N_i) = 0 \quad (1.3)$$

where the variation is taken with respect to changes in the gas and crystal composition. Since the system (crystal + gas) is closed, a change in crystal composition can only take place with a corresponding change in the gas phase. For example, decreasing the number of vacancies in the crystal by unity corresponds to a transfer of an A atom in the gas to a vacant site in the crystal. In general, changing the number of (i)-type defects involves a transfer of A and B atoms to or from the gas phase.

Equation 1.3 is, of course, valid for any arbitrary variation of crystal composition and can therefore be applied to any specific process of defect formation. For our purposes it is convenient and instructive to use (1.3) in conjunction with specifically defined variations that describe defect formation. Equation 1.3 will therefore be applied to (1.2) in the usual way to give

$$kT \left. \frac{\partial \ln w}{\partial N_{(i)}} \right|_P = \left. \frac{\partial G\{N_A, N_B, N_{(i)}\}}{\partial N_{(i)}} \right|_P + \left. \frac{\partial G^g(N_A{}^g, N_B{}^g)}{\partial N_{(i)}} \right|_P \quad (1.4)$$

with the understanding that the derivatives refer to changes in the statistical count and the free energies accompanying a particular process of defect formation.

The subscript P specifies the particular process for which the variation is applied.

7.2 THE STATISTICAL COUNT FOR SUBSTITUTIONAL DEFECTS

In this section the statistical count is derived for a dilute alloy containing substitutional defects only. We are given a lattice of L equivalent sites containing N_A atoms of type A, N_B atoms of type B, N_v monovacancies, N_{2v} divacancies, N_{2B} impurity pairs, and N_{Bv} impurity-vacancy complexes.

The alloy is dilute and, therefore,

$$N_B \ll N_A$$
$$N_v \ll N_A$$
$$N_{2v} \ll N_A \qquad (2.1)$$
$$N_{2B} \ll N_A$$
$$N_{Bv} \ll N_A$$

The total number of lattice sites is

$$L = N_A + N_B + N_v + 2N_{2v} + 2N_{2B} + 2N_{Bv} \qquad (2.2)$$

and the total numbers of B atoms and vacant sites is

$$N_B{}^T = N_B + 2N_{2B} + N_{Bv} \qquad (2.3)$$

$$N_v{}^T = N_v + 2N_{2v} + N_{Bv} \qquad (2.4)$$

As described in the previous section, each defect is defined as a small volume element such that if the volume elements of two defects overlap, the defects lose their identity and are transformed into a third defect. Let Z be the number of lattice sites in such a volume element for one-center defects (vacancy, impurity) and let Z' be the number of lattice sites in the volume element for two-center defects (divacancy, diimpurity, vacancy-impurity complex). In the simplest case, where the volume elements are defined by requiring that only nearest neighbors to vacant sites or impurity atoms are occupied by A atoms, $Z = z + 1$ and $Z' = z' + 2$ where z is the number of nearest neighbors to a lattice site, and z' is the number of nearest neighbors to a pair of adjacent sites.

The statistical count $w = w(N_A, N_B, N_v, N_{2v}, N_{2B}, N_{Bv})$ is the number of ways of distributing N_A A atoms, N_B B atoms, N_v monovacancies, N_{2v} divacancies, N_{2B} impurity pairs, and N_{Bv} impurity-vacancy complexes. A slight modification of the method of Lidiard and Howard* will first be used to get this number. In this procedure, we start with an empty lattice and first count the number of ways that impurity pairs can be placed on L sites, then the number of ways of placing the divacancies on the remaining sites, the number of ways of placing the impurity-vacancy complexes on the sites that are still empty, and finally the number of ways of filling in the lattice with A atoms, B atoms, and vacancies. Multiplying all these counts together gives the total statistical count for our system.

* Lidiard, A. B., "The Influence of Solutes on Self-Diffusion in Metals" *Phil. Mag.* **5**, 1171 (1960). Howard, R. E. and Lidiard, A. B. "Matter Transport in Solids" *Rep. Prog. Phys.* **27**, 161 (1964).

The number of ways of placing N_{2B} impurity pairs on L sites will be called w_{2B} and is obtained as follows:

The number of ways of placing the first pair on the lattice is just equal to the number of pairs that can be formed from L sites. Since the members of the pair are identical, this number is

$$\frac{zL}{2}$$

Now count the ways of placing a second impurity pair on the remaining sites. The first pair preempted Z' sites, so the number of ways the second pair can go on is given by

$$\frac{z}{2}(L - Z')$$

Continuing in this fashion, we get

$$\left(\frac{z}{2}\right)^{N_{2B}} \prod_{l=0}^{N_{2B}-1} (L - Z'l) \tag{2.5}$$

But (2.5) includes all distributions in which the pairs just change places with each other, and these cannot be allowed because the pairs are indistinguishable. Therefore, (2.5) must be divided by $N_{2B}!$ to obtain w_{2B}^{LH}, that is,

$$w_{2B}^{LH} = \frac{1}{N_{2B}!} \left(\frac{z}{2}\right)^{N_{2B}} \prod_{l=0}^{N_{2B}-1} (L - Z'l) \tag{2.6}$$

The superscript LH is used to indicate that the method of Lidiard and Howard is being used.

Now that the impurity pairs are in the lattice, count the number of ways of putting in divacancies. This is obtained in the same way as for w_{2B}^{LH}, except that the number of sites available to the divacancies is now $(L - \Delta_1)$ where

$$\Delta_1 = Z'N_{2B} \tag{2.7}$$

Therefore, w_{2v}^{LH}, the number of ways of placing divacancies in the lattice after the impurity pairs have been placed, is

$$w_{2v}^{LH} = \frac{1}{N_{2v}!} \left(\frac{z}{2}\right)^{N_{2v}} \prod_{m=0}^{N_{2v}-1} (L - \Delta_1 - Z'm) \tag{2.8}$$

In precisely the same way, the number of ways of placing N_{Bv} impurity-vacancy complexes after placing the impurity pairs and the divacancies is

$$w_{Bv}^{LH} = \frac{1}{N_{Bv}!} z^{N_{Bv}} \prod_{n=0}^{N_{Bv}-1} (L - \Delta_2 - Z'n) \tag{2.9}$$

where

$$\Delta_2 = Z'N_{2B} + Z'N_{2v} \tag{2.10}$$

The factor z appears in (2.9) rather than $z/2$ because the pair now consists of distinguishable entities for which the number of possible pairs per available lattice site is z.

There are now $(L - \Delta_3)$ available sites left, where

$$\Delta_3 = Z'N_{2B} + Z'N_{2v} + Z'N_{Bv} \qquad (2.11)$$

and we still need to distribute N_A A atoms, N_B B atoms, and N_v vacancies on them. The number of ways of arranging the B atoms on the remaining sites is just

$$w_B^{\text{LH}} = \frac{(L - \Delta_3)!}{N_B! (L - \Delta_3 - N_B)!} \qquad (2.12)$$

and, finally, the number of ways of placing N_v vacancies and N_A A atoms on the sites that are still empty is

$$w_v^{\text{LH}} = \frac{(L - \Delta_4)!}{N_v! (L - \Delta_4 - N_v)!} \qquad (2.13)$$

where

$$\Delta_4 = Z'N_{2B} + Z'N_{2v} + Z'N_{Bv} + ZN_B \qquad (2.14)$$

The total statistical count is then just the product of the individual counts and therefore the Lidiard-Howard method gives

$$w^{\text{LH}} = w_{2B}^{\text{LH}} w_{2v}^{\text{LH}} w_{Bv}^{\text{LH}} w_B^{\text{LH}} w_v^{\text{LH}} \qquad (2.15)$$

There are several comments to be made on this method. First, the answer depends on the order in which the defects are placed on the lattice. If, for example, the divacancies had been laid down first, and the impurity pairs next, w_{2v}^{LH} would not contain Δ_1 $(= Z'N_{2B})$ and w_{2B}^{LH} would contain $Z'N_{2v}$ instead of Δ_1. The next comment is that w_B^{LH} contains some configurations in which B atoms are adjacent to each other, and is not rigorously a count of the number of ways of placing "isolated" B atoms, that is, B atoms with their Z surrounding A atoms. Similar remarks hold for w_v^{LH}. However, w^{LH} is sufficiently accurate when the number of defects is small, and leads to defect concentration formulas that are correct within the approximation that $N(\text{defects}) \ll N_A$.

The products in (2.6), (2.8), and (2.9) can be simplified. They all have the form

$$P \equiv \prod_{j=0}^{N_d-1} (L - \Delta' - Z'j) \qquad (2.16)$$

or

$$\ln P = \sum_{j=0}^{N_d-1} \ln (L - \Delta' - Z'j) \qquad (2.17)$$

$$= \sum_{j=0}^{N_d-1} \ln L + \sum_{j=0}^{N_d-1} \ln \left(1 - \frac{\Delta' + Z'j}{L}\right) \qquad (2.18)$$

Since we restrict ourselves to defect concentrations such that $(\Delta' + Z'N_d) \ll L$, the logarithms in the second sum (2.18) can all be expanded to the first order. Thus, both sums in (2.18) can be evaluated to give

$$\ln P = N_d \ln L - N_a \frac{\Delta'}{L} - N_a \frac{Z'}{2L} \tag{2.19}$$

If defect concentrations, represented by Δ'/L, are, at most, of the order of 10^{-3}, and L is of the order of 10^{22}, then $\Delta'/2L$ can certainly be neglected relative to $\ln L$, within an accuracy of about 10^{-5} ($10^{-3}\%$). Accepting this error, and recognizing that the last term is completely negligible relative to the first, dropping the last two terms in (2.19) gives

$$P = L^{N_d} \tag{2.20}$$

Applying this formulae to the products in (2.6), (2.8), and (2.9) gives

$$w_{2B}^{LH} = \frac{1}{N_{2B}!} \left(\frac{zL}{2}\right)^{N_{2B}} \tag{2.21}$$

$$w_{2v}^{LH} = \frac{1}{N_{2v}!} \left(\frac{zL}{2}\right)^{N_{2v}} \tag{2.22}$$

$$w_{Bv}^{LH} = \frac{1}{N_{Bv}!} (zL)^{N_{Bv}} \tag{2.23}$$

The statistical counts for impurity atoms and vacancies given by (2.12) and (2.13) can also be simplified. Taking logarithms of (2.12) and applying Stirling's approximation gives

$$\ln w_B^{LH} = (L - \Delta_3) \ln (L - \Delta_3) - N_B \ln N_B$$
$$- (L - \Delta_3 - N_B) \ln (L - \Delta_3 - N_B) \tag{2.24}$$

Rearranging this equation, we get

$$\ln w_B^{LH} = -(L - \Delta_3) \ln \frac{(L - \Delta_3 - N_B)}{(L - \Delta_3)}$$
$$+ N_B \ln (L - \Delta_3 - N_B) - N_B \ln N_B$$

Now use the approximation $\ln (1 - x) = -x$ for small x to get

$$\ln w_B^{LH} = N_B + N_B \ln L - N_B \frac{(\Delta_3 + N_B)}{L} - N_B \ln N_B$$

or, neglecting the third term,

$$\ln w_B^{LH} = N_B + N_B \ln \frac{L}{N_B} \tag{2.25}$$

from which

$$w_B^{\mathrm{LH}} = e^{N_B} \left(\frac{L}{N_B} \right)^{N_B} \tag{2.26}$$

within the approximation that the number of defects is much less than the number of lattice sites.

Similarly, (2.14) reduces to

$$w_v^{\mathrm{LH}} = e^{N_v} \left(\frac{L}{N_v} \right)^{N_v} \tag{2.27}$$

Multiplying (2.21), (2.22), (2.23), (2.26), and (2.27) together gives the Lidiard-Howard count in the form

$$w^{\mathrm{LH}} = e^{N_{2B}} \left(\frac{zL}{2N_{2B}} \right)^{N_{2B}} e^{N_{2v}} \left(\frac{zL}{2N_{2v}} \right)^{N_{2v}} e^{N_{Bv}} \left(\frac{zL}{N_{Bv}} \right)^{N_{Bv}} e^{N_B} \left(\frac{L}{N_B} \right)^{N_B} e^{N_v} \left(\frac{L}{N_v} \right)^{N_v}$$

$$\tag{2.28}$$

This can be written in a more compact form by labeling the defects with a running index i so that $N_{(i)}$ is the number of defects of type (i) ($i = 2B$, $2v$, Bv, B, v) and defining $\omega_{(i)}$ to equal unity for one-center defects, z for two-center defects consisting of pairs of unlike species, and $z/2$ for defects consisting of pairs of identical species. Then, (2.28) can be written as

$$w^{\mathrm{LH}} = \prod_i e^{N_{(i)}} \left(\frac{\omega_{(i)}L}{N_{(i)}} \right)^{N_{(i)}} \tag{2.29}$$

$\omega_{(i)}$ can be thought of as a rotational factor given by the number of distinct rotations the defect can assume about its center.

The statistical count is now in a form that is independent of the order in which defects were placed on the lattice. This was accomplished by accepting the approximation that the number of defects is small relative to the number of lattice sites.

An alternate method of arriving at the statistical count is to recognize that it is the number of states for which the crystal has a given energy spectrum. Thus it can be computed by starting with a crystal having an allowed distribution of defects and counting the number of permutations among them that do not change the energy. This is done by counting the number of ways the defects can be moved about in the crystal without having any overlap of their core fields. The core fields contain Z and Z' sites respectively for one- and two-center defects, so the total number of sites preempted by the defects is

$$\Delta \equiv ZN_B + ZN_v + Z'N_{2B} + Z'N_{2v} + Z'N_{Bv} \tag{2.30}$$

Now consider a single impurity pair. If all other defects are held in fixed positions, the number of configurations that can be generated, without changing the crystal energy, by moving the pair is just the number of ways of distributing the pair on the available lattice sites. This is just

$$\frac{z}{2}(L - \Delta)$$

A second impurity pair also produces $z(L - \Delta)/2$ new configurations if it is moved through the crystal while all other defects are held fixed. Thus, the number of configurations generated by distributing all impurity pairs while other kinds of defects remain in a given distribution is

$$\left[\frac{z}{2}(L - \Delta)\right]^{N_{2B}}$$

This, however, clearly includes distributions that differ only by an interchange of impurity pairs and, therefore, must be divided by $N_{2B}!$ to give, for the number of configurations of indistinguishable pairs per configuration of all other kinds of defects,

$$\frac{1}{N_{2B}!}[\tfrac{z}{2}(L - \Delta)]^{N_{2B}}$$

There is obviously a factor of this kind for every type of defect, so the total statistical count is

$$w = \frac{\left[\dfrac{z}{2}(L - \Delta)\right]^{N_{2B}}\left[\dfrac{z}{2}(L - \Delta)\right]^{N_{2v}}[z(L - \Delta)]^{N_{Bv}}[L - \Delta]^{N_B}[L - \Delta]^{N_v}}{N_{2B}!\, N_{2v}!\, N_{Bv}!\, N_B!\, N_v!}$$

$$(2.31)$$

Note that this result does not depend on the order in which the counting was performed. Also, in counting the number of configurations arising from permutations of the vacancies and B atoms, no states in which pairs are formed were included. These two points make (2.31) conceptually preferable to the Lidiard-Howard count.

If Stirling's approximation for the factorials is used in the form

$$N_{2B}! = e^{-N_{2B}}N_{2B}^{N_{2B}} \qquad (2.32)$$

and so on and the rotational factor $\omega_{(i)}$ is again introduced, (2.31) becomes

$$w = \prod_{(i)} e^{N_{(i)}}\left[\frac{\omega_{(i)}(L - \Delta)}{N_{(i)}}\right]^{N_{(i)}} \qquad (2.33)$$

Comparison of this with (2.29) shows that it reduces to the Lidiard-Howard count when Δ is neglected relative to L.

Equation 2.33 is obviously a general result in that any defect, whether on a substitutional, interstitial, or mixed lattice, contributes a factor to the statistical count of the form

$$w_{(i)} = e^{N_{(i)}} \left[\frac{\omega_{(i)}(L - \Delta)}{N_{(i)}} \right]^{N_{(i)}} \tag{2.34}$$

In applying this formula, it is only necessary to determine the number of lattice sites available to the defect and the rotational weighting factor $\omega_{(i)}$.

The derivatives of the statistical count will be needed in the next section. They are given by

$$\frac{\partial \ln w_{(i)}}{\partial N_{(i)}} = \ln \frac{\omega_{(i)}(L - \Delta)}{N_{(i)}} \tag{2.35}$$

7.3 THE DEFECT-CONCENTRATION FORMULAS FOR SUBSTITUTIONAL DEFECTS

Now that the statistical count has been determined, the defect concentrations can be obtained from (1.4). In applying (1.4), it is only necessary to define the process for making the defect by transfers of atoms to or from the gas phase.

To create a vacancy, either an A atom or a B atom can be removed from its position in the crystal and transferred to the vapor.

If an A atom is transferred, then (1.4) is written as

$$kT \left. \frac{\partial \ln w}{\partial N_v} \right|_{A \to v} = \left. \frac{\partial G\{\ \}}{\partial N_v} \right|_{A \to v} + \left. \frac{\partial G^g}{\partial N_v} \right|_{A \to v} \tag{3.1}$$

where the arguments in the free energies have been omitted for convenience. The derivative of the statistical count for this process only involves the vacancy contribution, so that from (2.35)

$$\left. \frac{\partial \ln w}{\partial N_v} \right|_{A \to v} = \frac{\partial \ln w_v}{\partial N_v} = \ln \frac{(L - \Delta)}{N_v} \tag{3.2}$$

The first term on the right in (3.1) is the free-energy change on forming a vacancy by removing an A atom from the crystal and transferring it to the vapor phase. This free energy will be denoted by $G^f_{v(A)}$. The right-hand side of (3.1) will therefore be written as

$$\left. \frac{\partial G\{\ \}}{\partial N_v} \right|_{A \to v} + \left. \frac{\partial G^g}{\partial N_v} \right|_{A \to v} \equiv G^f_{v(A)} + \mu_A \tag{3.3}$$

where μ_A is the chemical potential of component A in the gas phase, that is,

$$\frac{\partial G^g}{\partial N_v}\bigg|_{A \to v} = \frac{\partial G^g}{\partial N_A} \equiv \mu_A$$

because the vacancy-formation process adds on A atom to the vapor at constant temperature and pressure. It is evident that $G^f_{v(A)} + \mu_A$ is the free energy of vacancy formation as it is usually defined. This can be seen by writing the derivative in (3.3) as

$$\frac{\partial G\{\ \}}{\partial N_v} = G\{N_A - 1, N_v + 1\} - G\{N_A, N_v\}$$

where $G\{N_A, N_v\}$ is the free energy of the crystal containing N_A A atoms and N_v vacancies before the transfer of an A atom to the gas, and $G\{N_A - 1, N_v + 1\}$ is the free energy of the crystal after the transfer that then contains one less A atom and one more vacancy. However, since at equilibrium μ_A is also the chemical potential of A in the crystal,

$$G\{N_A - 1, N_v + 1\} + \mu_A = G\{N_A, N_v + 1\}$$

which is the free energy of a crystal containing N_A atoms and $N_v + 1$ vacancies. Therefore

$$\frac{\partial G\{\ \}}{\partial N_v} + \mu_A = G\{N_A, N_v + 1\} - G\{N_A, N_v\}$$

This is just the vacancy-formation energy at constant number of atoms, as it is usually defined.

Combining (3.2) and (3.3) according to (3.1) gives the vacancy-concentration formulae as

$$\frac{N_v}{L - \Delta} = \lambda_A^{-1} e^{-G_{v(A)}{}^f/kT} \tag{3.4}$$

where λ_A is the absolute activity of component A defined by

$$\lambda_A \equiv e^{\mu_A/kT}$$

Now consider the process of forming a vacancy by removing a B atom from the crystal and transferring it to the vapor. This increases the number of vacancies and decreases the number of B atoms in the crystal, so changes in both the statistical count for the vacancies and for the B atoms must be accounted for. Therefore, for this process, (2.33) gives

$$\frac{\partial \ln w}{\partial N_v}\bigg|_{B \to v} = \frac{\partial \ln w_v}{\partial N_v} - \frac{\partial \ln w_B}{\partial N_B}$$

or

$$\frac{\partial \ln w}{\partial N_v}\bigg|_{B\to v} = \ln \frac{N_B}{N_v} \tag{3.5}$$

The change in the Gibbs free energy for this process is

$$\frac{\partial G\{\ \}}{\partial N_v}\bigg|_{B\to v} + \mu_B \equiv G^f_{v(B)} + \mu_B \tag{3.6}$$

since

$$\frac{\partial G^g}{\partial N_v}\bigg|_{B\to v} = \frac{\partial G^g}{\partial N_B} = \mu_B$$

where μ_B is the chemical potential of B.

For the process under consideration, (1.4) is

$$kT \frac{\partial \ln w}{\partial N_v}\bigg|_{B\to v} = \frac{\partial G\{\ \}}{\partial N_v}\bigg|_{B\to v} + \frac{\partial G^g}{\partial N_v}\bigg|_{B\to v} \tag{3.7}$$

so that (3.5), (3.6), and (3.7) give

$$\frac{N_v}{N_B} = \lambda_B^{-1} e^{-G^f_{v(B)}/kT} \tag{3.8}$$

where $\lambda_B = e^{\mu_B/kT}$ is the absolute activity of the impurity.

The definition of the vacancy-formation free energy in this case is completely analogous to that in (3.3), so $G^f_{v(B)} + \mu_B$ is the free energy of removing a B atom from the crystal and placing it on the surface, leaving behind a vacant site.

The equilibrium concentration of B atoms can be obtained by combining (3.4) and (3.8) to give

$$\frac{N_B}{L - \Delta} = \frac{\lambda_B}{\lambda_A} e^{-[G_{v(A)}^f - G_{v(B)}^f]/kT} \tag{3.9}$$

This equation determines the equilibrium solubility of B in A for a crystal in equilibrium with its vapor, and shows that there is a close connection between solubility and defect formation. Since we are restricting ourselves to dilute solutions, (3.9) shows that for an equilibrium alloy, $G^f_{v(A)}$ must always be sufficiently greater than $G^f_{v(B)}$ to satisfy the condition that $N_B \ll L$.

To obtain the equilibrium concentration of divacancies, consider the process in which two vacancies come together to form a divacancy, leaving the number of all other defects and the composition of the gas unchanged. In this case,

$$\frac{\partial \ln w}{\partial N_v}\bigg|_{2v\to v_2} = -2\frac{\partial \ln w_v}{\partial N_v} + \frac{\partial \ln w_{2v}}{\partial N_{2v}} \tag{3.10}$$

since the process destroys two vacancies and creates a divacancy. Only the free energy of the crystal changes in this process, so that the application of (1.4) gives

$$-2\frac{\partial \ln w_v}{\partial N_v} + \frac{\partial \ln w_{2v}}{\partial N_{2v}} = \frac{1}{kT}\frac{\partial G\{\ \}}{\partial N_{2v}}\bigg|_{2v \to v_2} \tag{3.11}$$

The right-hand side of (3.11) is the divacancy free energy of binding, because it represents the free-energy difference of a bound pair and a separated pair of vacancies, so it will be written as

$$G^{vv} \equiv \frac{\partial G\{\ \}}{\partial N_{2v}}\bigg|_{2v \to v_2} \tag{3.12}$$

Placing this in (3.12) and evaluating the derivatives of the statistical count gives

$$\frac{N_{2v}}{L-\Delta} = \frac{z}{2}\left(\frac{N_v}{L-\Delta}\right)^2 e^{G^{vv}/kT} \tag{3.13}$$

In a similar way, if we go through the process of forming a diimpurity from two isolated impurities, we obtain

$$\frac{N_{2B}}{L-\Delta} = \frac{z}{2}\left(\frac{N_B}{L-\Delta}\right)^2 e^{G^{BB}/kT} \tag{3.14}$$

where G^{BB} is the binding free energy for an impurity pair.

Finally, if the process of forming an impurity-vacancy complex by bringing together an isolated vacancy and an isolated B atom is applied to (1.4), the result is

$$\frac{N_{Bv}}{L-\Delta} = z\left(\frac{N_B}{L-\Delta}\right)\left(\frac{N_v}{L-\Delta}\right)e^{G^{Bv}/kT} \tag{3.15}$$

where G^{Bv} is the binding free energy for the impurity-vacancy complex.

All the equilibrium-defect concentrations have now been obtained. The numbers of one-center defects can be eliminated from the expressions for the two-center defects by using (3.4) and (3.9). A summary of the defect-concentration formula is given below, where we have written the atomic concentrations $N_{(i)}/(L-\Delta)$ as $C_{(i)}$ for convenience:

$$C_v = \lambda_A^{-1}e^{-G_{v(A)}^f/kT} \tag{3.16}$$

$$C_{2v} = \frac{z}{2}C_v^2 e^{G^{vv}/kT} \tag{3.17}$$

$$C_{2v} = \frac{z}{2}\lambda_A^{-2}e^{-G_{2v}^f/kT} \tag{3.18}$$

$$C_{Bv} = zC_BC_ve^{G^{Bv}/kT} \tag{3.19}$$

$$C_{2B} = \frac{z}{2}C_B{}^2e^{G^{BB}/kT} \tag{3.20}$$

$$C_B = \frac{\lambda_B}{\lambda_A}e^{-G_B{}^s/kT} \tag{3.21}$$

$$C_{Bv} = z\frac{\lambda_B}{\lambda_A{}^2}e^{-G_{Bv}{}^f/kT} \tag{3.22}$$

$$C_{2B} = \frac{z}{2}\left(\frac{\lambda_B}{\lambda_A}\right)^2 e^{-G_{2B}{}^f/kT} \tag{3.23}$$

In these equations, we have defined the following:
The free energy of divacancy formation:

$$G_{2v}{}^f = 2G_{v(A)}^f - G^{vv} \tag{3.24}$$

the free energy of impurity-atom dissolution:

$$G_B{}^s = G_{v(A)}^f - G_{v(B)}^f \tag{3.25}$$

the free energy of impurity-complex formation:

$$G_{Bv}{}^f = G_B{}^s + G_{v(A)}^f - G^{Bv} \tag{3.26}$$

the free energy of diimpurity formation:

$$G_{2B}{}^f = 2G_B{}^s - G^{BB} \tag{3.27}$$

Equations 3.16 to 3.23 were obtained for a dilute binary alloy in equilibrium with a vapor of its components. However, only the last three of these equations contain parameters that depend on the vapor-solid equilibrium. The first five equations describe internal-defect equilibria and are valid even if the crystal is not in equilibrium with its vapor, provided only that the crystal composition is a constant. In this case, of course, C_B is not given by (3.21), but is determined by the total impurity-concentration, C_{Bv} and C_{BB}. This partial equilibrium situation is quite frequent and is represented by a metal containing a small concentration of nonvolatile impurity whose vapor pressure is never high enough to alter the crystal composition. Equations 3.16 to 3.20 can therefore be taken to describe a system that is either only in internal equilibrium, or also in equilibrium with a vapor. If external equilibrium is in effect, the concentrations involving B atoms are not arbitrary, but are given by (3.21) to (3.23).

Note that the formation free energies were defined relative to the gas phase, since we are considering a crystal in equilibrium with its vapor. The vapor-crystal equilibrium is often neglected in defect studies, and the formation free energies are then defined by forming the defect by transferring atoms to or from the surface. The two definitions lead, of course, to the same defect-concentration formulae when only internal equilibria are considered. This is ensured by the presence of λ_A and λ_B in (3.16) to (3.20).

The total concentration of vacant sites and of B atoms is given by

$$C_v^{\ T} = C_v + 2C_{2v} + C_{Bv} \qquad (3.28)$$

$$C_B^{\ T} = C_B + 2C_{2B} + C_{Bv} \qquad (3.29)$$

or

$$C_v^{\ T} = C_v[1 + zC_v e^{G^{vv}/kT} + zC_B e^{G^{Bv}/kT}] \qquad (3.30)$$

$$C_B^{\ T} = C_B[1 + zC_B e^{G^{BB}/kT} + zC_v e^{G^{Bv}/kT}] \qquad (3.31)$$

It is sometimes necessary to compute C_B and C_{BB} from the total amount of B in the alloy, $C_B^{\ T}$. This is readily done by solving (3.31) to obtain

$$C_B = A\{[1 + BC_B^{\ T}]^{1/2} - 1\} \qquad (3.32)$$

where

$$A \equiv \frac{1 + zC_v e^{G^{Bv}/kT}}{2ze^{G^{BB}/kT}} \qquad (3.33)$$

and

$$B \equiv \frac{2}{A[1 + zC_v e^{G^{Bv}/kT}]} \qquad (3.34)$$

This gives C_B, and then C_{BB} can be obtained from (3.20). When the impurity concentration is low enough, C_B can be replaced by $C_B^{\ T}$ in the bracket of (3.31), provided the diimpurity binding energy is not too large. Then, solving for C_B gives

$$C_B \simeq C_B^{\ T}[1 + zC_B^{\ T} e^{G^{BB}/kT} + zC_v e^{G^{Bv}/kT}]^{-1} \qquad (3.35)$$

Once C_B is computed from (3.32) or (3.35), the other defect concentrations can be obtained from (3.16) to (3.20).

7.4 INTERNAL EQUILIBRIA FOR SUBSTITUTIONAL DEFECTS

For a monatomic metal with a nonvolatile impurity, the defect concentrations are governed by (3.16) to (3.20), along with (3.32) that relates the internal-defect equilibria to the total amount of impurity. From (3.16), (3.19), and (3.20) we have,

$$\frac{C_{2v}}{C_v^{\ 2}} = \frac{z}{2} e^{G^{vv}/kT} \qquad (4.1)$$

$$\frac{C_{Bv}}{C_B C_v} = ze^{G^{Bv}/kT} \qquad (4.2)$$

$$\frac{C_{2B}}{C_B^{\ 2}} = \frac{z}{2} e^{G^{BB}/kT} \qquad (4.3)$$

These equations have the form of equilibrium constants for chemical reactions. In fact, we can define a reaction in which two vacancies come together to form a divacancy, and its converse, in which a divacancy dissociates into two monovacancies. In chemical notation this reaction is represented by

$$V + V \rightleftarrows V_2 \qquad (4.4)$$

and the right-hand side of (4.1) is seen to be just the equilibrium constant for this reaction. Similarly two impurity atoms can form a diimpurity, or a diimpurity can dissociate. The corresponding reaction is

$$B + B \rightleftarrows B_2 \qquad (4.5)$$

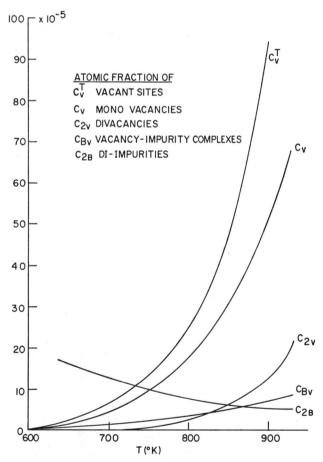

Fig. 7.2. Defect concentrations in aluminum containing 10^{-3} atomic fraction impurity.

with (4.3) as its equilibrium constant expression. The vacancy-impurity reaction is

$$B + V \rightleftarrows BV \qquad (4.6)$$

and its equilibrium constant is given by (4.2).

This scheme clearly shows the interdependence of the defect concentrations. Thus, adding B atoms to the crystal shifts the reaction (4.6) to the right according to (4.3), thereby increasing the number of impurity-vacancy complexes and the total number of vacant sites. Also, adding B atoms shifts the reaction (4.5) to the right, creating more impurity pairs. A rise in temperature, however, decreases the ratio of impurity pairs to single impurity atoms, since the equilibrium constant defined by (4.3) decreases with increasing temperature.

The addition of impurities has a significant effect on the defect equilibria. In Figure 7.2, the defect concentrations in aluminum are plotted for a total impurity concentration of 10^{-3} atomic fraction. For these calculations, the data in Table 6.1 were used, and it was assumed that both the diimpurity and vacancy-impurity binding energies were 0.2 eV, and that the binding entropies were zero. The figure shows that between 10 and 20% of the impurity is tied up in impurity pairs, and that between 3 and 12% are tied up in vacancy-impurity complexes, depending on the temperature. The impurity has also increased the total number of vacant sites. In fact, at low temperatures, the majority of vacant sites are in impurity-vacancy complexes. Even at high temperatures, the impurity has increased the vacancy concentration by over 10%.

The magnitude of the effect of impurities on the defect equilibria depends, of course, on the impurity concentration, but also on the value of the binding energies. These are not known with any accuracy, but they probably range from 0 to 0.5 eV.

7.5 QUENCHED-IN RESISTIVITY OF DILUTE BINARY ALLOYS

The electrical resistivity of a metal has its origin in deviations from the perfect crystal structure, which result in the electron-scattering mechanisms responsible for resistivity (see Chapter 4). Since point defects in metals scatter electrons, the electrical resistivity can be used to study defect properties.

The electrical resistivity is a function of temperature, impurity content, and defect concentrations. At normal temperatures, the effect of temperature is large since many lattice vibrations of all wavelengths are excited. At low temperatures, however, only a few phonons of long wavelength are active. Thus, the resistivity decreases with decreasing temperature. Because of the existence of impurities and defects, the resistivity retains a measurable value even at very low temperatures. It is this temperature independent intrinsic resistivity that is of interest for defect studies.

For purposes of illustration, consider a pure metal containing only vacancies and dislocations. We assume the dislocation density to be independent of temperature. Now consider the following experiment: the metal is maintained at some high temperature T_q long enough to insure that defect equilibrium is established. The temperature of the metal is then suddenly changed to a lower temperature T_f. The resistivity is then measured at liquid-helium temperatures. It will be assumed that T_f is low enough to immobilize the defects, and that the rate of quenching from T_q to T_f is so rapid that all defects present at T_q are frozen in at T_f. We can therefore write for the electrical resistivity of the quenched specimen

$$\rho(T_q \to T_f) = \rho(D) + \rho_v C_v(T_q) \tag{5.1}$$

where $\rho(D)$ is the resistivity arising from fixed defects such as dislocations, ρ_v is the specific resistivity per atomic fraction of vacancies, and C_v is the vacancy concentration. Phonon contributions to the resistivity do not appear since we are making the measurement at liquid-helium temperature. The vacancy concentration is characteristic of that at the initial quench temperature T_q. Now if another identical metal specimen is held at T_f long enough to insure point-defect equilibrium, its vacancy concentration will be $C_v(T_f)$, and the resistivity (measured at liquid-helium temperatures) will be

$$\rho(T_f) = \rho(D) + \rho_v C_v(T_f) \tag{5.2}$$

We assume that T_f is low enough so that vacancies do not anneal out while bringing the specimen to liquid-helium temperature. Taking the difference of (5.1) and (5.2) gives

$$(\Delta\rho)_{T_q} = \rho(T_q \to T_f) - \rho(T_f)$$
$$= \rho_v[C_v(T_q) - C_v(T_f)] \tag{5.3}$$

We will call $(\Delta\rho)_{T_q}$ the quenched-in resistivity. Note that T_f can be chosen so that $C_v(T_f)$ is negligible. The quenched-in resistivity is then proportional to the vacancy concentration at the quench temperature T_q, and substituting the vacancy-concentration formulas into (5.3) gives

$$(\Delta\rho)_{T_q} = \rho_v e^{-G_v{}^f/kT_q} \tag{5.4}$$

Now replace $G_v{}^f$ by $U_v{}^f - TS_v{}^f$ and take logarithms. The result is

$$\ln(\Delta\rho)_{T_q} = \ln \rho_v + \frac{S_v{}^f}{k} - \frac{U_v{}^f}{kT_q} \tag{5.5}$$

It is clear that the vacancy-formation energy can be obtained from experimental data by plotting $\ln (\Delta\rho)_{T_q}$ versus $1/T_q$ and computing the slope. In general, ρ_v is not known with accuracy, so S_v^f cannot be obtained from the intercept of the $\ln (\Delta\rho)_{T_q}$ versus $1/T_q$ plot. Thus, while the quenched-in resistivity can give defect-formation energies, it leads to unreliable results for total defect concentrations unless it is combined with some other type of data.

The idealized quenching experiment described above is difficult to realize in practice, the major problem being the inability to achieve an infinitely fast quench. Thus, in a real experiment the defect concentration after quenching is not characteristic of that at T_q since some defects are lost as the specimen cools. Nevertheless, by performing the experiments at several quenching rates and extrapolating the results to infinite quenching rate, significant data can be obtained. In fact, careful quenching studies have been performed on pure gold, and gold containing known amounts of impurity. It is therefore worthwhile to consider the case of the dilute binary alloy, using the results of Section 3.

For a dilute substitutional binary alloy as defined in Section 3, the point-defect contribution to the resistivity is the sum of contributions from vacancies, divacancies, solute atoms, vacancy-solute complexes, and solute-solute pairs. Thus, if we rewrite (5.3) for this case as

$$(\Delta\rho)_{T_q} = \rho(T_q \to T_f) - \rho(T_f) \tag{5.6}$$

then $\rho(T_q \to T_f)$, which is the resistivity of a specimen quenched from T_q to T_f and measured at liquid helium temperature, is

$$\rho(T_q \to T_f) = \rho_v C_v(T_q) + \rho_{2v} C_{2v}(T_q) + \rho_{Bv} C_{Bv}(T_q)$$
$$+ \rho_B C_B(T_q) + \rho_{2B} C_{2B}(T_q) \tag{5.7}$$

and $\rho(T_f)$, the resistivity of a specimen with defect concentrations characteristic of the final temperature, is

$$\rho(T_f) = \rho_v C_v(T_f) + \rho_{2v} C_{2v}(T_f) + \rho_{Bv} C_{Bv}(T_f) + \rho_B C_B(T_f) + \rho_{2B} C_{2B}(T_f) \tag{5.8}$$

In these equations, the defect concentrations are given by (3.16) to (3.20), and ρ_v, ρ_{2v}, ρ_{Bv}, ρ_B, and ρ_{2B} are the resistivities per atom fraction of vacancies, divacancies, vacancy-impurity complexes, impurity atoms, and impurity pairs, respectively.

It is obvious that for an impure metal, a plot of $\ln (\Delta\rho)T_q$ versus $1/T_q$ is not linear.

Careful measurements of the quenched-in resistivity have been made for gold containing tin and silver as impurities.* These experiments were

* Bass, J., "Quenched Resistance in Dilute Gold-Tin, Gold-Silver Alloys" *Phys. Rev.* **137**, A765 (1965).

analyzed* according to (5.6) to (5.8) in conjunction with (3.16) to (3.20). The results for gold with tin impurity are shown in Figure 7.3. Since a number of the parameters in the above equations were unknown, the results of Figure 7.3 had to be obtained by a computer programmed parametric

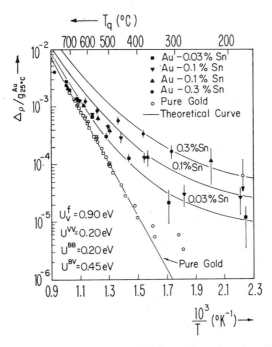

Fig. 7.3. Quenched-in resistivity in impure gold (experimental results of Bass).

curve-fitting process.

The deviation from a linear Arrhenius plot produced by a small amount of tin is remarkable. In fact, detailed analysis† shows that the large deviation from linearity at low temperature is the result of the existence of tin-tin pairs. If there were no tendency for pairs to form, the curves would be much closer to being linear.

The above results show that careful and detailed quenching experiments can give valuable results. The values for the binding energies that were consistent with experiment were $U^{vv} = 0.2$ eV, $U^{BB} = 0.2$ eV, and $U^{Bv} = 0.45$ eV.

* See reference 39 in Bibliography.
† See reference 39 in Bibliography.

7.6 SOME GENERAL THEORY INCLUDING INTERSTITIAL AND MIXED-LATTICE DEFECTS

In this section, the basic theory of defect concentrations in a dilute alloy will be summarized. The results previously obtained for substitutional defects will be generalized to include all the substitutional, interstitial, and mixed-lattice defects described in Section 1.

In Section 2, it was pointed out that any defect, whether of the substitutional, interstitial, or mixed-lattice type, contributes a factor given by (2.34) to the statistical count. Accordingly, for the general case we can write the number of ways of constructing the defect crystal as

$$w = \prod_i w_{(i)} = \prod_i e^{N_{(i)}} \left[\frac{\omega_{(i)}(L - \Delta)}{N_{(i)}} \right]^{N_{(i)}} \tag{6.1}$$

where the product is taken over all types of mono- and didefects. This result is restricted to lattices in which the number of substitutional sites is equal to the number of interstitial sites. However, it is readily generalized to lattices in which this is not true by letting L always be the number of substitutional sites and appropriately redefining $\omega_{(i)}$.

Examination of (3.16) to (3.23) shows that all the defect concentrations can be written in the following equivalent forms

$$C_{(i)} = \omega_{(i)}\lambda_{(i)}e^{-G_{(i)}^f/kT}$$
$$C_{(i)} = \omega_{(i)}e^{-G^{(i)}/kT} \tag{6.2}$$

These formulae are also correct for interstitial and mixed-lattice defects. The $G_{(i)}^f$ are defect free energies of formation referred to the vapor phase, while $G^{(i)}$ are defect formation free energies referred to the crystal surface. The $\omega_{(i)}$ are the rotational factors and $\lambda_{(i)}$ are products of absolute activities. The interpretations of the $\lambda_{(i)}$, $G_{(i)}^f$, $G^{(i)}$, and $\omega_{(i)}$ are obtained in precisely the same way for interstitial and mixed-lattice defects as for substitutional defects. Binding free energies also exist in the general case that are analogous to those already defined for substitutional two-center defects. These definitions are summarized in Table 7.1.

Not all of the defects listed in Table 7.1 will be important in all crystals. The concentration of each type of defect depends, of course, on their free energies of formation, and it is necessary to know these in order to decide which defects predominate in a particular system. For close-packed metals, the vacancy is the major intrinsic defect, and many impurities enter the lattice substitutionally. However, there is strong evidence that cadmium dissolves in lead by the so-called dissociative mechanism in which the cadmium exists at both substitutional and interstitial lattice sites. Furthermore, it appears

Table 7.1. Definitions of Defect-Formation Free Energies, Binding Free Energies, Rotational Factors, and Absolute Activity Factors

Defect	Formation Free Energy	Binding Free Energy	$\omega_{(i)}$	$\lambda_{(i)}$
V	$G^f_{v(A)}$	-	1	λ_A^{-1}
V₂	$G^f_{2v} = 2G^f_{v(A)} - G^{vv}$	$G^{vv} = 2G^f_{v(A)} - G^f_{2v}$	$z/2$	λ_A^{-2}
A'	G^f_I	-	1	λ_A
A'₂	$G^f_{2I} = 2G^f_I - G^{II}$	$G^{II} = 2G^f_I - G^f_{2I}$	$z/2$	λ_A^2
B	$G^s_B = G^f_{v(A)} - G^f_{v(B)}$	-	1	$\lambda_A^{-1}\lambda_B$
B₂	$G^f_{2B} = 2G^s_B - G^{BB}$	$G^{BB} = 2G^s_B - G^f_{2B}$	$z/2$	$\lambda_A^{-2}\lambda_B^2$
B'	$G^s_{B'}$	-	1	λ_B
B'₂	$G^f_{2B'} = 2G^s_{B'} - G^{B'B'}$	$G^{B'B'} = 2G^s_{B'} - G^f_{2B'}$	$z/2$	λ_B^2
BV	$G^f_{Bv} = G^s_B + G^f_{v(A)} - G^{Bv}$	$G^{Bv} = G^s_B + G^f_{v(A)} - G^f_{Bv}$	z	$\lambda_A^{-2}\lambda_B$
B'V	$G^f_{B'v} = G^s_{B'} + G^f_{v(A)} - G^{B'v}$	$G^{B'v} = G^s_{B'} + G^f_{v(A)} - G^f_{B'v}$	z	$\lambda_A^{-1}\lambda_B$
A'B'	$G^f_{IB'} = G^f_I + G^s_{B'} - G^{IB'}$	$G^{IB'} = G^f_I + G^s_B - G^f_{IB'}$	z	$\lambda_A\lambda_B$
A'V	$G^f_{Iv} = G^f_I + G^f_{v(A)} - G^{Iv}$	$G^{Iv} = G^f_{v(A)} + G^f_I - G^f_{Iv}$	z	1
A'B	$G^f_{IB} = G^f_I + G^s_B - G^{IB}$	$G^{IB} = G^f_I + G^s_B - G^f_{IB}$	z	λ_B
BB'	$G^f_{BB'} = G^s_B + G^s_{B'} - G^{BB'}$	$G^{BB'} = G^s_B + G^s_{B'} - G^f_{BB'}$	$z/2$	$\lambda_A^{-1}\lambda_B^2$

that there is a strong attraction between the interstitial cadmium impurity and a vacancy. In lead containing small amounts of impurity, therefore, the point defects that must be considered are V, V_2, B, B_2, B', B'_2, BV, $B'V$, and BB', in the notation of Table 7.1. Of these, it is probably safe to assume that B_2, B'_2, and BB' are in much smaller concentration than any of the others. It is of interest to write out the concentration formulae for the dissociative mechanism explicitly, since diffusion experiments exist that have been interpreted by this mechanism. Thus, using the notation of Table 7.2, (6.2) gives

$$C_v = \lambda_A^{-1} e^{-G_{v(A)}{}^f/kT} \tag{6.3}$$

$$C_{2v} = \frac{z}{2} \lambda_A^{-2} e^{-G_{2v}{}^f/kT} \tag{6.4}$$

$$C_B = \frac{\lambda_B}{\lambda_A} e^{-G_B{}^s/kT} \tag{6.5}$$

$$C_{B'} = \lambda_B e^{-G_{B'}{}^s/kT} \tag{6.6}$$

$$C_{Bv} = \frac{\lambda_B}{\lambda_A{}^2} z e^{-G_{Bv}{}^f/kT} \tag{6.7}$$

$$C_{B'v} = \frac{\lambda_B}{\lambda_A} z e^{-G_{B'v}{}^f/kT} \tag{6.8}$$

These equations represent the full equilibrium in which the impurity concentration is determined by the metal vapor in contact with the crystal. To obtain the internal equilibrium equations, consider the following reactions:

$$V + V \rightleftarrows V_2 \tag{6.9}$$

$$B + V \rightleftarrows BV \tag{6.10}$$

$$B' + V \rightleftarrows B'V \tag{6.11}$$

$$B \rightleftarrows B' + V \tag{6.12}$$

The first three of these reactions represent the usual didefect formation from the monodefects. The last reaction represents the transfer of the impurity among the substitutional and interstitial sites. Applying the law of mass action to the above reactions gives

$$\frac{C_{2v}}{C_v{}^2} = K_1 \tag{6.13}$$

$$\frac{C_{Bv}}{C_B C_v} = K_2 \tag{6.14}$$

$$\frac{C_{B'v}}{C_{B'} C_{v'}} = K_3 \tag{6.15}$$

$$\frac{C_{B'} C_v}{C_B} = K_4 \tag{6.16}$$

K_1, K_2, K_3, and K_4 are equilibrium constants that are readily obtained by substituting from (6.3) to (6.8) into the left-hand sides of (6.13) to (6.16). These equations, along with the vacancy-concentration formulas completely determine the conditions of internal equilibrium. Using Table 7.2 it is easy to show that the results are

$$C_v = e^{-G_{vA}{}^f/kT} \tag{6.17}$$

$$C_{2v} = \frac{z}{2} C_v{}^2 e^{G^{vv}/kT} \tag{6.18}$$

$$C_{Bv} = z C_B C_v e^{G^{Bv}/kT} \tag{6.19}$$

$$C_{B'v} = z C_{B'} C_v e^{G^{B'v}/kT} \tag{6.20}$$

$$\frac{C_{B'} C_v}{C_B} = \lambda_A e^{-[G^s_{B'} + G^f_{v(B)}]/kT} \tag{6.21}$$

The above equations are, of course, valid for both the full-scale equilibrium, and the case in which no metal atoms can be transferred to the vapor so that only internal equilibrium exists. In the latter case, the impurity concentration is a constant $C_B{}^T$, and we have

$$C_B{}^T = C_B + C_{B'} + C_{Bv} + C_{B'v} \tag{6.22}$$

7.7 THERMODYNAMICS OF THE GENERAL DILUTE ALLOY

The contribution of all the mono- and didefects to the thermodynamic functions can be obtained by a method similar to that used for vacancies in Chapter 6. This method consists of writing the free energy of the crystal as a sum of a configurationless part and the contribution from the configurational entropy. The defect-concentration formulae are then substituted into the configurational contribution and, finally, the configurationless free energy is expressed as a linear function of defect-formation energies.

The crystal free energy is

$$G_c(N_A, N_B, N_{(i)}) = G\{N_A, N_B, N_{(i)}\} - kT \ln w(N_A, N_B, N_{(i)}) \tag{7.1}$$

The configurationless free energy on the right-hand side of (7.1) is the same as that in (1.2) and is determined from the following partition function:

$$e^{-G\{N_A, N_B, N_{(i)}\}/kT} = \sum_{j,i} \exp\left\{-\frac{1}{kT}[E_j + PV_i]\right\} \tag{7.2}$$

where the sum is over all energy states and all possible volumes of the crystal

Now substitute (6.1) into (7.1) to get

$$G_c(N_A, N_B, N_{(i)}) = G\{N_A, N_B, N_{(i)}\}$$
$$- kT \sum_i N_{(i)} - kT \sum_i N_{(i)} \ln \left[\frac{\omega_{(i)}(L - \Delta)}{N_{(i)}} \right] \quad (7.3)$$

But from (6.2)

$$\ln \frac{\omega_{(i)}}{C_{(i)}} \equiv \ln \left[\frac{\omega_{(i)}(L - \Delta)}{N_{(i)}} \right] = \frac{G^f_{(i)}}{kT} - \ln \lambda_{(i)} \quad (7.4)$$

so (7.3) becomes

$$G_c(N_A, N_B, N_{(i)}) = G\{N_A, N_B, N_{(i)}\}$$
$$- \sum_{(i)} N_{(i)} G^f_{(i)} + kT \sum_i N_{(i)} \ln \lambda_i - kT \sum_i N_{(i)} \quad (7.5)$$

At this point, we wish to express the configurationless free energy as a linear function of the number of defects, choosing the perfect crystal containing only A atoms as the reference state. Starting with a perfect crystal (N_A atoms of type A), each vacancy that is formed increases the free energy by ($G^f_{v(A)} + \mu_A$) since the number of A atoms is not changed by vacancy formation, and $G^f_{v(A)}$ represents a free energy for removal of the A atom to the vapor phase. The free energy to add a B atom to the perfect crystal is ($G_B{}^s + \mu_A$) since $G_B{}^s$ is the free energy to exchange a B atom in the gas phase with an A atom in the crystal, and μ_A is the free energy required to maintain the number of A atoms constant. In a similar way, the change in free energy upon adding defects of all the types listed in Table 7.2 can be obtained from the definitions of the $G^f_{(i)}$. When this is done, and $G\{N_A, N_B, N_{(i)}\}$ is written as a linear function of these free-energy changes, it is found that the first three terms in (7.5) reduce to $G^0\{N_A\}$, the configurationless free energy of a perfect crystal containing N_A atoms of type A. The defect contributions to $G\{N_A, N_B, N_{(i)}\}$ are cancelled by the second and third terms in (7.5). Therefore, we can write (7.5) as

$$G_c(N_A, N_B, N_{(i)}) = G^0\{N_A\} - kT \sum_i N_{(i)} \quad (7.6)$$

where $G^0\{N_A\}$ is the free energy of the perfect crystal. This equation is just a generalization of (3.23) and (6.27) of Chapter 6.

The thermodynamic functions can now be obtained from (7.6) by application of the formulae (4.1) to (4.8) of Chapter 6, in the same way as for a crystal containing only vacancies and a crystal containing vacancies, divacancies, and interstitials (see Sections 4 and 6, Chapter 6). It is found that all defect contributions are additive and of similar form. The results are:

$$S_c = S^0 + \sum_i \frac{N_{(i)}}{T} [H^{(i)} + kT] \quad (7.7)$$

$$V_c = V^0 + \sum_i N_{(i)} V^{(i)} \quad (7.8)$$

$$U_c = U^0 + \sum_i N_{(i)} U^{(i)} \tag{7.9}$$

$$H_c = H^0 + \sum_i N_{(i)} H^{(i)} \tag{7.10}$$

$$C_P{}^c = C_P{}^0 + \sum_i N_{(i)} \left[C_P^{(i)} + \frac{H^{(i)^2}}{kT^2} \right] \tag{7.11}$$

$$C_V{}^c = C_V{}^0 + \sum_i N_{(i)} \left[C_V^{(i)} + \frac{H^{(i)^2}}{kT^2} \right] + (C_V{}^0 - C_P{}^0) \sum_i \frac{N_{(i)} V^{(i)}}{V_0}$$
$$\times \left[\frac{2\alpha^{(i)}}{\alpha_0} + \frac{2H^{(i)}}{\alpha^0 kT^2} - \frac{\beta^{(i)}}{\beta^0} - \frac{V^{(i)}}{\beta^0 kT} \right] \tag{7.12}$$

$$\alpha_c = \alpha^0 + \sum_i \frac{N_{(i)} V^{(i)}}{V_0} \left[\alpha^{(i)} - \alpha^0 + \frac{H^{(i)}}{kT^2} \right] \tag{7.13}$$

$$\beta_c = \beta^0 + \sum_i \frac{N_{(i)} V^{(i)}}{V_0} \left[\beta^{(i)} - \beta^0 + \frac{V^{(i)}}{kT} \right] \tag{7.14}$$

In these equations, the left-hand sides represent crystal properties obtained from $G_c(N_A, N_B, N_{(i)})$ by (4.1) to (4.8) of Chapter 6. We have omitted the argument $(N_A, N_B, N_{(i)})$ in these quantities for the sake of brevity. The thermodynamic functions with the superscript (i) are derived from the defect-formation free energies $G^{(i)}$ listed in the last column of Table 7.1 according to

$$S^{(i)} = -\left(\frac{\partial G^{(i)}}{\partial T} \right)_P \tag{7.15}$$

$$V^{(i)} = \left(\frac{\partial G^{(i)}}{\partial P} \right)_T \tag{7.16}$$

$$U^{(i)} = G^{(i)} + TS^{(i)} - PV^{(i)} \tag{7.17}$$

$$H^{(i)} = G^{(i)} + TS^{(i)} \tag{7.18}$$

$$C_P^{(i)} = \left(\frac{\partial H^{(i)}}{\partial T} \right)_P \tag{7.19}$$

$$C_V^{(i)} = \left(\frac{\partial U^{(i)}}{\partial T} \right)_V \tag{7.20}$$

$$\alpha^{(i)} = \frac{1}{V^{(i)}} \left(\frac{\partial V^{(i)}}{\partial T} \right)_P \tag{7.21}$$

$$\beta^{(i)} = -\frac{1}{V^{(i)}} \left(\frac{\partial V^{(i)}}{\partial P} \right)_T \tag{7.22}$$

In the equations for the thermal expansion and compressibility, quantities to the second order in the defect concentrations were neglected.

chapter 8

ATOMIC MIGRATION IN
SIMPLE CRYSTALS

8.1 THE EMPIRICAL LAWS OF DIFFUSION

The phenomenological description of diffusion is embodied in Fick's two laws, which are empirical statements that relate the diffusive flow of matter to concentration gradients. To illustrate these laws, consider a single phase dilute binary alloy in which the constituents are inhomogeneously distributed, and let C be the concentration of the minor constituent. The concentration is a function of position and time, and the concentration gradient induces a flow of matter. Fick's first law states that the flux of diffusing species in a given direction has a magnitude proportional to the concentration gradient in that direction. That is, if J_1 is the flux along the x_1 axis, then

$$J_1 = -D_1 \frac{\partial C}{\partial x_1} \tag{1.1}$$

where D_1 is a proportionality factor called the diffusion coefficient. The negative sign expresses the fact that diffusion occurs from regions of high to regions of low-concentration gradient. In fact, in crystals of low symmetry, it is found that the flux in the x_1 direction is also proportional to the gradients along the other coordinate axes. Thus, if (x_1, x_2, x_3) are the points in a Cartesian coordinate system, then the flux along x_1 is given by the generalized Fick's first law as

$$J_1 = -D_{11} \frac{\partial C}{\partial x_1} - D_{12} \frac{\partial C}{\partial x_2} - D_{13} \frac{\partial C}{\partial x_3} \tag{1.2}$$

Similarly, the flux along the x_2 and x_3 axes are

$$J_2 = -D_{21} \frac{\partial C}{\partial x_1} - D_{22} \frac{\partial C}{\partial x_2} - D_{23} \frac{\partial C}{\partial x_3} \qquad (1.3)$$

$$J_3 = -D_{31} \frac{\partial C}{\partial x_1} - D_{32} \frac{\partial C}{\partial x_2} - D_{33} \frac{\partial C}{\partial x_3} \qquad (1.4)$$

The D's are generalized diffusion coefficients coupling the flux in one direction with the concentration gradient in another. Thus $-D_{ij}$ is the flux in the ith direction arising from a unit-concentration gradient in the jth direction.

Fick's first law can be written in a compact form by using vector notation. If $\mathbf{i}_1$, $\mathbf{i}_2$, and $\mathbf{i}_3$ are three orthogonal unit vectors along the x_1, x_2, and x_3 directions, the flux vector is

$$\mathbf{J} \equiv J_1 \mathbf{i}_1 + J_2 \mathbf{i}_2 + J_3 \mathbf{i}_3 \qquad (1.5)$$

Also, we define the diffusion dyadic D by

$$\begin{aligned} \mathbf{D} = \; & D_{11} \mathbf{i}_1 \mathbf{i}_1 + D_{12} \mathbf{i}_1 \mathbf{i}_2 + D_{13} \mathbf{i}_1 \mathbf{i}_3 \\ & + D_{21} \mathbf{i}_2 \mathbf{i}_1 + D_{22} \mathbf{i}_2 \mathbf{i}_2 + D_{23} \mathbf{i}_2 \mathbf{i}_3 \\ & + D_{31} \mathbf{i}_3 \mathbf{i}_1 + D_{32} \mathbf{i}_3 \mathbf{i}_2 + D_{33} \mathbf{i}_3 \mathbf{i}_3 \end{aligned} \qquad (1.6)$$

Using the definitions (1.5) and (1.6), Fick's first law as represented by (1.2) to (1.4) can be condensed into the single form

$$\mathbf{J} = -\mathbf{D} \cdot \nabla C \qquad (1.7)$$

Fick's second law is obtained from (1.7) by use of the equation of continuity that states that

$$\frac{\partial C}{\partial t} + \nabla \cdot \mathbf{J} = 0 \qquad (1.8)$$

This is just an expression of the law of conservation of matter and states that any change of concentration in a volume element is a result of the difference in matter flow into and out of the volume element.

Combining (1.7) and (1.8) yields the generalized form of Fick's second law:

$$\frac{\partial C}{\partial t} = \nabla \cdot \mathbf{D} \cdot \nabla C \qquad (1.9)$$

The economy of the vector-dyadic notation is well illustrated by writing (1.9) in expanded form. If the diffusion coefficients are independent of concentration, (1.9) is

$$\frac{\partial C}{\partial t} = D_{11}\frac{\partial^2 C}{\partial x_1{}^2} + D_{12}\frac{\partial^2 C}{\partial x_1\,\partial x_2} + D_{13}\frac{\partial^2 C}{\partial x_1\,\partial x_3}$$

$$+ D_{21}\frac{\partial^2 C}{\partial x_2\,\partial x_1} + D_{22}\frac{\partial^2 C}{\partial x_2{}^2} + D_{23}\frac{\partial^2 C}{\partial x_2\,\partial x_3}$$

$$+ D_{31}\frac{\partial^2 C}{\partial x_3\,\partial x_1} + D_{32}\frac{\partial^2 C}{\partial x_3\,\partial x_2} + D_{33}\frac{\partial^2 C}{\partial x_3{}^2} \tag{1.10}$$

The above results are referred to an arbitrary coordinate system whose unit vectors are (i_1, i_2, i_3). However, it is always possible to choose a coordinate system such that the D_{ij} in (1.6) vanish unless $i = j$. We will assume that this has been done so the (1.6) becomes

$$D = D_1 i_1 i_1 + D_2 i_2 i_2 + D_3 i_3 i_3 \tag{1.11}$$

D_1, D_2, and D_3 are called the principal diffusion coefficients and the co-ordinate axes defined by i_1, i_2, and i_3 are called the principal diffusion axes.

In isotropic and cubic systems $D_1 = D_2 = D_3$, and Fick's laws take particularly simple forms, especially if the diffusion coefficient is constant. These forms are:

$$\mathbf{J} = -D\,\nabla C \tag{1.12}$$

$$\frac{\partial C}{\partial t} = D\,\nabla^2 C \tag{1.13}$$

The value of these results lies primarily in the fact that they can be used to compute the distribution of the diffusing species in space and time. Thus, if the initial boundary conditions are known, (1.13) can be solved for $C(\mathbf{r}, t)$. Also, it is possible to establish experimental conditions corresponding to particularly simple solutions of (1.13), thereby making it possible to determine the diffusion coefficient from experiment.

The purpose of theory at this point is to provide a derivation of Fick's laws from the statistical-kinetic theory of matter, and to relate the diffusion coefficients to the atomistic mechanisms responsible for diffusion. This will be done in the following sections. First, it will be shown that the diffusion coefficient is proportional to the mean square of atomic displacements per unit time by the method of transition probabilities. Then the mean square displacement will be related to defect concentrations and atomic jump frequency. Finally, the jump frequency will be treated by statistical mechanics and expressed in terms of a migration energy and crystal vibration frequencies. The detailed treatment will be restricted to simple interstitial and tracer diffusion in metals, but will be sufficient to illustrate the general principles of the statistical-mechanical theory of diffusion.

8.2 TRANSITION PROBABILITIES AND FICK'S LAWS

Consider a volume element $d\mathbf{r}$ in the crystal centered about a point defined by the position vector $\mathbf{r}$. We wish to count the number of particles of the diffusing species entering and leaving this volume per unit time. To do this define a conditional transition probability density per unit time $\wedge(\mathbf{r} \mid \mathbf{R})$ such that $\wedge(\mathbf{r} \mid \mathbf{R}) d\mathbf{R} dt$ is the probability that, if a particle is at $\mathbf{r}$, it will jump a distance $\mathbf{R}$ into the volume element $d\mathbf{R}$ during a time dt. Thus, since $C(\mathbf{r}, t) d\mathbf{r}$ is the number of particles in $d\mathbf{r}$, then the number of particles leaving $d\mathbf{r}$ in time dt is

$$\int_{\mathbf{R}} C(\mathbf{r}, t) \wedge (\mathbf{r} \mid \mathbf{R}) \, d\mathbf{R} \, d\mathbf{r} \, dt \qquad (2.1)$$

where the integral is taken over all possible migration vectors $\mathbf{R}$.

Now consider a volume element about a point $(\mathbf{r} - \mathbf{R})$. The number of particles moving from this volume element in time dt to the original volume element $d\mathbf{r}$ is

$$C(\mathbf{r} - \mathbf{R}, t) \wedge (\mathbf{r} - \mathbf{R} \mid \mathbf{R}) \, d\mathbf{R} \, d\mathbf{r} \, dt \qquad (2.2)$$

and integrating this over all $\mathbf{R}$ gives the number of particles entering the volume element $d\mathbf{r}$ in time dt as

$$\int_{\mathbf{R}} C(\mathbf{r} - \mathbf{R}, t) \wedge (\mathbf{r} - \mathbf{R} \mid \mathbf{R}) \, d\mathbf{R} \, d\mathbf{r} \, dt \qquad (2.3)$$

Subtracting (2.1) from (2.3) and dividing through by $d\mathbf{r} \, dt$ gives the net rate of increase of the concentration at $\mathbf{r}$:

$$\frac{\partial C}{\partial t} = \int_{\mathbf{R}} [C(\mathbf{r} - \mathbf{R}, t) \wedge (\mathbf{r} - \mathbf{R} \mid \mathbf{R}) - C(\mathbf{r}, t) \wedge (\mathbf{r} \mid \mathbf{R})] \, d\mathbf{R} \qquad (2.4)$$

Fick's second law can be obtained from (2.4) by expanding the first term in the integrand in a Taylor expansion and retaining only the first two terms in the series. That is, we take

$$C(\mathbf{r} - \mathbf{R}, t) \wedge (\mathbf{r} - \mathbf{R} \mid \mathbf{R})$$
$$= C(\mathbf{r}, t) \wedge (\mathbf{r} \mid \mathbf{R}) - \wedge (\mathbf{r} \mid \mathbf{R})\mathbf{R} \cdot \nabla C + \tfrac{1}{2} \wedge (\mathbf{r} \mid \mathbf{R})(\mathbf{R} \cdot \nabla)^2 C \qquad (2.5)$$

This presumes that the transition probability is very small for large $\mathbf{R}$, a condition certainly satisfied in crystals. Also, we have taken $\wedge(\mathbf{r}, \mathbf{R})$ to be independent of $\mathbf{r}$. This will be true in the cases we are considering. Substituting (2.5) into (2.4) gives

$$\frac{\partial C}{\partial t} = - \int_{\mathbf{R}} \wedge (\mathbf{r} \mid \mathbf{R})\mathbf{R} \cdot \nabla C \, d\mathbf{R} + \tfrac{1}{2} \int_{\mathbf{R}} \wedge (\mathbf{r} \mid \mathbf{R})(\mathbf{R} \cdot \nabla)^2 C \, d\mathbf{R} \qquad (2.6)$$

Remember that the derivatives of the concentration entering into (2.6) are evaluated at **r** and are independent of **R**. Also, we make the fundamental assumption that the probability of a particle migration **R** is equal to that for a migration −**R**. This means that we are taking individual migration events to be completely random. In this case, the first integral on the right of (2.6) vanishes, and

$$\frac{\partial C}{\partial t} = \frac{1}{2} \int_{\mathbf{R}} \Lambda(\mathbf{r} \mid \mathbf{R})(\mathbf{R} \cdot \nabla)^2 C \, d\mathbf{R} \tag{2.7}$$

It must be pointed out that there are many diffusion phenomena of interest in which the transition probability is not isotropic and, therefore, the first integral in (2.6) does not vanish. These include diffusion in strain fields, gravitational fields, electric fields, and temperature gradients. By taking $\Lambda(\mathbf{r} \mid \mathbf{R})$ to be isotropic, we exclude these cases from current consideration.

If the components of **R** are X_1, X_2, and X_3, and x_1, x_2, and x_3 are the components of **r**, then

$$(\mathbf{R} \cdot \nabla)^2 C = \sum_{i,j} X_i X_j \frac{\partial^2 C}{\partial x_i \, \partial x_j} \tag{2.8}$$

and (2.7) can be written as

$$\frac{\partial C}{\partial t} = \frac{1}{2} \sum_{i,j} \langle X_i X_j \rangle \frac{\partial^2 C}{\partial x_i \, \partial x_j} \tag{2.9}$$

where

$$\langle X_i X_j \rangle \equiv \int_{\mathbf{R}} X_i X_j \, \Lambda \, (\mathbf{r} \mid \mathbf{R}) \, d\mathbf{R} \tag{2.10}$$

are the mean values of the quadratic migration distance products $X_i X_j$, the averages being taken over all possible migration distances.

Comparing (2.9) with (1.10), we see that the diffusion coefficients are defined by

$$D_{ij} = \tfrac{1}{2} \langle X_i X_j \rangle \tag{2.11}$$

If the coordinate axes (x_1, x_2, x_3) are chosen to be the principal axes of diffusion, the mixed products are zero, and (2.9) becomes

$$\frac{\partial C}{\partial t} = \frac{1}{2} \sum_i \langle X_i^2 \rangle \frac{\partial^2 C}{\partial x_i^2} \tag{2.12}$$

This is Fick's second law, with the principal diffusion coefficients D_i, being given by

$$D_i = \tfrac{1}{2} \langle X_i^2 \rangle \tag{2.13}$$

Equation 2.13 is a fundamental relation that connects the diffusion coefficient to the mean square particle displacement per unit time. It is this result that enables the macroscopic diffusion coefficient to be understood in terms of atomistic mechanisms.

When all the D_i are equal (as in isotropic systems or cubic crystals) (2.12) becomes

$$\frac{\partial C}{\partial t} = \frac{\langle X^2 \rangle}{2} \nabla^2 C \qquad (2.14)$$

where $\langle X^2 \rangle = \langle X_1^2 \rangle = \langle X_2^2 \rangle = \langle X_3^2 \rangle$.

The diffusion coefficient is often expressed in terms of $\langle R^2 \rangle$ rather than $\langle X^2 \rangle$. Since

$$\langle R^2 \rangle = \langle X_1^2 \rangle + \langle X_2^2 \rangle + \langle X_3^2 \rangle \qquad (2.15)$$

we have, for an isotropic system,

$$\langle R^2 \rangle = 3 \langle X^2 \rangle \qquad (2.16)$$

and we write (2.14) as

$$\frac{\partial C}{\partial t} = D \nabla^2 C \qquad (2.17)$$

where

$$D = \tfrac{1}{6} \langle R^2 \rangle \qquad (2.18)$$

The above results show that, to understand the diffusion coefficient at an atomistic level, the mean square atomic displacement per unit time must be computed.

Note that comparing (2.17) to the equation of continuity gives Fick's first law as

$$\mathbf{J} = -D \, \nabla C \qquad (2.19)$$

8.3   ATOMIC JUMPS AND THE DIFFUSION COEFFICIENT

Diffusion in crystals takes place by discrete atomic jumps in which an atom moves from one lattice position to another. The simplest diffusion system consists of a dilute interstitial impurity in a cubic metal, as exemplified by carbon in iron. In such a system, an impurity atom in an interstitial position spends most of its time executing small vibratory motions about the mean position. Occasionally, however, the impurity atom acquires a large amount of energy as the result of a local thermal fluctuation, and jumps to a neighboring interstitial site. Consequently, the impurity wanders through the crystal in a tortuous path that is the sum of a large number of random jumps.

The motion of a vacancy in a cubic crystal takes place in a similar fashion, the elementary vacancy jump consisting of the movement of an atom adjacent to the vacancy into the vacant site. In tracer diffusion by a vacancy mechanism, the tracer atom moves by jumping into an adjacent vacancy. In all of these cases, the length of the displacement vector over a period of time is the sum of elementary jump vectors of equal length. Thus, if $\mathbf{R}$ is the overall displacement of a diffusing entity per unit time, then

$$\mathbf{R} = \sum_{i=1}^{\Gamma} \mathbf{r}_i \tag{3.1}$$

where $\mathbf{r}_i$ is the ith elementary jump vector and Γ is the number of atomic jumps per unit time.

Squaring (3.1) gives

$$R^2 = \sum_{i,j}^{\Gamma} \mathbf{r}_i \cdot \mathbf{r}_j \tag{3.2}$$

Now rewrite this by separating the terms for which $i = j$ from those for which $i \neq j$:

$$R^2 = \sum_{i=1}^{\Gamma} r_i^2 + \sum_{i \neq j}^{\Gamma} \mathbf{r}_i \cdot \mathbf{r}_j \tag{3.3}$$

The second term in (3.3) can be written in terms of partial sums as

$$\sum_{i \neq j}^{\Gamma} \mathbf{r}_i \cdot \mathbf{r}_j = 2 \sum_{i=1}^{\Gamma-1} \mathbf{r}_i \cdot \mathbf{r}_{i+1} + 2 \sum_{i=1}^{\Gamma-2} \mathbf{r}_i \cdot \mathbf{r}_{i+2} + \cdots \tag{3.4}$$

so that (3.3) becomes

$$\mathbf{R}^2 = \sum_{i=1}^{\Gamma} r_i^2 + 2 \sum_{j=1}^{\Gamma-1} \sum_{i}^{\Gamma-j} \mathbf{r}_i \cdot \mathbf{r}_{i+j} \tag{3.5}$$

In a cubic crystal all the r_i are equal. Also

$$\mathbf{r}_i \cdot \mathbf{r}_{i+j} = r^2 \cos \theta_{i,i+j} \tag{3.6}$$

where $\theta_{i,i+j}$ is the angle between the ith and $(i+j)$th jump; so (3.5) can be written as

$$R^2 = \Gamma r^2 + 2r^2 \sum_{j=1}^{\Gamma-1} \sum_{i=1}^{\Gamma-j} \cos \theta_{i,i+j} \tag{3.7}$$

In this equation, R is the magnitude of the displacement of a single atom in unit time. To obtain the mean square displacement, it is only necessary to average (3.7) over a large number of atoms, each atom making Γ jumps per unit time. That is,

$$\langle R^2 \rangle = \Gamma r^2 f \tag{3.8}$$

where

$$f \equiv 1 + \frac{2}{\Gamma} \sum_{j=1}^{\Gamma-1} \sum_{i=1}^{\Gamma-j} \langle \cos \theta_{i,i+j} \rangle \tag{3.9}$$

f is called the correlation factor. It is determined by the crystal structure and the diffusion mechanism. In general, computation of f is difficult, but for simple diffusion mechanisms, f is a constant near unity.

Now make use of the following identity from spherical trigonometry:

$$\cos \theta_{i,i+2} = \cos \theta_{i,i+1} \cos \theta_{i+1,i+2} + \sin \theta_{i,i+1} \sin \theta_{i+1,i+2} \cos \alpha \quad (3.10)$$

In this formula, α is the angle between a plane defined by r_i and r_{i+1}, and a plane defined by r_{i+1} and r_{i+2}. For every angle α between two such planes in cubic crystals, there is an angle $(180° - \alpha)$ between two other such planes. Also, for every angle $\theta_{i+1,i+2}$ between successive jump vectors, there is an angle $(180° - \theta_{i+1,i+2})$. Therefore, on taking averages over all possible jumps, the last term in (3.10) vanishes and we have

$$\langle \cos \theta_{i,i+2} \rangle = \langle \cos \theta_{i,i+1} \cos \theta_{i+1,i+2} \rangle \quad (3.11)$$

But each of the angles on the right-hand side of (3.11) connects successive jumps, and successive jumps are independent. Therefore

$$\langle \cos \theta_{i,i+2} \rangle = \langle \cos \theta_{i,i+1} \rangle \langle \cos \theta_{i+1,i+2} \rangle$$

$$= \langle \cos \theta \rangle^2 \quad (3.12)$$

where θ is the angle between successive jumps. Similarly,

$$\langle \cos \theta_{i,i+j} \rangle = \langle \cos \theta \rangle^j \quad (3.13)$$

Substitution of (3.13) into (3.9) gives

$$f = 1 + 2 \sum_{j=1}^{\Gamma-1} \left(1 - \frac{j}{\Gamma}\right) \langle \cos \theta \rangle^j \quad (3.14)$$

Γ is always very large, so the sum in (3.14) can be taken from unity to infinity. The term containing j/Γ then goes to zero, and (3.14) becomes

$$f = 1 + 2 \sum_{j=1}^{\infty} \langle \cos \theta \rangle^j \quad (3.15)$$

But, from the formula for the sum of a geometric series,

$$\sum_{j=1}^{\infty} \langle \cos \theta \rangle^j = \frac{1}{1 - \langle \cos \theta \rangle} - 1 \quad (3.16)$$

so (3.15) is

$$f = \frac{1 + \langle \cos \theta \rangle}{1 - \langle \cos \theta \rangle} \quad (3.17)$$

For diffusion of an interstitial impurity in a cubic crystal, it is clear that $\langle \cos \theta \rangle = 0$ since for every possible atomic jump in a given direction, there is also a possible jump in the opposite direction. In this case, therefore, $f = 1$, and diffusion is said to occur by an uncorrelated random flight of the impurity atoms. Physically, this means that successive jumps of an interstitial impurity are completely independent. This is a consequence of the fact that the jump frequency is much less than the vibration frequency of the crystal normal modes. Thus an impurity atom exists at its interstitial site long enough to lose all memory of its previous jump. Clearly, if we regard a vacancy as a diffusing entity, its migration is also uncorrelated, and $f = 1$ for vacancy diffusion.

Diffusion of a substitutional tracer atom by a vacancy mechanism is another matter, however. After a tracer atom jumps into a vacant site, the vacancy is still next to the tracer atom. Furthermore, the time it takes for the vacancy to move away from the tracer is comparable to the time it takes for the tracer to move back into the vacancy. Therefore, the probability that a tracer will move back to a site it has left is greater than the probability it will move to some other site. This means that the jump probabilities are not equal for all directions; the jumps are correlated because the direction of a jump depends on the direction of the preceding jump. Thus $\langle \cos \theta \rangle$ is not zero. Detailed calculation shows that for tracer diffusion by a vacancy mechanism, $f = 0.72$ for the BCC structure and $f = 0.78$ for the FCC structure. We will consider only such simple cases in which f is a constant near unity.

Combining (3.8) with (2.18) shows that the diffusion coefficient is given by

$$D = \frac{f}{6} r^2 \Gamma \tag{3.18}$$

For the case of interstitial diffusion, $f = 1$, $\Gamma = \Gamma_I$ is the jump frequency of interstitial atoms, and $r = r_I$ is the distance between interstitial sites. Therefore, the diffusion coefficient for interstitials is

$$D_I = \frac{r_I^2 \Gamma_I}{6} \quad \text{(interstitial diffusion)} \tag{3.19}$$

A similar formula holds for the diffusion of vacancies

$$D_v = \frac{r_L^2 \Gamma_v}{6} \quad \text{(vacancy diffusion)} \tag{3.20}$$

In this equation, r_L is the distance between two nearest-neighbor lattice points and Γ_v is the jump frequency of a vacancy. Note that Γ_v is really the jump frequency for an atom into an adjacent vacancy.

For self-diffusion by a vacancy mechanism $\Gamma = \Gamma_v C_v$ where C_v is the atomic fraction of vacant sites. This is so because Γ is the actual number of jumps an atom makes per second. Of course, the atom cannot move unless a vacancy is next to it, so the jump frequency for an atom into a vacancy, Γ_v, must be multiplied by C_v, the probability that a site is vacant. Thus, for self diffusion,

$$D_s = \frac{fr_L^2 \Gamma_v C_v}{6} \tag{3.21}$$

It is found experimentally that diffusion coefficients in crystals have an Arrhenius-type dependence on temperature and pressure of the form:

$$D = D_0 e^{-(Q^* + PV^*)/kT} \tag{3.22}$$

where D_0, V^*, and Q^* are constants. A major purpose of the application of statistical mechanics to diffusion is the derivation of this experimental result, and the interpretation of the preexponential factor D_0, and the volume and heat of activation V^* and Q^*. All that remains to be done to achieve this goal is to develop the theory of the jump frequency.

8.4 THE JUMP FREQUENCY IN ONE DIMENSION

During an atomic jump, an atom passes from one equilibrium position to another. It is clear that in the process of doing this, the atom meets strong repulsive forces from its neighbors, so that it must surmount an energy barrier. The migrating atom only occasionally acquires enough energy to climb this barrier. This energy is a local thermal fluctuation consisting of a coming together of phonons of sufficient energy and directionality to move the atom over the barrier.

To introduce the method of dealing with this situation, we consider a one-dimensional analogue of the jump process as shown in Figure 8.1. This

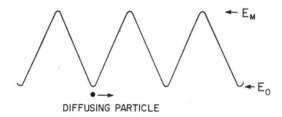

DIFFUSING PARTICLE

Fig. 8.1. Diffusion of a particle in a one-dimensional periodic potential.

represents a single particle of mass m moving in one dimension in a one-dimensional periodic potential. The single particle represents the diffusing atom, and the periodic potential is the analogue of the interaction of this atom with the rest of the crystal. The equilibrium positions of the particle are at the minima of the periodic potential, and the difference between the maximum energy E_M and the minimum energy E_0 is the activation barrier. $(E_M - E_0)$ is taken to be large enough so that most of the time the particle vibrates about the bottom of the well in accord with Hooke's law. When the particle acquires an energy equal to or greater than $(E_M - E_0)$, by thermal interaction with its surroundings, it climbs out of the well and moves over the barrier to enter a new equilibrium position adjacent to its old one. We want to know how often the particle will do this. The answer will be sought by using classical statistical mechanics.

Assume that the top of each barrier is flat enough so that a very small, but nonzero, distance *l* can be defined over each maximum such that the potential energy is very nearly constant over this distance. When crossing through this distance *l*, the particle will have a mean velocity $\bar{v}$ given by classical statistical mechanics as

$$\bar{v} = \frac{\displaystyle\int_0^\infty v e^{-mv^2/2kT}\, dv}{\displaystyle\int_0^\infty e^{-mv^2/2kT}\, dv} \tag{4.1}$$

Performing the integrations gives

$$\bar{v} = \sqrt{\frac{kT}{2\pi m}} \tag{4.2}$$

Out of some long time interval τ, the particle spends most of its time near the bottom of a well, and a small amount of time in one of the regions *l* at the top of a barrier. Let τ_B be the time it spends near an energy minimum and $\tau(l)$ be the time it spends in one of the regions of the barrier tops. Also, let $\bar{\tau}$ be the average time it takes the particle to pass through the distance *l*. Then $\tau(l)/\bar{\tau}$ is just the total number of crossings the particle makes from one well to another during the total time τ. Dividing this number by τ gives the number of jumps per unit time, which is just the definition of the jump frequency Γ. Therefore

$$\Gamma = \frac{\tau(l)}{\bar{\tau}\tau} \tag{4.3}$$

But the mean velocity $\bar{v}$ is

$$\bar{v} = \frac{l}{\bar{\tau}} \tag{4.4}$$

so (4.3) becomes

$$\Gamma = \frac{\bar{v}}{l}\frac{\tau(l)}{\tau} \tag{4.5}$$

Furthermore, τ is very nearly equal to τ_B, since a jump is a relatively rare occurrence. Then, using this fact and (4.2), (4.5) becomes

$$\Gamma = \sqrt{\frac{kT}{2\pi m}} \frac{1}{l} \frac{\tau(l)}{\tau_B} \tag{4.6}$$

The ratio $\tau(l)/\tau_B$ can be obtained from the basic axiom of statistical mechanics that states that ensemble averages are equal to time averages, so that the time a system spends in any group of states is proportional to the partition function for those states. Therefore

$$\frac{\tau(l)}{\tau_B} = \frac{\sum\limits^{(l)} e^{-E(l)/kT}}{\sum\limits^{(B)} e^{-E_B/kT}} \tag{4.7}$$

where the numerator is the partition function when the particle is in the region l at the top of the barrier, and the denominator is the partition function when the particle is in the region around the bottom of the well. In the classical limit, the partition sums become integrals over coordinates and momenta (see Chapter 1, Section 16). The integrals over the momentum are the same for both partition functions, so (4.7) becomes, in the classical limit,

$$\frac{\tau(l)}{\tau_B} = \frac{\int_{x_M-l/2}^{x_M+l/2} e^{-\varphi(x)/kT} \, dx}{\int_{-\infty}^{\infty} e^{-\varphi(x)/kT} \, dx} \tag{4.8}$$

where x_M is the position of the maximum.

$\varphi(x)$ is the energy of the particle measured as a function of distance from the position of an energy minimum. In the region l, this energy is very nearly constant, that is,

$$\varphi(x) = E_M \quad \text{(in region } l) \tag{4.9}$$

Thus, the integral in the numerator of (4.8) becomes

$$le^{-E_M/kT} \tag{4.10}$$

In the region near the bottom of the well, the particle is harmonically bound to the energy minimum so that

$$\varphi(x) = E_0 + \frac{B}{2} x^2 \tag{4.11}$$

where B is the Hooke's law force constant.

The integral in the denominator of (4.8) should be taken over the bottom of the well and part way up its walls. However, the exponential factor rapidly approaches zero as $\varphi(x)$ becomes large, so a negligible error is introduced if the integration is taken from $-\infty$ to ∞. Using (4.11), the denominator in (4.8) becomes

$$e^{-E_0/kT}\int_{-\infty}^{\infty} e^{-Bx^2/2kT}\, dx = e^{-E_0/kT}\sqrt{\frac{2\pi kT}{B}} \qquad (4.12)$$

Substitution of (4.10) and (4.12) into (4.8) gives

$$\frac{\tau(l)}{\tau_B} = e^{-(E_M-E_0)/kT} l \sqrt{\frac{B}{2\pi kT}} \qquad (4.13)$$

Combining this with (4.6), we get the jump-frequency formula

$$\Gamma = \frac{1}{2\pi}\sqrt{\frac{B}{m}}\, e^{-(E_M-E_0)/kT} \qquad (4.14)$$

But the preexponential factor is just the vibration frequency ν of the particle at the bottom of the well, so (4.14) can be put in the more usual form

$$\Gamma = \nu e^{-E^m/kT} \qquad (4.15)$$
where
$$E^m \equiv E_M - E_0 \qquad (4.16)$$

is called the activation energy for migration.

The physical interpretation of (4.15) is straightforward. The particle vibrates near the bottom of a well ν times per second. In order to climb out of the well, it must acquire an energy equal to or greater than the well height E^m. The probability that the particle can do this is given by the Boltzmann factor. The product of this probability with the vibration frequency gives the jump frequency.

This one-dimensional example illustrates the application of statistical mechanics to the theory of the jump frequency. In real crystals, of course, atomic migration is a many-body process involving the motion of many atoms. The migrating atom acquires its energy from the motion of other atoms; also, as the migrating atom approaches the energy barrier, it interacts with other atoms that move as a result of the interaction. In the next section we take up the jump frequency in a many-body system.

8.5 MANY-BODY THEORY OF THE JUMP FREQUENCY

The many-body theory of the jump frequency can be developed as a generalization of the method of the previous section. The system under consideration will be a single vacancy in an otherwise perfect monatomic crystal, and the temperature will be taken to be high enough so that the semiclassical description of Chapter 1, Section 16 can be used when needed.

In accord with the semiclassical description, the system will be represented by a point in configuration phase space. This space is defined by the co-ordinates of the system, so that at any particular time the configuration phase point has coordinates $(q_1, q_2, q_3, \ldots)$. Most of the time, the atoms vibrate about their mean positions, and the harmonic approximation can be used to express the potential energy of the crystal as

$$\varphi(q) = \varphi_0 + \frac{1}{2} \sum_{i=1}^{3N} m\omega_i^2 q_i^2 \qquad (5.1)$$

where ω_i is the angular frequency of the ith normal mode, q_i is the ith normal mode coordinate, and m is the atomic mass. φ_0 is the energy of the system when all atoms are at their equilibrium positions.

Occasionally, however, an atom next to the vacancy will acquire enough energy to jump into it. In the process of jumping, the atom will pass an energy barrier separating the initial and final equilibrium positions. The midpoint is a critical position; if the atom reaches this midpoint with a non-zero velocity it will move into the vacancy, leaving a vacancy behind, and an atomic jump will have occurred. The coordinates of all the atoms when the migrating atom is at the critical position will be called the activated state for atomic migration. Actually, there is not just one activated state; there is an entire ensemble of them since many "midpoint configurations" are con-sistent with the requirement that the migrating atom be in a proper position with sufficient velocity to move into the vacancy. This ensemble of activated states is just a subensemble of the complete ensemble of the system.

It can be expected that the activated states are all close together in the same sense that the normal states are all close together. The activated state of lowest energy will be that in which the migrating atom is midway between its initial and final position, and all other atoms are at rest at equilibrium positions. Of course, these equilibrium positions will be different than those for the normal states because the migrating atom at the top of the energy barrier interacts strongly with the surrounding atoms and pushes them to new positions.

In terms of configuration phase space, the migration process can be described as follows: the phase point representing the system spends most of its time executing small motions around a region whose potential energy is given by (5.1). That is, it moves through the ensemble of normal states. Occasionally, however, the phase point leaves that $3N$-dimensional well to jump into an adjacent similar well. In doing this, it passes through a region of configuration-phase space representing the subensemble of activated states. If we let (q_1, q_2, q_3) be the coordinates of the migrating atom, and choose the coordinate system such that q_1 is along a line joining the initial and final positions, it is clear that the configuration region corresponding to the activated states will have a very small extension along the q_1-axis. Call this extension δ, and let $\bar{v}_1$ be the average velocity of the migrating atom along the q_1-axis when it is in the activated region. Then the mean time that the phase point spends in the activated region during a migration from one well to the next is

$$\bar{\tau} = \frac{\delta}{\bar{v}_1} \tag{5.2}$$

Now let $\tau(\delta)$ be the total time the phase point spends in the activated region out of a total time τ. Then $\tau(\delta)/\bar{\tau}$ is the number of times an atom crosses the activated region, and dividing this by the total time τ gives the jump frequency for a vacancy

$$\Gamma_v = \frac{\tau(\delta)}{\bar{\tau}\tau} \tag{5.3}$$

or, using (5.2),

$$\Gamma_v = \frac{\bar{v}_1}{\delta}\frac{\tau(\delta)}{\tau} \tag{5.4}$$

Replacing $\bar{v}_1$ by $\bar{p}_1/m$, where $\bar{p}_1$ is the average momentum of the moving atom along the migration direction in the activated region gives

$$\Gamma_v = \frac{\bar{p}_1}{m\delta}\frac{\tau(\delta)}{\tau} \tag{5.5}$$

The average momentum is readily obtained from (17.23) of Chapter 1 as

$$\bar{p}_1 = \frac{\int_0^\infty p_1 e^{-p_1^2/2mkT}\,dp_1 \int \exp\left(-\sum_{i=2}^{3N} p_i^2/2mkT\right)dp_2\cdots dp_{3N}}{(2\pi mkT)^{3N/2}}$$

or, performing the integrations,

$$\bar{p}_1 = m\left(\frac{kT}{2\pi m}\right)^{1/2} \tag{5.6}$$

where the integration on dp_1 was taken from zero to infinity because we are averaging only over the positive momenta that take the moving atom across the barrier. Equation 5.4 now reads

$$\Gamma_v = \left(\frac{kT}{2\pi m}\right)^{1/2}\frac{1}{\delta}\frac{\tau(\delta)}{\tau} \tag{5.7}$$

Just as in the one-dimensional case, we now invoke the equivalence of time and ensemble averages, and replace the ratio of the times in (5.7) by a ratio of partition functions. Now, however, the pressure canonical ensemble will be introduced for this purpose so that the final results can be expressed in terms of Gibbs free energies, thereby facilitating the treatment of the effect of pressure on diffusion. Thus

$$\frac{\tau(\delta)}{\tau} = \frac{Z_P^{\ddagger}}{Z_P} \tag{5.8}$$

where $Z_P^{\ddagger}$ is the pressure canonical partition function for the activated region and Z_P is the pressure canonical partition function for the normal crystal. These are given by

$$Z_P = e^{-G/kT} = \sum_{E,V} e^{-(E+PV)/kT} \tag{5.9}$$

$$Z_P^{\ddagger} = e^{-G^{\ddagger}/kT} = \sum_{E,V}^{\ddagger} e^{-(E+PV)/kT} \tag{5.10}$$

The sum in (5.9) is over all volumes and energy states of the normal crystal, but the sum in (5.10) is taken only over the activated states. G is the Gibbs free energy of the normal crystal, and $G^{\ddagger}$ is the Gibbs free energy of the activated state as defined by (5.10). It is convenient to separate the pressure term from these free energies:

$$G = A + PV \tag{5.11}$$

$$G^{\ddagger} = A^{\ddagger} + PV^{\ddagger} \tag{5.12}$$

where A, V, and $A^{\ddagger}$, $V^{\ddagger}$ are the Helmholtz free energies and volume of the normal crystal and the activated state, respectively.

Using (5.7) to (5.12) the vacancy jump frequency now reads

$$\Gamma_v = \left(\frac{kT}{2\pi m}\right)^{1/2} \frac{1}{\delta} e^{-PV_v{}^m/kT} \frac{e^{-A^{\ddagger}/kT}}{e^{-A/kT}} \tag{5.13}$$

where

$$V_v{}^m \equiv V^{\ddagger} - V \tag{5.14}$$

is the volume of vacancy migration.

Equation 5.13 can be put into a simpler and more useful form by evaluating the exponentials of the free energies using canonical ensemble theory. This states that

$$e^{-A/kT} = \sum_i e^{-E_i/kT} \tag{5.15}$$

$$e^{-A^{\ddagger}/kT} = \sum_i^{\ddagger} e^{-E_i/kT} \tag{5.16}$$

The sum in (5.15) is taken over all states in the normal crystal, while the sum in (5.16) is taken over all activated states. In the semiclassical approximation,

(5.15) becomes

$$e^{-A/kT} = \left(\frac{2\pi mkT}{h^2}\right)^{3N/2} \int e^{-\varphi(q)/kT}\, dq \tag{5.17}$$

the integration being taken over all coordinates. Assuming the crystal is harmonic, the quadratic potential of (5.1) can be substituted into the integral in (5.17) to give

$$\int e^{-\varphi(q)/kT}\, dq = e^{-\varphi_0/kT} \int \cdots \int e^{-m\omega_1^2 q_1^2/2kT} e^{-m\omega_2^2 q_2^2/2kT} \cdots dq_1\, dq_2 \cdots \tag{5.18}$$

The multidimensional integral in (5.18) separates into a product of $3N$ one-dimensional integrals, all of the form

$$\int_{-\infty}^{\infty} e^{-m\omega^2 q^2/2kT}\, dq = \left(\frac{2\pi kT}{m\omega^2}\right)^{1/2} \tag{5.19}$$

Thus

$$\int e^{-\varphi(q)/kT}\, dq = \prod_{i=1}^{3N} \left(\frac{2\pi kT}{m\omega_i^2}\right)^{1/2} e^{-\varphi_0/kT} \tag{5.20}$$

and substituting this into (5.17) gives

$$e^{-A/kT} = \left(\frac{2\pi kT}{h}\right)^{3N} \prod_{i=1}^{3N} \frac{1}{\omega_i}\, e^{-\varphi_0/kT} \tag{5.21}$$

Equation 5.16 can be treated in a similar way. In the semiclassical approximation this is

$$e^{-A^{\ddagger}/kT} = \left(\frac{2\pi mkT}{h^2}\right)^{3N/2} \int^{\ddagger} e^{-\varphi^{\ddagger}(q)/kT}\, dq \tag{5.22}$$

where the integration is now carried out over all coordinates in the activated region and $\varphi^{\ddagger}(q)$ is the potential energy in that region. Written out explicitly, the integral in (5.22) is

$$\int^{\ddagger} e^{-\varphi^{\ddagger}(q)/kT}\, dq = \int_{x_1-\delta/2}^{x_1+\delta/2} \int_{-\infty}^{\infty} \overset{\cdots}{(3N-1)} \int_{-\infty}^{\infty} e^{-\varphi(q_1, q_2 \cdots q_{3N})/kT} dq_1\, dq_2 \cdots dq_{3N} \tag{5.23}$$

The integral over dq_1 is taken only over a length δ since this was defined as the thickness of the activated region along the direction of migration. x_1 is the value of q_1 at the minimum activated configuration. Since δ is small and, indeed, we will take the limit as $\delta \to 0$, $\varphi^{\ddagger}(q)$ will be taken as independent of q_1 over the length δ. The integration over q_1 is therefore readily performed and (5.23) becomes

$$\int^{\ddagger} e^{-\varphi^{\ddagger}(q)/kT} \, dq = \delta \int_{-\infty}^{\infty} \overset{\cdots}{(3N-1)} \int_{-\infty}^{\infty} e^{-\varphi^{\ddagger}(x_1, q_2 \cdots q_{3N})/kT} dq_2 \, dq_3 \cdots dq_{3N} \quad (5.24)$$

The harmonic approximation will again be used to express the potential energy as a function of coordinates in the activated region as

$$\varphi^{\ddagger}(x_1, q_2 \cdots q_{3N}) = \varphi_0^{\ddagger} + \tfrac{1}{2} \sum_{i=1}^{3N-1} m\omega_i^{\ddagger 2} q_i^{2} \quad (5.25)$$

where $\varphi_0^{\ddagger}$ is the potential energy when the migrating atom is midway between its initial and final positions and all other atoms are at the mean positions in the activated region. The $\omega_i^{\ddagger}$ are normal mode frequencies of the activated state. Using (5.25) in (5.24) and evaluating the integrals gives

$$\int^{\ddagger} e^{-\varphi^{\ddagger}(q)/kT} \, dq = \delta e^{-\varphi_0^{\ddagger}/kT} \prod_{i=1}^{3N-1} \left(\frac{2\pi kT}{m\omega_i^{\ddagger 2}} \right)^{1/2} \quad (5.26)$$

so that (5.22) becomes

$$e^{-A^{\ddagger}/kT} = \delta e^{-\varphi_0^{\ddagger}/kT} \left(\frac{2\pi kT}{h} \right)^{3N} \left(\frac{m}{2\pi kT} \right)^{1/2} \prod_{i=1}^{3N-1} \frac{1}{\omega_i^{\ddagger}} \quad (5.27)$$

Now combine (5.28), (5.21), and (5.13). The result is

$$\Gamma_v = e^{-(E_v{}^m + PV_v{}^m)/kT} \frac{\displaystyle\prod_{i=1}^{3N} \nu_i}{\displaystyle\prod_{i=1}^{3N-1} \nu_i^{\ddagger}} \quad (5.28)$$

where

$$E_v{}^m \equiv \varphi_0^{\ddagger} - \varphi_0 \quad (5.29)$$

is the vacancy-migration energy, and the angular frequencies have been replaced by the ordinary frequencies $\nu = \omega/2\pi$.

Another form of (5.28) can be written by making use of the identities

$$\prod_{i=1}^{3N} \nu_i = \left(\frac{kT}{h} \right)^{3N} \exp \left(\frac{1}{k} \sum_{i=1}^{3N} k \ln \frac{h\nu_i}{kT} \right) \quad (5.30)$$

$$\prod_{i=1}^{3N-1} \nu_i^{\ddagger} = \left(\frac{kT}{h} \right)^{3N-1} \exp \left(\frac{1}{k} \sum_{i=1}^{3N-1} k \ln \frac{h\nu_i^{\ddagger}}{kT} \right) \quad (5.31)$$

The reason for doing this is that it can easily be shown (see, for example, Chapter 5, Section 5) that the high-temperature approximation for the entropy of a harmonic system is

$$S = -k \sum_i \ln \frac{h\nu_i}{kT} \tag{5.32}$$

Thus, (5.30) and (5.31) give

$$\prod_{i=1}^{3N} \nu_i = \left(\frac{kT}{h}\right)^{3N} e^{-S/k} \tag{5.33}$$

$$\prod_{i=1}^{3N-1} \nu_i^{\ddagger} = \left(\frac{kT}{h}\right)^{3N-1} e^{-S^{\ddagger}/k} \tag{5.34}$$

where S and $S^{\ddagger}$ are the vibrational entropies of the normal and activated states, respectively. Equation 5.28, therefore, becomes

$$\Gamma_v = \left(\frac{kT}{h}\right) e^{-(E_v{}^m + PV_v{}^m - TS_v{}^m)/kT} \tag{5.35}$$

where

$$S_v{}^m \equiv S^{\ddagger} - S \tag{5.36}$$

is the entropy of vacancy migration. The free energy of vacancy migration is defined by

$$G_v{}^m \equiv E_v{}^m + PV_v{}^m - TS_v{}^m \tag{5.37}$$

so an alternative form for the jump frequency is

$$\Gamma_v = \frac{kT}{h} e^{-G_v{}^m/kT} \tag{5.38}$$

This derivation of the jump-frequency formula was carried out for vacancy motion for the sake of definiteness. But obviously, similar results can be obtained in a similar way for other diffusion mechanisms. In particular, the jump frequency for an interstitial is

$$\Gamma_I = \frac{kT}{h} e^{-G_I{}^m/kT} \tag{5.39}$$

$G_I{}^m$ being the Gibbs free energy of motion for an interstitial, defined in complete analogy with $G_v{}^m$.

It is important to note that the free energies in (5.38) and (5.39) are not free-energy differences in the usual sense. G is the Gibbs free energy of the crystal, and has its usual meaning. However, $G^{\ddagger}$ is defined in terms of restricted partition sums. Furthermore, $G^{\ddagger}$ refers to a system having one less degree of freedom than the crystal in its normal state. This missing degree of freedom is responsible for the factor kT/h in the jump-frequency formulas.

Often, it is the temperature dependence of the jump frequency that is of major interest. It is then more convenient to retain the jump frequency in the form (5.28) rather than (5.38), since $G_v{}^m$ is linear in temperature. The appearance of the formulae is simplified by defining an effective frequency $\tilde{\nu}_v$ by

$$\tilde{\nu}_v = \frac{\prod\limits_{i=1}^{3N} \nu_i}{\prod\limits_{i=1}^{3N-1} \nu_i^{\ddagger}} \tag{5.40}$$

so that (5.28) becomes

$$\Gamma_v = \tilde{\nu}_v e^{-(E_v{}^m + PV_v{}^m)/kT} \tag{5.41}$$

Similarly, for interstitial diffusion, we can write

$$\Gamma_I = \tilde{\nu}_I e^{-(E_I{}^m + PV_I{}^m)/kT} \tag{5.42}$$

where $E_I{}^m$ and $V_I{}^m$ are the energy and volume of interstitial migration, respectively, and $\tilde{\nu}_I$ is the effective frequency for interstitial migration defined in analogy with (5.40).

It must be stressed that this treatment of the jump frequency is based on equilibrium statistical mechanics. The activated state is one of the states in the ensemble representing the complete crystal equilibrium, and in computing the jump frequency, we have just counted the frequency with which the system moves from one set of states to another in the ensemble. In applying this theory to diffusion in a concentration gradient, it is assumed that the deviation from equilibrium caused by the nonzero gradients is not sufficient to affect seriously the equations derived on an equilibrium basis. The theory presented here, therefore, does not touch the fundamental question of irreversible processes, and in those cases in which irreversibility is of prime importance, such as diffusion in a thermal gradient, the present theory can be expected to fail. In many cases, however, particularly when the diffusing material is present in small quantities, and when the system is near equilibrium, the theory is quite satisfactory.

Migration energies in metals are often of the order of 1 eV and effective frequencies are usually of the order of 10^{13} sec^{-1}. This means that an interstitial or a vacancy makes about 10^8 jumps per second at 1000°K. This number decreases rapidly with decreasing temperature.

8.6   THE DIFFUSION COEFFICIENT

The results obtained so far are sufficient to give the diffusion coefficient as an explicit function of temperature and pressure. Substitution of (5.41) and

Table 8.1. Self-Diffusion Parameters in Metals

Metal	D_o cm^2/sec	Q cal/mol
Aluminum	0.035	28,750
Chromium	0.20	73,700
Cobalt	0.83	67,700
Copper	0.62	49,560
Germanium	7.8	68,500
Gold	0.091	41,700
∂-Iron	2.0	57,300
Lead	1.37	26,060
Lithium	0.39	13,490
Molybdenum	0.1	92,200
Nickel	1.9	68,000
Niobium	1.1	96,000
Palladium	0.21	63,600
Platinum	0.33	68,200
Potassium	0.31	9,750
Silicon	1800	110,000
Silver	0.44	44,270
Sodium	0.145	10,090
Tantalum	0.124	98,700
β -Thallium	0.7	20,000
α -Thorium	1.2	76,600
Tungsten	42.8	153,000

From a compilation by N. L. Peterson, *Solid State Physics*, Vol. 22, F. Seitz, D. Turnbull, and H. Ehrenreich, Eds., Academic Press, New York, 1968.

(5.42) into (3.19), (3.20), and (3.21) gives

$$D_I = \tfrac{1}{6}\tilde{\nu}_I r_I^2 e^{-(E_I{}^m + PV_I{}^m)/kT} \tag{6.1}$$

$$D_v = \tfrac{1}{6}\tilde{\nu}_v r_L^2 e^{-(E_v{}^m + PV_v{}^m)/kT} \tag{6.2}$$

$$D_S = \tfrac{1}{6}\tilde{\nu}_v f r_L^2 e^{-(E_v{}^m + PV_v{}^m)/kT} C_v \tag{6.3}$$

From the results of Chapter 5 and 6, the atomic fraction of vacancies is

$$C_v = e^{-(E_v{}^f + PV_v{}^f)/kT} \tag{6.4}$$

Thus the diffusion coefficients (6.1) to (6.3) can all be written in the form

$$D = D_0 e^{-(Q^* + PV^*)/kT} \tag{6.5}$$

where the preexponential factors, the heats of activation, and volumes of activation are given by:
Interstitial diffusion:

$$\begin{aligned}
D_0 &= \tilde{\nu}_I r_I^2/6 \\
Q^* &= E_I{}^m \\
V^* &= V_I{}^m
\end{aligned} \tag{6.6}$$

Vacancy diffusion:

$$\begin{aligned}
D_0 &= \tilde{\nu}_v r_L^2/6 \\
Q^* &= E_v{}^m \\
V^* &= V_v{}^m
\end{aligned} \tag{6.7}$$

Self diffusion:

$$\begin{aligned}
D_0 &= \tilde{\nu}_v f r_L^2/6 \\
Q^* &= E_v{}^m + E_v{}^f \\
V^* &= V_v{}^m + V_v{}^f
\end{aligned} \tag{6.8}$$

This formulation of the diffusion coefficient has been successful in treating diffusion data in a large variety of systems. The energy of activation can be determined by diffusion experiments as a function of temperature at a fixed pressure since then the slope of a plot of $\ln D$ versus $1/T$ gives Q^*. The volume of activation is determined from a plot of $\ln D$ versus P using data obtained at various pressures but at the same temperature. Precise diffusion experiments are extremely difficult to perform, but a number of systems have been carefully studied. Table 8.1 gives some experimentally determined values of the diffusion parameters.

chapter 9

DEFECTS AND DIFFUSION IN
ORDER–DISORDER ALLOYS

9.1 KINETIC PROCESSES IN ORDER-DISORDER ALLOYS

Since the degree of order of an order-disorder alloy varies with temperature, a mechanism must exist that permits ordering or disordering to take place. The degree of order is defined by the position of atoms on different types of lattice sites; therefore, a change in order must be the result of the diffusive motion of atoms in the crystal. Suppose, for example, that the temperature T_1 of an AB alloy is so low that the alloy is completely ordered, and suppose that the temperature is suddenly changed to some value T_2 above the critical temperature. Then the atoms will have to move so as to convert the ordered atomic arrangement to a disordered one. That is, A and B atoms execute jumps from site to site, and the degree of order decreases continuously to zero. For common order-disorder alloys, the atomic jumps most probably take place by a vacancy mechanism. It will be assumed that this is the case for the alloys considered in this chapter.

In the kinetics of ordering, it is clear that the important factors are the number of vacancies and the atomic jump frequencies. As shown in previous chapters, these quantities are sensitive functions of the temperature and depend on interaction energies in the crystal. The vacancy concentration is determined by the vacancy-formation energy. But in an AB alloy, the energy required to remove an atom from a lattice site obviously depends on the kinds of atoms surrounding that site. On the average, the site occupation is described by the long-range-order parameter. Therefore, the vacancy concentration will depend on the degree of order. Similarly, the jump frequency is a function of the degree of order, because the energy of migration depends on the environment of the jumping atom.

The energies of migration and vacancy formation are both functions of the degree of order and, therefore, so is the activation energy of diffusion. Thus if the diffusion coefficients for the alloy constituents are measured at various temperatures, deviations from the Arrhenius form can be expected that are related to the degree of order. We will see later that experiments confirm this expectation.

The rate of change of the order parameter itself, in a nonequilibrium alloy, is a complex function of the order parameter. The change in order is clearly a kind of superposition of elementary diffusion processes that depend on order.

9.2 DESCRIPTION OF ORDER IN AN ALLOY CONTAINING VACANCIES

Consider a two-component order-disorder crystal containing vacancies and equal numbers of A atoms and B atoms. The crystal lattice is divided into two sublattices labeled α and β as described in Chapter 5. We adopt the following notation:

$$N_A = \text{total number of } A \text{ atoms}$$
$$N_B = \text{total number of } B \text{ atoms}$$
$$N = \text{total number of atoms}$$
$$L_\alpha = \text{number of } \alpha \text{ sites}$$
$$L_\beta = \text{number of } \beta \text{ sites}$$
$$L = \text{total number of sites}$$
$$N_{A\alpha} = \text{number of } A \text{ atoms on } \alpha \text{ sites}$$
$$N_{A\beta} = \text{number of } A \text{ atoms on } \beta \text{ sites}$$
$$N_{B\alpha} = \text{number of } B \text{ atoms on } \alpha \text{ sites}$$
$$N_{B\beta} = \text{number of } B \text{ atoms on } \beta \text{ sites}$$
$$N_{V\alpha} = \text{number of vacancies on } \alpha \text{ sites}$$
$$N_{V\beta} = \text{number of vacancies on } \beta \text{ sites}$$
$$N_V = \text{total number of vacancies}$$

These quantities, of course, are not independent but satisfy the following obvious relations:

$$N_A + N_B + N_V = L \qquad (2.1)$$

$$N_{A\alpha} + N_{B\alpha} + N_{V\alpha} = L_\alpha \qquad (2.2)$$

$$N_{A\beta} + N_{B\beta} + N_{V\beta} = L_\beta \qquad (2.3)$$

$$N_{A\alpha} + N_{A\beta} = N_A \qquad (2.4)$$

$$N_{B\alpha} + N_{B\beta} = N_B \qquad (2.5)$$

$$N_A + N_B = N \qquad (2.6)$$

$$L_\alpha + L_\beta = L \qquad (2.7)$$

$$N_{V\alpha} + N_{V\beta} = N_V \qquad (2.8)$$

$$N_A = N_B = \frac{N}{2} \qquad (2.9)$$

Since $N_{V\alpha}$ and $N_{V\beta}$ are the total number of vacancies on the complete sublattices, we should be able to use the long-range-order parameter to relate the vacancy concentration to the degree of order. For a 50-50 AB alloy, the long-range-order parameter is defined by (Chapter 5)

$$R = \frac{N_{A\alpha}/L_\alpha - L_\alpha/L}{1 - L_\alpha/L} \qquad (2.10)$$

For a perfect crystal, R is symmetric with respect to an interchange of the subscripts, $(A\alpha \rightleftarrows B\beta)$. That is, the long-range-order parameter can be defined either in terms of occupation of the α sublattice by A atoms, or the occupation of the β sublattice by B atoms. The reason for this is that an A atom can be transferred from the α to the β lattice only by simultaneously transferring a B atom from the β to the α lattice. Therefore, for a perfect crystal,

$$N_{A\beta} = N_{B\alpha} \qquad (2.11)$$

$$N_{A\alpha} = N_{B\beta} \qquad (2.12)$$

If R is to be symmetric for a crystal containing vacancies, it is necessary that (2.11) and (2.12) be valid for this case also. That this is so can be shown by starting with a crystal containing the equilibrium number of vacancies, but having all A atoms on α sites and all B atoms on β sites. Now interchange A and B atoms until the equilibrium values of $N_{A\alpha}$, $N_{A\beta}$, $N_{B\alpha}$, $N_{B\beta}$ are attained. Since the number of vacant sites in the α and β sublattices remain constant throughout this process, everytime an A atom is transferred from the α to the β lattice, a B atom must be transferred from β to α. Thus, at the end of this process, (2.11) and (2.12) will be valid and the crystal will be at equilibrium. Since the equilibrium state is independent of the method of arriving at it, the validity of (2.12) and (2.13) is established.

The conditions (2.11) and (2.12) are not sufficient to guarantee the symmetry of R with respect to the interchange $(A\alpha \rightleftarrows B\beta)$, because the number of α sites does not necessarily equal the number of β sites, and the desired symmetry requires that $L_\alpha = L_\beta$. If we use (2.2), (2.3), (2.11), and (2.12), then (2.10) can be converted to

$$R = \frac{(N_{B\beta}/L_\beta) - (L_\beta/L) + (N_{V\beta}/L_\beta) - (N_{V\alpha}/L_\alpha)}{1 - L_\beta/L} \tag{2.13}$$

Comparing this to (2.10) shows that R has the desired symmetry only if the vacancy concentrations on the two sublattices are equal. To determine whether this is true, it is necessary to solve the statistical-mechanical problem for $N_{V\alpha}$ and $N_{V\beta}$. In doing so, it will be shown that, at least within the Bragg-Williams approximation, the sublattices do have equal vacancy concentrations so that a symmetric long-range-order parameter can indeed be defined.

Although we anticipate that the long-range-order parameter will be given by (2.10) even for a crystal containing vacancies, it is not necessary to assume this at this point. All that is needed is to recognize that the degree of order is a measure of the number of the two types of atoms on the two sublattices. Then we can define a function $g(R)$ in complete analogy with (5.2) of Chapter 5, that gives the number of ways of arranging $N_{A\alpha}$ A atoms, $N_{B\alpha}$ B atoms, and $N_{V\alpha}$ vacancies on α sites; and $N_{A\beta}$ A atoms, $N_{B\beta}$ B atoms, and $N_{V\beta}$ vacancies on β sites:

$$g(R) = \frac{L_\alpha!}{N_{A\alpha}! \, N_{B\alpha}! \, N_{V\alpha}!} \frac{L_\beta!}{N_{A\beta}! \, N_{B\beta}! \, N_{V\beta}!} \tag{2.14}$$

Just as in Chapter 5, $g(R)$ is the number of states consistent with a given degree of order. Therefore, $\overline{W}(R)$, the arithmetic average of the energy of the crystal is

$$\overline{W}(R) = \frac{1}{g(R)} \sum_k W_k \tag{2.15}$$

It is possible to apply the Kirkwood method, just as we did for the perfect crystal. However, since we will stay in the Bragg-Williams approximation, we will write the partition function in the Bragg-Williams form immediately as

$$Z_c = g(R)e^{-\overline{W}(R)/kT} \tag{2.16}$$

The Helmholtz free energy is, as usual,

$$A_c = -kT \ln Z_c \tag{2.17}$$

or, using (2.16),

$$A_c = -kT \ln g(R) + \overline{W}(R) \tag{2.18}$$

These results are sufficient to give the vacancy-concentration formulae within the limits of the Bragg-Williams approximation.

9.3   THE AVERAGE VACANCY CONCENTRATION IN AN *AB* ALLOY

The equilibrium number of vacancies can be obtained by minimizing the Helmholtz free energy which, in the Bragg-Williams approximation, is given by (2.18). The free energy is a function of the number of vacancies as well as the degree of order because W_k is determined by the number of AA, BB, and AB pairs, and these numbers are a function of both the degree of order and the vacancy concentration. The equilibrium state of the crystal is therefore determined by

$$\frac{\partial A_c}{\partial R} = 0 \tag{3.1}$$

$$\frac{\partial A_c}{\partial N_{V\alpha}} = 0 \tag{3.2}$$

$$\frac{\partial A_c}{\partial N_{V\beta}} = 0 \tag{3.3}$$

The first of these equations is the ordinary equilibrium condition for the long-range-order parameter and will not be considered any further. It will be assumed that the vacancy concentration is too small to seriously affect the temperature dependence of R. From (2.18), (3.1), and (3.2) become

$$\frac{\partial \ln g(R)}{\partial N_{V\alpha}} = \frac{1}{kT} \frac{\partial \overline{W}(R)}{\partial N_{V\alpha}} \tag{3.4}$$

$$\frac{\partial \ln g(R)}{\partial N_{V\beta}} = \frac{1}{kT} \frac{\partial \overline{W}(R)}{\partial N_{V\beta}} \tag{3.5}$$

The left-hand sides of these equations can be expressed directly in terms of $N_{V\alpha}$ and $N_{V\beta}$ from (2.14). Using Stirling's approximation and performing the differentiations, (3.4) and (3.5) become

$$\ln \frac{N_{V\alpha}}{L_\alpha} = -\frac{1}{kT} \frac{\partial \overline{W}(R)}{\partial N_{V\alpha}} \tag{3.6}$$

$$\ln \frac{N_{V\beta}}{L_\beta} = -\frac{1}{kT} \frac{\partial \overline{W}(R)}{\partial N_{V\beta}} \tag{3.7}$$

The average energy is given in terms of the number of pairs by averaging (4.8) in Chapter 5, which is valid whether vacancies are present or not:

$$\bar{W} = -v_{AA}\bar{Q}_{AA} - v_{BB}\bar{Q}_{BB} - v_{AB}[\bar{Q}_{AB} + \bar{Q}_{BA}] \qquad (3.8)$$

Differentiation of (3.8) gives the right-hand sides of (3.6) and (3.7):

$$\frac{\partial \bar{W}}{\partial N_{V\alpha}} = -v_{AA}\frac{\partial \bar{Q}_{AA}}{\partial N_{V\alpha}} - v_{BB}\frac{\partial \bar{Q}_{BB}}{\partial N_{V\alpha}} - v_{AB}\left[\frac{\partial \bar{Q}_{AB}}{\partial N_{V\alpha}} + \frac{\partial \bar{Q}_{BA}}{\partial N_{V\alpha}}\right] \qquad (3.9)$$

$$\frac{\partial \bar{W}}{\partial N_{V\beta}} = -v_{AA}\frac{\partial \bar{Q}_{AA}}{\partial N_{V\beta}} - v_{BB}\frac{\partial \bar{Q}_{BB}}{\partial N_{V\beta}} - v_{AB}\left[\frac{\partial \bar{Q}_{AB}}{\partial N_{V\beta}} + \frac{\partial \bar{Q}_{BA}}{\partial N_{V\beta}}\right] \qquad (3.10)$$

The derivatives of the average number of pairs with respect to the number of vacancies can be computed by counting the average number of bonds of various types that are broken when a vacancy is formed. For example, to obtain $\partial \bar{Q}_{AA}/\partial N_{V\alpha}$, create a vacancy on an α site. The probability that this site was occupied by an A atom is $N_{A\alpha}/L_\alpha$, and the probability that a particular adjacent β site contains an A atom is $N_{A\beta}/L_\beta$. Therefore, the average number of AA pairs that were destroyed by removing the A atom from the α site is

$$z\,\frac{N_{A\alpha}}{L_\alpha}\frac{N_{A\beta}}{L_\beta}$$

where z is the coordination number.

In the vacancy-formation process, the energy of vacancy formation is defined by removing the atom from the interior and placing it on the surface. It will now be assumed that when the atom is placed on the surface, it is in contact (on the average) with just one-half of the number of the atoms of both types that surrounded it in the interior of the crystal. This is consistent with the definition of the formation energy in monatomic systems, and amounts to choosing the surface in such a way that the degree of order is not altered by creating a vacancy. The average number of AA pairs destroyed on forming a vacancy is therefore one-half of the number computed above:

$$z\,\frac{N_{A\alpha}N_{A\beta}}{2L_\alpha L_\beta}$$

so that if $\Delta N_{\alpha V}$ vacancies are created on α sites, the change in the average number of AA pairs, $\Delta \bar{Q}_{AA}$ is

$$\Delta \bar{Q}_{AA} = -\frac{zN_{A\alpha}N_{A\beta}}{2L_\alpha L_\beta}N_{V\alpha} \qquad (3.11)$$

from which

$$\frac{\partial \bar{Q}_{AA}}{\partial N_{V\alpha}} = - \frac{zN_{A\alpha}N_{A\beta}}{2L_\alpha L_\beta} \tag{3.12}$$

In a similar way, the other derivatives in (3.9) and (3.10) are found to be

$$\frac{\partial \bar{Q}_{AA}}{\partial N_{V\beta}} = - \frac{zN_{A\alpha}N_{A\beta}}{2L_\alpha L_\beta} \tag{3.13}$$

$$\frac{\partial \bar{Q}_{BB}}{\partial N_{V\alpha}} = \frac{\partial \bar{Q}_{BB}}{\partial N_{V\beta}} = - \frac{zN_{B\alpha}N_{B\beta}}{2L_\alpha L_\beta} \tag{3.14}$$

$$\frac{\partial \bar{Q}_{AB}}{\partial N_{V\alpha}} = \frac{\partial \bar{Q}_{AB}}{\partial N_{V\beta}} = - \frac{zN_{A\alpha}N_{B\beta}}{2L_\alpha L_\beta} \tag{3.15}$$

$$\frac{\partial \bar{Q}_{BA}}{\partial N_{V\alpha}} = \frac{\partial \bar{Q}_{BA}}{\partial N_{V\beta}} = - \frac{zN_{B\alpha}N_{A\beta}}{2L_\alpha L_\beta} \tag{3.16}$$

Now the derivatives given by (3.13) to (3.16) can be substituted into (3.9) and (3.10) with the result

$$\frac{\partial \bar{W}}{\partial N_{V\alpha}} = \frac{\partial \bar{W}}{\partial N_{V\beta}} = \frac{z}{2L_\alpha L_\beta}$$

$$\times [v_{AA}N_{A\alpha}N_{A\beta} + v_{BB}N_{B\alpha}N_{B\beta} + v_{AB}(N_{A\alpha}N_{B\beta} + N_{B\alpha}N_{A\beta})] \tag{3.17}$$

Thus the right-hand sides of (3.6) and (3.7) are the same and therefore

$$\frac{N_{V\alpha}}{L_\alpha} = \frac{N_{V\beta}}{L_\beta} \tag{3.18}$$

In accord with the discussion in the previous section, it follows that the long-range-order parameter can be defined by (2.10) for an alloy containing vacancies. The long-range-order parameter is then connected to the number of atoms of the two types on the two sublattices by (5.3) and (5.4) of Chapter 5:

$$N_{A\alpha} = N_{B\beta} = \frac{L}{4}(1 + R) \tag{3.19}$$

$$N_{A\beta} = N_{B\alpha} = \frac{L}{4}(1 - R) \tag{3.20}$$

The vacancy concentration can now be expressed in terms of the long-range-order parameter by substituting (3.19) and (3.20) in (3.17), and substituting the result into (3.6) and (3.7). Performing a little algebraic simplification gives

$$C_V \equiv \frac{N_{V\alpha}}{L_\alpha} = \frac{N_{V\beta}}{L_\beta} = e^{-(E_V^0 + 1/4zvR^2)/kT} \qquad (3.21)$$

where E_V^0 is the energy of vacancy formation at zero degree of order, given by

$$E_V^0 \equiv \frac{z}{8}(v_{AA} + v_{BB} + 2v_{AB}) \qquad (3.22)$$

and v is the ordering energy.

The energy of vacancy formation is a quadratic function of R that is a minimum when the alloy is completely disordered and increases with increasing order. This is a physically reasonable result, since to form a vacancy in a completely ordered crystal only AB bonds are broken; while to form a vacancy in a disordered alloy, AA and BB bonds must also be broken. Since $v_{AB} > \frac{1}{2}(v_{AA} + v_{BB})$, it follows that the formation energy increases with increasing order.

Clearly, a log n_V versus $1/T$ plot will not have the linear Arrhenius form, because R is a function of temperature. We can deduce from (3.21) that the slope of the plot will increase with increasing temperature until the critical temperature is reached, above which the plot will be linear.

9.4 THE JUMP FREQUENCY IN AN *AB* ALLOY

The jump frequency of a particular atom in an alloy depends on the atom's local environment. If each A atom is labeled by an index j, the average jump frequency of the A atoms is defined by

$$\Gamma_A = \frac{1}{N_A} \sum_j \tilde{\nu}_A(j) e^{-E_A^m(j)/kT} \qquad (4.1)$$

where $\tilde{\nu}_A(j)$ is the effective vibration frequency for an A atom on the jth site next to a vacancy, and $E_A^m(j)$ is the migration energy for an A atom on the jth site moving by a vacancy mechanism (see Section 5, Chapter 8).

The primary source of the differences among the jump frequencies of different atoms is the migration energy. All $\tilde{\nu}_A(j)$ will therefore be taken as equal to a single constant ν_A. This is consistent with our general approach to order-disorder alloys in which variations of the lattice vibrations with order are ignored.

Now let $\bar{E}_A^m$ be the mean migration energy:

$$\bar{E}_A^m \equiv \frac{1}{N_A} \sum_j E_A^m(j) \qquad (4.2)$$

and expand each exponential in (4.1) about this mean in a Taylor series. The result is

$$\Gamma_A = v_A e^{-\bar{E}_A{}^m/kT}\left[1 - \frac{1}{N_A kT}\sum_j [\bar{E}_A{}^m - E_A{}^m(j)] \right.$$

$$\left. + \frac{1}{2N_A(kT)^2}\sum_j [\bar{E}_A{}^m - E_A{}^m(j)]^2 + \cdots \right] \quad (4.3)$$

This shows that the jump frequency can be expanded in moments of the energy in a manner similar to the Kirkwood theory. If only the first term in (4.3) is retained, we have the equivalent of the Bragg-Williams theory. We will therefore take Γ_A to be

$$\Gamma_A = v_A e^{-\bar{E}_A{}^m/kT} \quad (4.4)$$

Similarly, the average jump frequency of B atoms, to the same approximation, is

$$\Gamma_B = v_B e^{-\bar{E}_B{}^m/kT} \quad (4.5)$$

To get these jump frequencies as functions of order, we need to relate the average migration energies to the site occupation numbers. This can be done by writing the average migration energy as averages over the two sublattices. For A atoms this gives

$$\bar{E}_A{}^m = \frac{1}{N_A}[N_{A\alpha}\bar{E}_{A\alpha}{}^m + N_{A\beta}\bar{E}_{A\beta}{}^m] \quad (4.6)$$

where $\bar{E}_{A\alpha}{}^m$ and $\bar{E}_{A\beta}{}^m$ are the average migration energies for A atoms on the α and β sublattices, respectively.

Now consider an A atom on an α site adjacent to a vacancy. It is surrounded by $z\beta$ sites, $(z-1)$ of which contain atoms. The probability that one of these $(z-1)\beta$ sites contains an A atom is $N_{A\beta}/L_\beta$, while the probability that one of these sites contains a B atom is $N_{B\beta}/L_\beta$. The average energy of the atom is therefore

$$-(z-1)\frac{N_{A\beta}}{L_\beta}v_{AA} - (z-1)\frac{N_{B\beta}}{L_\beta}v_{AB} \quad (4.7)$$

An expression similar to this, but referring to an average A atom in a saddle-point position, is needed in order to compute $\bar{E}_{A\alpha}{}^m$. The precise form of the saddle-point energy depends on the crystal structure and the specific configuration of the saddle point. For example, in a body-centered cubic structure such as β-brass, the A atom passes through two groups of three atoms each as it moves from its initial to a nearest-neighbor position. The three atoms in each group are at the corners of two equilateral triangles, each triangle being perpendicular to the line joining the initial and final site. The first triangle is one-third the way along this line, while the second triangle is two-thirds the way, and rotated relative to the first by sixty degrees. The atoms in the first triangle are first neighbors to the initial site and second neighbors to the vacancy, while the reverse is true for the atoms in the second triangle. The migration-energy barrier therefore has a double hump. If the minimum between the humps is deep, the migration energy will be determined by the triangle with the highest maximum. However, if the minimum is shallow relative to kT, the activated position can be taken midway between the humps. The situation can be somewhat complicated, and involves factors about which we have little precise information. We can, however, define $z_\alpha^\ddagger$ and $z_\beta^\ddagger$ to be the number of atoms on α sites and on β sites, respectively, that surround the activated position. Then, following the procedure used to get (4.7), the energy at the saddle point for the motion of an A atom out of an α site is

$$-v_{AA}^\ddagger \left(z_\alpha^\ddagger \frac{N_{A\alpha}}{L_\alpha} + z_\beta^\ddagger \frac{N_{A\beta}}{L_\beta} \right) - v_{AB}^\ddagger \left(z_\alpha^\ddagger \frac{N_{B\alpha}}{L_\alpha} + z_\beta^\ddagger \frac{N_{B\beta}}{L_\beta} \right) \qquad (4.8)$$

where $v_{AA}^\ddagger$ and $v_{AB}^\ddagger$ are the interaction energies between AA and AB pairs, respectively, when the A atom is at the saddle point. We have assumed that these interaction energies are the same for all AA pairs and for all AB pairs at the saddle point. This is not necessarily true, in general, but the form of our final results will not be affected by this assumption.

The migration energy for an A atom on an α site is just the difference of (4.8) and (4.7):

$$\bar{E}_{A\alpha}{}^m = -v_{AA}^\ddagger \left(z_\alpha^\ddagger \frac{N_{A\alpha}}{L_\alpha} + z_\beta^\ddagger \frac{N_{A\beta}}{L_\beta} \right) - v_{AB}^\ddagger \left(z_\alpha^\ddagger \frac{N_{B\alpha}}{L_\alpha} + z_\beta^\ddagger \frac{N_{B\beta}}{L_\beta} \right)$$

$$+ (z - 1) \frac{N_{A\beta}}{L_\beta} v_{AA} + (z - 1) \frac{N_{B\beta}}{L_\beta} v_{AB} \qquad (4.9)$$

This can now be expressed in terms of the long-range-order parameter. Multiplying (4.9) by $N_{A\alpha}/N_A$ and using (3.19) and (3.20), we obtain

$$\frac{N_{A\alpha}\bar{E}_{A\alpha}{}^m}{N_A} = \tfrac{1}{4}\{(z - 1)(1 - R^2)v_{AA} + (z - 1)(1 + R^2)v_{AB}$$

$$- [z_\alpha^\ddagger(1 + R)^2 + z_\beta^\ddagger(1 - R^2)]v_{AA}^\ddagger - [z_\alpha^\ddagger(1 - R^2) + z_\beta^\ddagger(1 + R)^2]v_{AB}^\ddagger\} \qquad (4.10)$$

If we perform a similar calculation for the migration energy of an A atom out of a β site to get $N_{A\beta}\bar{E}_{A\beta}{}^m/N_A$ and add the result to (4.10), then (4.6) gives

$$
\begin{aligned}
\bar{E}_A{}^m = {}&\tfrac{1}{2}(z-1)[(1-R^2)v_{AA} + (1+R^2)v_{AB}] \\
&- \tfrac{1}{4}v_{AA}^{\ddagger}[z_\alpha^{\ddagger}(1+R)^2 + z_\beta^{\ddagger}(1-R)^2] \\
&- \tfrac{1}{4}v_{AA}^{\ddagger}[z_\beta^{\ddagger}(1-R)^2 + z_\alpha'^{\ddagger}(1-R^2)] \\
&- \tfrac{1}{4}v_{AB}^{\ddagger}[z_\alpha^{\ddagger}(1-R^2) + z_\beta^{\ddagger}(1+R)^2] \\
&- \tfrac{1}{4}v_{AB}^{\ddagger}[z_\beta'^{\ddagger}(1-R^2) + z_\alpha'^{\ddagger}(1-R)^2]
\end{aligned}
\tag{4.11}
$$

In this equation, $z_\alpha'^{\ddagger}$ and $z_\beta'^{\ddagger}$ are the saddle-point configuration numbers when the diffusing atom starts from a site on the β sublattice. Since the number of α sites surrounding an activated atom coming from a β site is equal to the number of β sites surrounding an activated atom coming from an α site, we have that

$$
\begin{aligned}
z_\alpha'^{\ddagger} &= z_\beta^{\ddagger} \\
z_\beta'^{\ddagger} &= z_\beta^{\ddagger}
\end{aligned}
\tag{4.12}
$$

Equation 4.11 can be written in the form

$$
\bar{E}_A{}^m = \bar{E}_A{}^m(0) + \Delta_A R^2
\tag{4.13}
$$

where

$$
\bar{E}_A{}^m(0) = \tfrac{1}{2}(z-1)(v_{AA} + v_{AB}) - \tfrac{1}{2}(z_\alpha^{\ddagger} + z_\beta^{\ddagger})(v_{AA}^{\ddagger} + v_{AB}^{\ddagger})
\tag{4.14}
$$

and

$$
\Delta_A = \tfrac{1}{2}(z-1)(v_{AB} - v_{AA}) - \tfrac{1}{2}(z_\alpha^{\ddagger} - z_\beta^{\ddagger})(v_{AA}^{\ddagger} - v_{AB}^{\ddagger})
\tag{4.15}
$$

The calculation of the average migration energy can be computed in the same way, with the result that

$$
\bar{E}_B{}^m = \bar{E}_B{}^m(0) + \Delta_B R^2
\tag{4.16}
$$

where

$$
\bar{E}_B{}^m(0) = \tfrac{1}{2}(z-1)(v_{BB} + v_{AB}) - \tfrac{1}{2}(z_\alpha^{\ddagger} + z_\beta^{\ddagger})(v_{BB}^{\ddagger} + v_{AB}^{\ddagger})
\tag{4.17}
$$

and

$$
\Delta_B = \tfrac{1}{2}(z-1)(v_{AB} - v_{BB}) - \tfrac{1}{2}(z^{\ddagger} - z_\alpha^{\ddagger})(v_{BB}^{\ddagger} - v_{AB}^{\ddagger})
\tag{4.18}
$$

Using (4.13) and (4.16) in (4.4) and (4.5) gives the average jump frequencies for A and B atoms as

$$
\Gamma_A = \nu_A e^{-[\bar{E}_A{}^m(0) + \Delta_A R^2]/kT}
\tag{4.19}
$$

$$
\Gamma_B = \nu_B e^{-[\bar{E}_B{}^m(0) + \Delta_B R^2]/kT}
\tag{4.20}
$$

The jump frequency therefore has the same kind of an exponential R^2 dependence as the vacancy concentration. Note that the jump frequencies of the two kinds of atoms are different.

9.5   THE DIFFUSION COEFFICIENT IN β-BRASS

Now that the effect of order on the vacancy concentration and the jump frequency have been studied, the results can be used to analyze the diffusion coefficient in an AB alloy. According to the results of Chapter 8, the diffusion coefficient for a species moving by a vacancy mechanism is

$$D = \frac{fr_L^2 \Gamma_V C_V}{6} \tag{5.1}$$

where r_L is the jump distance, C_V is the probability that a site is vacant, and Γ_V is the jump frequency for a vacancy mechanism, and f is the correlation factor. In an alloy, sites with different environments will contribute differently to the diffusion coefficient. This could be studied by taking the average of the product of the jump frequency and vacancy concentration over all sites and doing a moment expansion in the Kirkwood manner. In treating the jump frequency and vacancy concentration separately, however, we have consistently stayed within the Bragg-Williams approximation by ignoring higher moments. We will do this also for the diffusion coefficient by simply using the average jump frequency and average vacancy concentration. Thus, the diffusion coefficients for atoms of type A and of type B will be written as

$$D_A = \frac{fr_L^2}{6} \Gamma_A C_V \tag{5.2}$$

$$D_B = \frac{fr_L^2}{6} \Gamma_B C_V \tag{5.3}$$

where C_V is given by (3.21), and the jump frequencies are given by (4.19) and (4.20). Thus, the diffusion coefficients take the form

$$D_A = D_0^A e^{-Q_A(R)/kT} \tag{5.4}$$
$$D_B = D_0^B e^{-Q_B(R)/kT} \tag{5.5}$$

where D_0^A and D_0^B are constants given by

$$D_0^A = \tfrac{1}{6} fr_L^2 \nu_A \tag{5.6}$$

$$D_0^B = \tfrac{1}{6} fr_L^2 \nu_B \tag{5.7}$$

and the heats of activation $Q_A(R)$ and $Q_B(R)$ are defined by

$$Q_A(R) = \bar{E}_A^m(0) + E_V^0 + R^2\left(\Delta_A + \frac{zv}{4}\right) \tag{5.8}$$

$$Q_B(R) = \bar{E}_B^m(0) + E_V^0 + R^2\left(\Delta_B + \frac{zv}{4}\right) \tag{5.9}$$

These equations can be put in a more convenient form by defining the following quantities:

$$\varphi = \frac{2kT}{zv} = \frac{T}{T_c} \tag{5.10}$$

$$\alpha_A \equiv -\frac{\Delta Q_A}{Q_A(0)} \tag{5.11}$$

$$\alpha_B \equiv \frac{\Delta Q_B}{Q_B(0)} \tag{5.12}$$

$$Q_A(0) \equiv \bar{E}_A{}^m(0) + E_V{}^0 \tag{5.13}$$

$$Q_B(0) \equiv \bar{E}_B{}^m(0) + E_V{}^0 \tag{5.14}$$

$$\Delta Q_A \equiv \Delta_A + \frac{zv}{4} \tag{5.15}$$

$$\Delta Q_B \equiv \Delta_B + \frac{zv}{4} \tag{5.16}$$

Then, using (5.7) to (5.15), (5.3) and (5.4) become

$$D_A = D_0{}^A e^{-2Q_A(0)(1+\alpha_A R^2)/zv\varphi} \tag{5.17}$$

$$D_B = D_0{}^B e^{-2Q_B(0)(1+\alpha_B R^2)/zv\varphi} \tag{5.18}$$

Careful measurements of the diffusion of copper and zinc in β-brass have been made* that can be used to test the validity of these results. To do this it is necessary to know the constants $D_0{}^A$, $D_0{}^B$, $Q_A(0)$, and $Q_B(0)$. These constants can be obtained from the experimental data for temperatures above the critical temperature, where $R = 0$. The results of a least squares analysis of the log D versus $1/T$ plots above the critical temperature gives the following results:

$$\ln D_{Zn} = -\frac{5.39}{\varphi} - 2.633 \tag{5.19}$$

$$\ln D_{Cu} = -\frac{6.71}{\varphi} - 1.785 \tag{5.20}$$

so that below the critical temperature, (5.16) and (5.17) give, for β-brass,

$$\ln D_{Zn} = -5.39 \frac{T_c}{T}(1 + \alpha_{Zn} R^2) - 2.633 \tag{5.21}$$

$$\ln D_{Cu} = -6.71 \frac{T_c}{T}(1 + \alpha_{Cu} R^2) - 1.785 \tag{5.22}$$

* Kuper, A. B., Lazarus, D., Manning, J. R., and Tomizuka, C. T., "Diffusion in Ordered and Disordered Copper-Zinc," *Phys. Rev.* **104**, 1536 (1956).

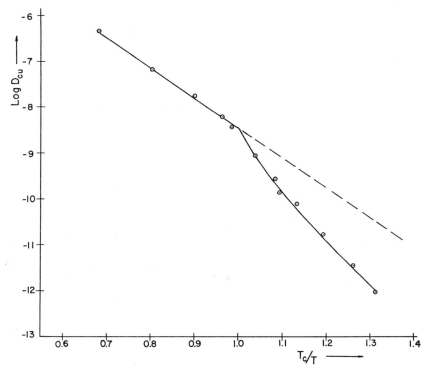

Fig. 9.1. Diffusion coefficient of copper in β-brass.

All that is necessary to compare these equations to experiment is to compute
the constants α_{Zn} and α_{Cu}. This has been done* by curve-fitting the experimental points to these equations. The results are shown in Figure 9.1 where
the line represents the theoretical equation, and the circles the experimental
points for copper. In performing the computation it is necessary to know
R as a function of T. This information was obtained from the experimental
data of Chipman and Warren.†

The values of the α's determined from the curve-fitting procedure are:

$$\alpha_{Zn} = 0.369 \tag{5.23}$$

$$\alpha_{Cu} = 0.204 \tag{5.24}$$

It is of interest to try to separate the effects of migration and formation
energies. This can be done because an experimental value of the migration

* Girifalco, L. A., "Vacancy Concentration and Diffusion in Order-Disorder Alloys,"
J. Phys. Chem. Solids **24,** 323 (1964).

† Chipman, D. and Warren, B. E., "X-Ray Measurement of Long Range Order in
β-Brass," *J. Appl. Phys.*, **21, 696** (1950).

energy of zinc is available. Work on the internal friction of β-brass strongly suggests that the migration energy of zinc in ordered material is 15 kcal/mole.* The activation energy for diffusion at $R = 1$ is

$$Q_{Zn}(1) = \bar{E}_{Zn}{}^m(1) + \bar{E}_V(1) \qquad (5.25)$$

where $\bar{E}_{Zn}{}^m(1)$ and $\bar{E}_V(1)$ are the migration energy for zinc, and the vacancy formation energy in β-brass, with $\bar{E}_{Zn}{}^m(1) = 15$ kcal/mole. Thus, we have that $\bar{E}_V(1) = 10$ kcal/mole, because the activation energy for zinc diffusion at perfect order is 25 kcal/mole. Furthermore, since $Q_{Cu}(1) = 27.38$ kcal/mole, and

$$Q_{Cu}(1) = \bar{E}_{Cu}{}^m(1) + \bar{E}_V(1) \qquad (5.26)$$

we have that $\bar{E}_{Cu}{}^m = 17.4$ kcal/mole. Now we can compute $E_V{}^0$ since

$$\bar{E}_V(1) = E_V{}^0 + \frac{zv}{4} \qquad (5.27)$$

From the critical temperature of β-brass ($T_c = 468°C$) the last term in (5.27) is 0.735 kcal/mol, so $E_V{}^0$ is readily computed. Finally, using the experimentally determined values of the activation energies for diffusion at zero order, the migration energies at zero order are obtained. The results are summarized in Table 9.1 and have the following analytical representation as

Table 9.1. Energy Parameters for Self-Diffusion in β-Brass (kcal/mole)

	Q_{Zn}	$\bar{E}^m_{Zn}$	Q_{Cu}	$\bar{E}^m_{Cu}$	E_V
S = 1	25.03	15	27.38	17.4	10
S = O	18.73	9.04	22.74	13.5	9.27

functions of R:

$$Q_{Zn}(R) = 18.73(1 + 0.369R^2)$$
$$Q_{Cu}(R) = 22.74(1 + 0.204R^2)$$
$$\bar{E}_{Zn}{}^m(R) = 9.04(1 + 0.660R^2) \qquad (5.28)$$
$$\bar{E}_{Cu}{}^m(R) = 13.5(1 + 0.291R^2)$$
$$E_V(R) = 9.27(1 + 0.0787R^2)$$

where all the energies are in kcal/mole.

* Clarebrough, L. M., "Internal Friction of β-Brass," *Acta Met.*, **5**, 413 (1957).

We see that the simple approach presented here is rather successful. However, careful examination of the diffusion data (Figure 9.1) shows that the diffusion coefficients begin to deviate from the Arrhenius linearity somewhat above the critical temperature, whereas our theory states that deviations should begin at the critical temperature. The observed deviation above T_c is the result of short-range ordering, which is neglected in the treatment given here.

appendix I

COMBINATORIAL PROBLEMS

IN STATISTICAL MECHANICS

A. ENSEMBLE STATISTICS

In the ensemble statistics, it is necessary to compute the number of complexions of the ensemble for a given distribution. A distribution is defined by a set of numbers $(N_1, N_2, \ldots, N_j, \ldots)$ such that N_j is the number of member systems in the jth state. The number of complexions for a given distribution is the number of *states of the ensemble* consistent with the specification of the number of member systems in the various *system states*. Now the members of the ensemble are macroscopic systems and are therefore distinguishable from one another. Therefore, we need to count the number of ways of arranging X systems, such that N_1 are in state 1, N_2 in state $2, \ldots, N_j$ in state $j \cdots$ and so on with

$$\sum_j N_j = X \tag{I.1}$$

Clearly, this is just the same as the number of ways of putting X marbles in boxes such that N_1 are in the first box, N_2 in the second box, and so on, without regard to the order of arrangement of marbles within a particular box. Let us denote this number by W.

Note that if we were arranging the marbles in such a way that the ordering of marbles in a particular box must be taken into account, then the total number of possible arrangements would be $X!$. For this case each box would have a set of subcompartments, and a permutation of marbles among the subcompartments would be counted as a new arrangement. If we put the marbles in their positions one at a time, the first could be placed in X ways; X now being the total number of available subcompartments as well as the number of marbles. The second could be placed in $(X - 1)$ ways, since one place is already filled, the third in $(X - 2)$ ways, and so on. The total number of ways of arranging X marbles in the boxes when the order of positioning them is accounted for is therefore the product $X(X - 1) \times (X - 2) \cdots = X!$ But we are not really interested in this problem; we want W, in which the arrangement within boxes is neglected. Obviously W does not take all conceivable permutations into account and is therefore smaller than $X!$. If, however, we multiply W by the number of ways of ordering the N_1 marbles in the first box, the N_2 marbles in the second box, and so on, we will obviously get $X!$, the total number of possible permutations. That is,

$$W N_1! \, N_2! \cdots = X!$$

or

$$W = \frac{X!}{\prod\limits_{j} N_j!} \tag{I.2}$$

since $N_j!$ is the number of permutations within the jth box. This is just the number of complexions of the ensemble for a given distribution.

B. MAXWELL-BOLTZMANN STATISTICS

In the particle statistics, all particle states are divided into groups such that the jth group consists of states with nearly the same energy $\epsilon_j \pm \Delta\epsilon$, where $\Delta\epsilon \ll \epsilon_j$. We want to calculate the total number of states of the system of N particles when N_1 particles are in the first group, N_2 particles in the second group, and so on. In the Maxwell-Boltzmann case, the particles are distinguishable and there is no limit to the number of particles in each group, and we denote the desired number of complexions by W_{MB}.

First, we calculate the number of ways of putting the N particles in groups such that there are N_j in the jth group. The result has just the same form as for the ensemble statistics:

$$\frac{N!}{\prod\limits_{j} N_j!} \tag{I.3}$$

However, there is a difference between our present problem and the ensemble problem. In the ensemble statistics, all states in a j group were identical. In the present problem, all states in a j group are different although they have very nearly the same energy. Also, there is no limit to the number of particles that can go into any particular state in a given j group. Therefore, the above expression (I.3) must be multiplied by the number of ways of arranging N_j particles among the number of states in each j group. If the jth group has ω_j states, then N_j particles can be put in these states in $\omega_j{}^{N_j}$ ways. This is so, because one particle can be put in any of ω_j states; a second particle can also be placed in ω_j ways since any number of particles can occupy a particular state. Continuing the process, N_j particles can be placed in $\omega_j{}^{N_j}$ ways. To get W_{MB} we therefore multiply (I.3) by $\omega_1{}^{N_1}$, $\omega_2{}^{N_2}$, and so on, that is,

$$W_{\mathrm{MB}} = N! \prod_j \frac{\omega_j{}^{N_j}}{N_j!} \tag{I.4}$$

C. FERMI-DIRAC STATISTICS

Again we have N particles to be distributed among particle states that are divided into groups such that the jth group, with energy $\epsilon_j \pm \Delta\epsilon$, contains N_j particles, and we want to compute the number of ways of doing this. But now the particles are not distinguishable and there can be at most one particle in a given state.

First, let us confine our attention to one j-group that has ω_j states, and let us calculate the number of ways of putting in N_j particles. If the particles were distinguishable, this would be just $\omega_j!/(\omega_j - N_j)!$ because the first particle could be placed in ω_j ways, the second in $(\omega_j - 1)$ ways, and so on, and the last in $(\omega_j - N_j + 1)$ ways, and the product of all these gives

$$\omega_j(\omega_j - 1)(\omega_j - 2) \cdots (\omega_j - N_j + 1) = \frac{\omega_j.}{(\omega_j - N_j)!} \tag{I.5}$$

But the particles are *not* distinguishable and the above expression must be corrected to get the number we are after. This correction consists of dividing (I.5) by $N_j!$, because that many permutations of the particles among the states count as the same arrangement when the particles are indistinguishable. The number of ways of putting N_j particles in the j-group for the Fermi-Dirac case is therefore

$$\frac{\omega_j!}{N_j!\,(\omega_j - N_j)!} \tag{I.6}$$

To obtain the total number of complexions for all j-groups, we just take the product of (I.6) over all j. There is no need to worry about permutations from one j-group to another because the particles are indistinguishable and the N_j are fixed so that any interchange of particles between j-groups does not give a new arrangement. The desired number, W_{FD} is therefore

$$W_{\mathrm{FD}} = \prod_j \frac{\omega_j!}{N_j!\,(\omega_j - N_j)!} \tag{I.7}$$

D. BOSE-EINSTEIN STATISTICS

We want the number of ways of distributing N particles among states so that N_j are in the jth group that has ω_j particle states. The particles are indistinguishable and there is no limit to the number of particles that can be in any state. Just as in the Fermi-Dirac case, we can compute the number of complexions for one j-group and then take the product over all j. This procedure is valid because of the particle indistinguishability.

Now pick out a j-group. Each possible distribution of the N_j particles among the ω_j states can be schematically represented by the following scheme:

$$|3, 7|\ 5\ |\cdot|\ 1, 2, 9\ |23|\cdots$$
$$\omega_j =\quad 1\quad 2\quad 3\quad\ 4\qquad 5\quad 6\cdots$$

The vertical lines separate the particle states and the numbers label the particles in each state. Thus the distribution represented by the above arrangement is the one in which particles 3 and 7 are in state 1, particle 5 is in state 2, there are no particles in state 3, particles 1, 2, and 9 are in state 4, particle 23 is in state 5, and so on. Clearly then, if the particles were distinguishable, the number of complexions for the j-group would be the number of ways of permuting the integers 1 to N_j and the vertical lines separating the states. There are a total of ω_j vertical lines, but in our scheme, the first one must be kept in place so that we do not have any particles not in some state. Therefore we can permute only $(\omega_j - 1)$ lines. The number of permutations of $(N_j + \omega_j - 1)$ objects (integers plus lines) among ordered positions is $(N_j + \omega_j - 1)!$. This is greater than the number of j-complexions (arrangements of the particles in the states of the j-group) for two reasons. First, the permutation of a set of symbols consisting of a line and the integers following it with another such set does not alter the combined state of the j-group. There are $(\omega_j - 1)$ such sets of symbols so we must divide out $(\omega_j - 1)!$. Second, permutations of the N_j particles with each other leaves everything unchanged because they are indistinguishable so we must divide through by $N_j!$. The number of ways of putting the N_j particles in the ω_j states is therefore

$$\frac{(N_j + \omega_j - 1)!}{N_j!\,(\omega_j - 1)!}$$

Taking the product over all j gives the number of complexions for the Bose-Einstein case, that is,

$$W_{\mathrm{BE}} = \prod_j \frac{(N_j + \omega_j - 1)!}{N_j!\,(\omega_j - 1)!} \tag{I.8}$$

appendix II

THE METHOD OF

UNDETERMINED MULTIPLIERS

Lagrange's method of undetermined multipliers is used frequently in physical problems to determine the maxima or minima of certain functions subject to subsidiary conditions. In this appendix, a brief description of the mathematics involved is presented without encumbering the analysis with statistical mechanics.

We are given a known function F of a number of variables $y_1, y_2, \ldots$, and so on

$$F = F(y_1, y_2 \cdots) \tag{II.1}$$

The variables $y_1, y_2, \ldots$ are themselves functions of a set of parameters $x_1, x_2, \ldots$, and so on, that is,

$$y_1 = y_1(x_1, x_2 \cdots)$$
$$y_2 = y_2(x_1, x_2 \cdots)$$
$$\begin{array}{cc} \cdot & \cdot \\ \cdot & \cdot \\ \cdot & \cdot \end{array} \tag{II.2}$$

Now we want to find the functional form of equations (II.2) that gives F a stationary value (makes F maximum or minimum).* Throughout the

---

* In the theory of the canonical ensemble, for example, F corresponds to $\ln W\{n_i\}$, $y_1, y_2 \cdots$ correspond to $n_1, n_2 \cdots$ and $x_1, x_2 \cdots$ correspond to $E_1, E_2 \cdots$.

search for this functional form, the x's are taken to be given and to remain constant.

For $F(y_1, y_2 \cdots)$ to be stationary, we only need any variation in F, resulting from a variation in the y's, to be zero. That is,

$$\delta F = \frac{\partial F}{\partial y_1} \delta y_1 + \frac{\partial F}{\partial y_2} \delta y_2 + \cdots = \sum_{i=1}^{r} \frac{\partial F}{\partial y_i} \delta y_i = 0 \tag{II.3}$$

where r is the total number of y's. If the y_i are all independent, each coefficient of δy_i must be zero because the variations δy_i are completely arbitrary. The problem is then solved. All we need to do is set each partial derivative equal to zero. We are much more interested in the case in which the y_i are not completely independent, but in which some functions of the y_i and x_i exist that must also be stationary. Let us assume that we have two such functions G_1 and G_2:

$$G_1 = G_1(y_1, y_2 \cdots \mid x_1, x_2 \cdots) \tag{II.4}$$

$$G_2 = G_2(y_1, y_2 \cdots \mid x_1, x_2 \cdots) \tag{II.5}$$

whose form is known, and whose variation is zero with respect to a variation in the y's.

$$\delta G_1 = \sum_{i=1}^{r} \frac{\partial G_1}{\partial y_i} \delta y_i = 0 \tag{II.6}$$

$$\delta G_2 = \sum_{i=1}^{r} \frac{\partial G_2}{\partial y_i} \delta y_i = 0 \tag{II.7}$$

The y_i are no longer independent because of the relations (II.4) and (II.5). Since the variations in G_1 and G_2 are zero, G_1 and G_2 are constants and we have two relations among the y_i. But we can take this lack of complete independence into account in the following manner: Because of the above equations, we have

$$a_1 \, \delta G_1 = 0$$
$$a_2 \, \delta G_2 = 0 \tag{II.8}$$

and therefore, adding these to $\delta F = 0$,

$$\delta F + a_1 \, \delta G_1 + a_2 \, \delta G_2 = 0 \tag{II.9}$$

for *any* constants a_1 and a_2. This means that we can combine (II.3), (II.6), and (II.7) according to (II.9) to get

$$\sum_{i=1}^{r} \left[\frac{\partial F}{\delta y_i} + a_1 \frac{\partial G_1}{\partial y_i} + a_2 \frac{\partial G_2}{\partial y_i} \right] \delta y_i = 0 \tag{II.10}$$

and this equation will be true regardless of the values of a_1 and a_2. Now let us choose a_1 and a_2 to have values such that they satisfy the simultaneous equations

$$\left(\frac{\partial F}{\partial y_1}\right)_0 + a_1\left(\frac{\partial G_1}{\partial y_1}\right)_0 + a_2\left(\frac{\partial G_2}{\partial y_1}\right)_0 = 0 \qquad (\text{II}.11)$$

$$\left(\frac{\partial F}{\partial y_2}\right)_0 + a_1\left(\frac{\partial G_1}{\partial y_2}\right)_0 + a_2\left(\frac{\partial G_2}{\partial y_2}\right)_0 = 0 \qquad (\text{II}.12)$$

the derivatives being evaluated at the values of the y_i that make F a minimum (or maximum). Once this is done, the first two terms in (II.10) reduce to zero and y_1 and y_2 are fixed. Equation II.10 then becomes

$$\sum_{i=3}^{r} \left[\frac{\partial F}{\partial y_i} + a_1\frac{\partial G_1}{\partial y_i} + a_2\frac{\partial G_2}{\partial y_i}\right] \delta y_i = 0 \qquad (\text{II}.13)$$

The derivatives, of course, all being evaluated at values of y_i for which F is stationary. But now all the remaining y_i are independent and the corresponding δy_i are arbitrary, because there were only two conditions on the y_i and two y_i have been fixed. This means that all the coefficient of δy_i in (II.13) must be zero and we have for all i,

$$\frac{\partial F}{\partial y_i} + a_1\frac{\partial G_1}{\partial y_i} + a_2\frac{\partial G_2}{\partial y_i} = 0 \qquad (\text{II}.14)$$

This is the solution to our problem. Since F, G_1, and G_2 are known functions of the y_i, and since G_1 and G_2 contain the parameters x_j, (II.14) can be solved for y_i in terms of the x_j. These solutions still contain a_1 and a_2, but these multipliers can now be removed by substituting the solutions $y_i = y_i(x_j, a_1, a_2)$ into (II.4) and (II.5) and solving for a_1 and a_2. In actual applications to physical problems, the undetermined multipliers have important physical interpretations, so that they are usually retained in the functional form of the y_i.

The generalization to any number of subsidiary conditions with corresponding multipliers a_1, a_2, a_3, . . . , and so on is obvious.

appendix III

STIRLING'S APPROXIMATION

For large, positive integers N, Stirling's approximation states that $N!$ is approximately given by

$$\ln N! \simeq N \ln N - N \tag{III.1}$$

This relation can be proven by using the relation between a sum and an integral. Since $\ln x$ is a monatomic increasing function of x, we have that

$$\sum_{j=1}^{N} \ln j < \int_0^N \ln x \, dx < \sum_{j=1}^{N} \ln (j + 1) \tag{III.2}$$

These inequalities can be clearly made evident by graphing the function $\ln x$, and comparing it to the summed areas of the unit stepwise divisions representing the sums in (III.2). From (III.2), it follows that

$$\int_0^{N-1} \ln x \, dx < \sum_{j=1}^{N} \ln j < \int_0^N \ln x \, dx \tag{III.3}$$

Performing the integrals, and recognizing that the sum in the middle is $\ln N!$, we get

$$(N - 1)[\ln (N - 1) - 1] < \ln N! < N[\ln N - 1] \tag{III.4}$$

For large N,

$$\ln (N - 1) = \ln N \left(1 - \frac{1}{N}\right) = \ln N + \ln \left(1 - \frac{1}{N}\right)$$

so that the outside members of the inequality (III.4) differ by a quantity of order $\ln N$, so that

$$\ln N! = N \ln N - N + 0 \, (\ln N) \tag{III.5}$$

where $0 \, (\ln N)$ is the term of order $\ln N$.   For large N, therefore, we recover (III.1).

The following table shows that Stirling's approximation is a good one for remarkably small values of N.

N	$\ln N!$	$N \ln N - N$
50	148	146
100	363	360
200	864	860
300	1415	1411
400	2000	1997
500	2611	2607
600	3242	3238

appendix IV

SUMS AND INTEGRALS

In this appendix we evaluate certain sums and integrals that are useful in statistical mechanics. The sums are:

$$S_1 \equiv \sum_{j=0}^{\infty} x^j = \frac{1}{1-x} \, ; \qquad x < 1 \tag{IV.1}$$

$$S_2 \equiv \sum_{j=0}^{\infty} j x^j = \frac{x}{(1-x)^2} \, ; \qquad x < 1 \tag{IV.2}$$

$$S_3 \equiv \sum_{j=1}^{\infty} \frac{(-1)^{j-1}}{j^2} = \frac{\pi^2}{12} \tag{IV.3}$$

$$S_4 \equiv \sum_{j=1}^{\infty} \frac{1}{j^2} = \frac{\pi^2}{6} \tag{IV.4}$$

$$S_5 \equiv \sum_{j=1}^{\infty} \frac{1}{j^4} = \frac{\pi^4}{90} \tag{IV.5}$$

To prove (IV.1), consider the partial sums

$$S_1(n) \equiv \sum_{j=0}^{n} x^j = 1 + x + x^2 + \cdots + x^n \tag{IV.6}$$

Multiply this by x to get

$$x S_1(n) = x \sum_{j=0}^{n} x^j = x + x^2 + x^3 + \cdots + x^{n+1} \tag{IV.7}$$

Now subtract (IV.7) from (IV.6), and solve for $S_1(n)$. The result is

$$S_1(n) = \sum_{j=0}^{n} x^j = \frac{1 - x^{n+1}}{1 - x} \tag{IV.8}$$

Since $x < 1$, taking the limit of (IV.8) as $n \to \infty$ gives (IV.1). To obtain (IV.2), differentiate (IV.1) with respect to x to get

$$\frac{dS_1}{dx} = \sum_{j=0}^{\infty} j x^{j-1} = \frac{1}{(1 - x)^2} \tag{IV.9}$$

from which (IV.2) follows immediately.

(IV.3) and (IV.4) can be obtained by expanding the function $f(x) = x^2$ in a Fourier series in the interval $-\pi \le x \le \pi$. The result is

$$x^2 = \frac{\pi^2}{3} + 4 \sum_{j=1}^{\infty} \frac{(-1)^j}{j^2} \cos jx \tag{IV.10}$$

Letting $x = 0$ in this equation gives (IV.3), and letting $x = \pi$ gives (IV.4).

A similar procedure works for (IV.5). Expand $f(x) = x^4$ in a Fourier series to get

$$x^4 = \frac{\pi^4}{5} + 8\pi^2 \sum_{j=1}^{\infty} \frac{(-1)^j}{j^2} \cos jx - 48 \sum_{j=1}^{\infty} \frac{(-1)^j}{j^4} \cos jx \tag{IV.11}$$

Letting $x = \pi$, (IV.11) becomes

$$\pi^4 = \frac{\pi^4}{5} + 8\pi^2 \sum_{j=1}^{\infty} \frac{1}{j^2} - 48 \sum_{j=1}^{\infty} \frac{1}{j^4} \tag{IV.12}$$

Replacing the first sum on the right by $\pi^2/6$ according to (IV.4) and solving for the second sum gives (IV.5).

We now want to prove the following:

$$A_1 \equiv \int_{-\infty}^{\infty} e^{-ax^2} dx = \left(\frac{\pi}{a}\right)^{1/2} \tag{IV.13}$$

$$A_2 \equiv \int_{0}^{\infty} x^2 e^{-ax^2} dx = \frac{1}{4}\left(\frac{\pi}{a^3}\right)^{1/2} \tag{IV.14}$$

$$A_3 \equiv \int_{0}^{\infty} x^{1/2} e^{-ax} dx = \frac{1}{2}\left(\frac{\pi}{a^3}\right)^{1/2} \tag{IV.15}$$

$$A_4 \equiv \int_{0}^{\infty} x^{3/2} e^{-ax} dx = \frac{3}{4}\left(\frac{\pi}{a^5}\right)^{1/2} \tag{IV.16}$$

$$A_5 \equiv \int_{0}^{\infty} x^{2n} e^{-jx} dx = \frac{2n!}{j^{2n+1}} ; \qquad n, j = \text{positive integers} \tag{IV.17}$$

(IV.13) is obtained by first changing variables to $u \equiv x\sqrt{a}$ to get

$$A_1 = \frac{1}{\sqrt{a}} \int_{-\infty}^{\infty} e^{-u^2} \, du \qquad \text{(IV.18)}$$

which can be squared to give

$$A_1{}^2 = \frac{1}{a} \int_{-\infty}^{\infty} \int_{-\infty}^{\infty} e^{-(u^2+v^2)} \, du \, dv \qquad \text{(IV.19)}$$

Now transform to polar coordinates so that

$$r^2 \equiv u^2 + v^2$$
$$du \, dv = 2\pi r \, dr$$

which transforms (IV.19) into

$$A_1{}^2 = \frac{2\pi}{a} \int_{0}^{\infty} e^{-r^2} r \, dr \qquad \text{(IV.20)}$$

Now, letting $y \equiv r^2$, this gives

$$A_1{}^2 = \frac{\pi}{a} \int_{0}^{\infty} e^{-y} \, dy = \frac{\pi}{a} \qquad \text{(IV.21)}$$

from which (IV.13) follows.

(IV.14) is easily obtained from (IV.13) by differentiating with respect to a:

$$\frac{dA_1}{da} = -\int_{-\infty}^{\infty} x^2 e^{-ax^2} \, dx = -\frac{1}{2}\left(\frac{\pi}{a^3}\right)^{1/2} \qquad \text{(IV.22)}$$

which is just (IV.14), since the integrand is an even function.

To prove (IV.15), transform the integration variable to $u \equiv x^{1/2}$, so that

$$A_3 = 2 \int_{0}^{\infty} u^2 e^{-au^2} \, du \qquad \text{(IV.23)}$$

Comparing this with (IV.14) shows that (IV.15) is correct.

Differentiation of (IV.15) gives

$$\frac{dA_3}{da} = -\int_{0}^{\infty} x^{3/2} e^{-ax} \, dx = -\frac{3}{4}\left(\frac{\pi}{a^5}\right)^{1/2} \qquad \text{(IV.24)}$$

which proves (IV.16).

The proof of (IV.17) starts with the variable transformation $y = jx$ so that

$$A_5 = \frac{1}{j^{2n+1}} \int_{0}^{\infty} y^{2n} e^{-y} \, dy \qquad \text{(IV.25)}$$

Equation IV.I7 then follows from a successive integration by parts; that is,

$$\int_0^\infty y^{2n}e^{-y}\,dy = -y^{2n}e^{-y}\Big|_0^\infty + 2n\int_0^\infty y^{2n-1}e^{-y}\,dy$$

$$= 2n\int_0^\infty y^{2n-1}e^{-y}\,dy$$

$$= 2n(2n-1)\int_0^\infty y^{2n-2}e^{-y}\,dy\cdots$$

$$= 2n!\int_0^\infty e^{-y}\,dy = 2n!$$

Putting this result in (IV.25), we get (IV.17).

Two integrals in the Debye theory of crystals are:

$$A_6 \equiv \int_0^\infty \frac{x^3\,dx}{e^x - 1} = \frac{\pi^4}{15} \tag{IV.26}$$

$$A_7 = \int_0^\infty \frac{x^4 e^{-x}}{(1 - e^{-x})^2}\,dx = \frac{4\pi^4}{15} \tag{IV.27}$$

Equation IV.26 can be obtained by using the series expansion

$$(e^x - 1)^{-1} = e^{-x} + e^{-2x} + e^{-3x} + \cdots \tag{IV.28}$$

so that

$$A_6 = \sum_{n=1}^\infty \int_0^\infty x^3 e^{-nx}\,dx \tag{IV.29}$$

Now let $y = nx$, so that (IV.29) can be written as

$$A_6 = \sum_{n=1}^\infty \frac{1}{n4}\int_0^\infty y^3 e^{-y}\,dy \tag{IV.30}$$

The integral can be evaluated by successive integration by parts:

$$\int_0^\infty y^3 e^{-y}\,dy = -y^3 e^{-y}\Big|_0^\infty + 3\int_0^\infty y^2 e^{-y}\,dy$$

$$= 3\int_0^\infty y^2 e^{-y}\,dy$$

$$= 6\int_0^\infty y e^{-y}\,dy$$

$$= 6\int_0^\infty e^{-y}\,dy = 6$$

Thus (1V.30) becomes

$$A_6 = 6\sum_{n=1}^\infty \frac{1}{n^4} \tag{IV.31}$$

The sum has been evaluated previously and is given by (IV.5). Using this result in (IV.31) we recover (IV.26).

Equation IV.27 can be obtained by integration of (IV.26) by parts. Thus:

$$\frac{\pi^4}{15} = \int_0^\infty \frac{x^3 \, dx}{e^x - 1} = \frac{x^4}{4(e^x - 1)}\Bigg|_0^\infty + \frac{1}{4}\int_0^\infty \frac{x^4 e^x}{(e^x - 1)^2} \, dx$$

$$= \frac{1}{4}\int_0^\infty \frac{x^4 e^{-x}}{(1 - e^{-x})^2} \, dx$$

which reduces to (IV.27).

appendix V

FERMI INTEGRALS

If $g(\epsilon)$ is a monotonically increasing function of ϵ whose value is zero when ϵ is zero, and $f(\epsilon)$ is the Fermi function defined by

$$f(\epsilon) = \frac{1}{e^{(\epsilon-\mu)/kT} + 1} \tag{V.1}$$

then the Fermi integral is defined by

$$I = \int_0^\infty g(\epsilon)f(\epsilon)\, d\epsilon$$

or $\qquad\qquad\qquad\qquad\qquad\qquad\qquad\qquad\qquad\qquad$ (V.2)

$$I = -\int_0^\infty F(\epsilon)\frac{\partial f}{\partial \epsilon}\, d\epsilon$$

The second of the equations in (V.2) is obtained from the first by an integration by parts and $F(\epsilon)$ is defined by

$$F(\epsilon) = \int_0^\epsilon g(x)\, dx \tag{V.3}$$

Since $\partial f/\partial \epsilon$ is practically zero except near $\epsilon = \mu$, a rapidly converging series expansion for I can be obtained by expanding $F(\epsilon)$ in a Taylor series about μ.

$$F(\epsilon) = \sum_{r=0}^\infty \frac{1}{r!}(\epsilon - \mu)^r F_r(\mu) \tag{V.4}$$

$F_r(\mu)$ being the rth derivative of $F(\epsilon)$ evaluated at $\epsilon = \mu$. That is,

$$F_r(\mu) = \left(\frac{d^r F}{d\epsilon^r}\right)_{\epsilon=\mu} \tag{V.5}$$

Equation V.2 can now be written as

$$I = -\sum_{r=0}^{\infty} \frac{F_r(\mu)}{r!} I_r \qquad \text{(V.6)}$$

I_r being defined by

$$I_r = \int_0^{\infty} (\epsilon - \mu)^r \frac{\partial f}{\partial \epsilon} d\epsilon \qquad \text{(V.7)}$$

Now let us work on this integral. From the definition of the Fermi function, the derivative is

$$\frac{\partial f}{\partial \epsilon} = -\frac{e^{(\epsilon-\mu)/kT}}{kT[1 + e^{(\epsilon-\mu)/kT}]^2}$$

Define a variable, z, by

$$z = \frac{(\epsilon - \mu)}{kT} \qquad \text{(V.8)}$$

so that

$$\frac{\partial f}{\partial \epsilon} = -\frac{e^z}{kT(1 + e^z)^2} \qquad \text{(V.9)}$$

and (V.7) becomes

$$I_r = -(kT)^r \int_{-\mu/kT}^{\infty} \frac{z^r e^z}{(1 + e^z)^2} dz \qquad \text{(V.10)}$$

First, we note that for all reasonable temperatures, μ/kT is large, and since the integrand in (V.10) is appreciably different from zero only near $z = 0$, the lower limit can be replaced by $-\infty$ with practically no loss in accuracy. Next we note that if r is odd, the integrand in (V.10) is antisymmetric, while if n is even it is symmetric. This is so because the derivative of the Fermi function is symmetric about z

$$\frac{e^z}{(1 + e^z)^2} = \frac{e^{-z}}{(1 + e^{-z})^2} \qquad \text{(V.11)}$$

as can be verified by a little algebra. This means that the integrals defined as I_r all vanish if r is odd, whereas if r is even so that $r = 2n$, where n is any integer, we can write

$$I_{2n} = -(kT)^{2n} \int_{-\infty}^{\infty} \frac{z^{2n} e^z}{(1 + e^z)^2} dz$$

$$I_{2n} = -2(kT)^{2n} \int_0^{\infty} \frac{z^{2n} e^z}{(1 + e^z)^2} dz \qquad \text{(V.12)}$$

$$I_{2n} = -2(kT)^{2n} \int_0^{\infty} \frac{z^{2n} e^{-z}}{(1 + e^{-z})^2} dz$$

The integral in the last of these equations can be related to known functions by expanding the denominator and integrating term by term. According to the binomial theorem

$$(1 + e^{-z})^{-2} = \sum_{j=1}^{\infty} (-1)^{j-1} j e^{-(j-1)z} \tag{V.13}$$

so that (V.12) becomes

$$I_{2n} = -2(kT)^{2n} \sum_{j=1}^{\infty} (-1)^{j-1} j \int_0^{\infty} z^{2n} e^{-jz} \, dz \tag{V.14}$$

The integrals in this equation are standard forms given by (IV.17)

$$\int_0^{\infty} z^2 n e^{-jz} \, dz = \frac{(2n)!}{j^{2n+1}} \tag{V.15}$$

and therefore (V.14) reduces to

$$I_{2n} = -2(kT)^{2n}(2n)! \sum_{j=1}^{\infty} \frac{(-1)^{j-1}}{j^{2n}} \tag{V.16}$$

This is what we were after, because the sums can be evaluated. The sum with $n = 1$ has already been given in Appendix IV by (IV.3) and has the value $\pi^2/12$. Putting this in (V.16) for $n = 1$, we get

$$I_2 = -\tfrac{1}{3}(\pi kT)^2 \tag{V.17}$$

Thus, we can write (V.6) to an approximation including the terms for $r = 0$ and $r = 2$ as

$$I = -F_0(\mu)I_0 + F_2(\mu) \frac{(\pi kT)^2}{6} \tag{V.18}$$

This is ordinarily sufficiently accurate. A more general development starts with recognizing that the sums in (V.16) are closely related to the Riemann zeta function, which has been extensively studied and whose values are known. The relation is

$$\sum_{j=1}^{\infty} \frac{(-1)^{j-1}}{j^{2n}} = \xi(2n)(1 - 2^{1-2n}) \tag{V.19}$$

For integral values of the argument, which is all that concerns us here, the zeta function $\xi(2n)$ has the values

$$\xi(2n) = 2^{2n-1} \frac{\pi^{2n}}{(2n)!} |B_{2n}| \tag{V.20}$$

where the B_n are the Bernoulli numbers, the first four of which are

$$B_2 = \tfrac{1}{6}; \qquad B_4 = -\tfrac{1}{30}; \qquad B_6 = 0.02381; \qquad B_8 = -\tfrac{1}{30} \tag{V.21}$$

Substituting (V.19) and (V.20) into (V.16),

$$I_{2n} = -(2\pi kT)^{2n} |B_{2n}| (1 - 2^{1-2n}) \tag{V.22}$$

Now go back to (V.6). We found that for odd values of r, I_r was zero, so we only need to substitute the I_{2n} from (V.22). If the first term in the series in (V.6) is written separately, then

$$I = -F_0(\mu) \int_0^\infty \frac{\partial f}{\partial \epsilon} d\epsilon + \sum_{n=1}^\infty \frac{1}{(2n)!} F_{2n}(\mu)(2\pi kT)^{2n} B_n (1 - 2^{1-2n}) \tag{V.23}$$

or, since

$$\int_0^\infty \frac{\partial f}{\partial \epsilon} d\epsilon = f(\epsilon) \Big|_0^\infty = -1$$

$$I = F_0(\mu) + \sum_{n=1}^\infty \frac{1}{(2n)!} F_{2n}(\mu)(2\pi kT)^{2n} B_n (1 - 2^{1-2n}) \tag{V.24}$$

$F_0(\mu)$ is, of course, just the value of $F(\epsilon)$ at $\epsilon = \mu$. Equation V.24 is the equation we were looking for. It is an expression for the Fermi integral as a sum whose first term is large compared to all the others, and whose second term is a rapidly converging series. In general, it is sufficiently accurate to retain only one or two terms in the series expansion. In truncated form, we write (V.24) as

$$I = F_0(\mu) + \frac{(2\pi kT)^2}{24} F_2(\mu) + 9.92 \times 10^{-4}(2\pi kT)^4 F_4(\mu) \tag{V.25}$$

Even this is more accurate than is usually required; taking only the first two terms is nearly always sufficient.

appendix VI

FERROMAGNETISM AND ORDER–DISORDER THEORY

In the Ising model of a ferromagnetic crystal, an electron spin is assigned to each lattice site. The spin can take on either of two values, ± 1, and the ferromagnetic energy is written as

$$E\{S_i\} = -J \sum_{\langle ij \rangle} S_i S_j \qquad \text{(VI.1)}$$

where S_i and S_j can assume the values ± 1, $E\{S_i\}$ is the energy of the crystal for a given distribution of spins on the lattice and J is a constant (the exchange integral). The summation is taken over all nearest neighbor pairs in the crystal. Equation VI.1 states that nearest neighbors with parallel spins have an energy of interaction of $-J$, while nearest neighbors with antiparallel spins have an energy J. Thus, parallel spins are energetically favored.

The statistical-mechanical problem is to determine the thermodynamic state of the system and the spontaneous magnetization given (VI.1).

Let

N = total number of atoms (lattice sites) in the crystal
N_+ = number of sites with spin up
N_- = number of sites with spin down

Clearly
$$N = N_+ + N_- \qquad \text{(VI.2)}$$

The spontaneous magnetization M is defined as the excess number of up spins:

$$M = N_+ - N_-$$

or

$$\frac{M}{N} = \frac{N_+ - N_-}{N} \tag{VI.3}$$

Equation VI.1 is formally identical to (4.20) of Chapter 5 except for a constant term that can be defined away to zero by a proper choice of reference states. Therefore, all the statistical mechanics based on (VI.1) will have the same structure as that for the 50-50 AB order-disorder alloy. It only remains to identify the role of the spontaneous magnetization. In fact, if N_+ is taken to correspond to the number of sites "rightly" occupied in the AB alloy (A on α or B on β) and N_- the number of sites "wrongly" occupied (A on β or B on α), then M/N corresponds to the long-range-order parameter R. This can be seen by going to (5.3) and (5.4) in Chapter 5 and combining them to obtain

$$R = \frac{(N_{A\alpha} + N_{B\beta})}{N} - \frac{(N_{A\beta} + N_{B\alpha})}{N} \tag{VI.4}$$

We therefore conclude that the theory of the Ising model of a ferromagnetic can be obtained directly from that of the AB order-disorder alloy by making the following replacements:

$$\frac{v}{2} \to J$$

$$N_{A\alpha} + N_{B\beta} \to N_+$$

$$N_{A\beta} + N_{B\alpha} \to N_-$$

$$R \to \frac{M}{N}$$

Of course, the zero of energy must be redefined by taking $Qv_{AB}/2 = 0$ in (4.20) of Chapter 5.

appendix VII

THE EINSTEIN FUNCTION

$$E(x) = \frac{x^2 e^{-x}}{(1 + e^{-x})^2}$$

x	1/x	E(x)	x	1/x	E(x)
0.05	20.000000	0.999792	0.95	1.052632	0.928068
0.10	10.000000	0.999167	1.00	1.000000	0.920674
0.15	6.666667	0.998127	1.05	0.952381	0.912976
0.20	5.000000	0.996673	1.10	0.909091	0.904986
0.25	4.000000	0.994808	1.15	0.869565	0.896714
0.30	3.333333	0.992534	1.20	0.833333	0.888170
0.35	2.857143	0.989854	1.25	0.800000	0.879366
0.40	2.500000	0.986773	1.30	0.769231	0.870314
0.45	2.222222	0.983294	1.35	0.740741	0.861024
0.50	2.000000	0.979425	1.40	0.714286	0.851509
0.55	1.818182	0.975168	1.45	0.689655	0.841780
0.60	1.666667	0.970532	1.50	0.666667	0.831849
0.65	1.538462	0.965523	1.55	0.645161	0.821728
0.70	1.428571	0.960148	1.60	0.625000	0.811429
0.75	1.333333	0.954415	1.65	0.606061	0.800964
0.80	1.250000	0.948331	1.70	0.588235	0.790345
0.85	1.176471	0.941906	1.75	0.571429	0.779584
0.90	1.111111	0.935148	1.80	0.555556	0.768693

$$E_{(x)} = \frac{x^2 e^{-x}}{(1 + e^{-x})^2}$$

x	1/x	E(x)	x	1/x	E(x)
1.85	0.540541	0.757684	2.80	0.357143	0.540486
1.90	0.526316	0.746568	2.85	0.350877	0.529304
1.95	0.512821	0.735356	2.90	0.344828	0.518203
2.00	0.500000	0.724062	2.95	0.338983	0.507189
2.05	0.487805	0.712694	3.00	0.333333	0.496269
2.10	0.476190	0.701266	3.05	0.327869	0.485448
2.15	0.465116	0.689787	3.10	0.322581	0.474732
2.20	0.454545	0.678269	3.15	0.317460	0.464125
2.25	0.444444	0.666721	3.20	0.312500	0.453633
2.30	0.434783	0.655154	3.25	0.307692	0.443260
2.35	0.425532	0.643578	3.30	0.303030	0.433010
2.40	0.416667	0.632002	3.35	0.298507	0.422887
2.45	0.408163	0.620347	3.40	0.294118	0.412894
2.50	0.400000	0.608890	3.45	0.289855	0.403036
2.55	0.392157	0.597372	3.50	0.285714	0.393313
2.60	0.384615	0.585890	3.55	0.281690	0.383731
2.65	0.377358	0.574452	3.60	0.277778	0.374290
2.70	0.370370	0.563068	3.65	0.273973	0.364993
2.75	0.363636	0.551743	3.70	0.270270	0.355843

$$E(x) = \frac{x^2 e^{-x}}{(1 + e^{-x})^2}$$

x	1/x	E(x)	x	1/x	E(x)
3.75	0.266667	0.346840	4.75	0.210526	0.198626
3.80	0.263158	0.337987	4.80	0.208333	0.192773
3.85	0.259740	0.329284	4.85	0.206186	0.187060
3.90	0.256410	0.320733	4.90	0.204082	0.181485
3.95	0.253165	0.312333	4.95	0.202020	0.176046
4.00	0.250000	0.304087	5.00	0.200000	0.170742
4.05	0.246914	0.295995	5.05	0.198020	0.165570
4.10	0.243902	0.288055	5.10	0.196078	0.160528
4.15	0.240964	0.280270	5.15	0.194175	0.155614
4.20	0.238095	0.272637	5.20	0.192308	0.150827
4.25	0.235294	0.265158	5.25	0.190476	0.146165
4.30	0.232558	0.257832	5.30	0.188679	0.141624
4.35	0.229885	0.250658	5.35	0.186916	0.137204
4.40	0.227273	0.243635	5.40	0.185185	0.132901
4.45	0.224719	0.236763	5.45	0.183486	0.128715
4.50	0.222222	0.230040	5.50	0.181818	0.124642
4.55	0.219780	0.223465	5.55	0.180180	0.120680
4.60	0.217391	0.217038	5.60	0.178571	0.116827
4.65	0.215054	0.210757	5.65	0.176991	0.113082
4.70	0.212766	0.204620	5.70	0.175439	0.109442
			5.75	0.173913	0.105904

$$E(x) = \frac{x^2 e^{-x}}{(1 + e^{-x})^2}$$

x	1/x	E(x)	x	1/x	E(x)
5.80	0.172414	0.102466	6.75	0.148148	0.053473
5.85	0.170940	0.099127	6.80	0.147059	0.051616
5.90	0.169492	0.095885	6.85	0.145985	0.049818
5.95	0.168067	0.092736	6.90	0.144928	0.048078
6.00	0.166667	0.089679	6.95	0.143885	0.046393
6.05	0.165289	0.086712	7.00	0.142857	0.044764
6.10	0.163934	0.083833	7.05	0.141844	0.043187
6.15	0.162602	0.081039	7.10	0.140845	0.041662
6.20	0.161290	0.078329	7.15	0.139860	0.040187
6.25	0.160000	0.075700	7.20	0.138889	0.038761
6.30	0.158730	0.073151	7.25	0.137931	0.037382
6.35	0.157480	0.070680	7.30	0.136986	0.036048
6.40	0.156250	0.068284	7.35	0.136054	0.034759
6.45	0.155039	0.065962	7.40	0.135135	0.033513
6.50	0.153846	0.063712	7.45	0.134228	0.032309
6.55	0.152672	0.061531	7.50	0.133333	0.031145
6.60	0.151515	0.059419	7.55	0.132450	0.030021
6.65	0.150376	0.057373	7.60	0.131579	0.028935
6.70	0.149254	0.055392	7.65	0.130719	0.027886

$$E(x) = \frac{x^2 e^{-x}}{(1 + e^{-x})^2}$$

x	1/x	E(x)	x	1/x	E(x)
7.70	0.129870	0.026872	8.75	0.114286	0.012136
7.75	0.129032	0.025894	8.80	0.113636	0.011676
7.80	0.128205	0.024949	8.85	0.112994	0.011233
7.85	0.127389	0.024036	8.90	0.112360	0.01806
7.90	0.126582	0.023155	8.95	0.111732	0.010395
7.95	0.125786	0.022305	9.00	0.111111	0.009999
8.00	0.125000	0.021484	9.05	0.110497	0.009617
8.05	0.124224	0.020692	9.10	0.109890	0.009249
8.10	0.123457	0.019927	9.15	0.109290	0.008895
8.15	0.122699	0.019190	9.20	0.108696	0.008554
8.20	0.121951	0.018478	9.25	0.108108	0.008225
8.25	0.121212	0.017791	9.30	0.107527	0.007909
8.30	0.120482	0.017129	9.35	0.106952	0.007604
8.35	0.119760	0.016490	9.40	0.106383	0.007311
8.40	0.119048	0.015874	9.45	0.105820	0.007028
8.45	0.118343	0.015280	9.50	0.105263	0.006756
8.50	0.117647	0.014707	9.55	0.104712	0.006495
8.55	0.116959	0.014154	9.60	0.104167	0.006243
8.60	0.116279	0.013621	9.65	0.103627	0.006000
8.65	0.115607	0.013108	9.70	0.103093	0.005767
8.70	0.114943	0.012613			

$$E(x) = \frac{x^2 e^{-x}}{(1 + e^{-x})^2}$$

x	1/x	E(x)	x	1/x	E(x)
9.75	0.102564	0.005542	10.75	0.093023	0.002478
9.80	0.102041	0.005326	10.80	0.092593	0.002379
9.85	0.101523	0.005118	10.85	0.092166	0.002284
9.90	0.101010	0.004918	10.90	0.091743	0.002193
9.95	0.100503	0.004726	10.95	0.091324	0.002105
10.00	0.100000	0.004540	11.00	0.090909	0.002021
10.05	0.099502	0.004362	11.05	0.090498	0.001940
10.10	0.099010	0.004191	11.10	0.090090	0.001862
10.15	0.098522	0.004026	11.15	0.089686	0.001787
10.20	0.098039	0.003867	11.20	0.089286	0.001715
10.25	0.097561	0.003715	11.25	0.088889	0.001646
10 30	0.097087	0.003568	11.30	0.088496	0.001580
10.35	0.096618	0.003427	11.35	0.088106	0.001516
10.40	0.096154	0.003292	11.40	0.087719	0.001455
10.45	0.095694	0.003161	11.45	0.087336	0.001396
10.50	0.095238	0.003036	11.50	0.086957	0.001340
10.55	0.094787	0.002916	11.55	0.086580	0.001285
10.60	0.094340	0.002800	11.60	0.086207	0.001233
10.65	0.093897	0.002688	11.65	0.085837	0.001183
10.70	0.093458	0.002581	11.70	0.085470	0.001135

$$E(x) = \frac{x^2 e^{-x}}{(1 + e^{-x})^2}$$

x	1/x	E(x)	x	1/x	E(x)
11.75	0.085106	0.001089	12.75	0.078431	0.000472
11.80	0.084746	0.001045	12.80	0.078125	0.000452
11.85	0.084388	0.001002	12.85	0.077821	0.000434
11.90	0.084034	0.000962	12.90	0.077519	0.000416
11.95	0.083682	0.000922	12.95	0.077220	0.000398
12.00	0.083333	0.000885	13.00	0.076923	0.000382
12.05	0.082988	0.000849	13.05	0.076628	0.000366
12.10	0.082645	0.000814	13.10	0.076336	0.000351
12.15	0.082305	0.000781	13.15	0.076046	0.000336
12.20	0.081967	0.000749	13.20	0.075758	0.000322
12.25	0.081633	0.000718	13.25	0.075472	0.000309
12.30	0.081301	0.000689	13.30	0.075188	0.000296
12.35	0.080972	0.000660	13.35	0.074906	0.000284
12.40	0.080645	0.000633	13.40	0.074627	0.000272
12.45	0.080321	0.000607	13.45	0.074349	0.000261
12.50	0.080000	0.000582	13.50	0.074074	0.000250
12.55	0.079681	0.000558	13.55	0.073801	0.000239
12.60	0.079365	0.000535	13.60	0.073529	0.000229
12.65	0.079051	0.000513	13.65	0.073260	0.000220
12.70	0.078740	0.000492	13.70	0.072993	0.000211

$$E(x) = \frac{x^2 e^{-x}}{(1 + e^{-x})^2}$$

x	1/x	E(x)	x	1/x	E(x)
13.75	0.072727	0.000202	14.75	0.067797	0.000085
13.80	0.072464	0.000193	14.80	0.067568	0.000082
13.85	0.072202	0.000185	14.85	0.067340	0.000078
13.90	0.071942	0.000178	14.90	0.067114	0.000075
13.95	0.071685	0.000170	14.95	0.066890	0.000072
14.00	0.071429	0.000163	15.00	0.066667	0.000069
14.05	0.071174	0.000156	15.05	0.066445	0.000066
14.10	0.070922	0.000150	15.10	0.066225	0.000063
14.15	0.070671	0.000143	15.15	0.066007	0.000060
14.20	0.070423	0.000137	15.20	0.065789	0.000058
14.25	0.070175	0.000132	15.25	0.065574	0.000055
14.30	0.069930	0.000126	15.30	0.065359	0.000053
14.35	0.069686	0.000121	15.35	0.065147	0.000051
14.40	0.069444	0.000116	15.40	0.064935	0.000049
14.45	0.069204	0.000111	15.45	0.064725	0.000047
14.50	0.068966	0.000106	15.50	0.064516	0.000045
14.55	0.068729	0.000102	15.55	0.064309	0.000043
14.60	0.068493	0.000097	15.60	0.064103	0.000041
14.65	0.068259	0.000093	15.65	0.063898	0.000039
14.70	0.068027	0.000089	15.70	0.063694	0.000037

appendix VIII

DEBYE ENERGY AND HEAT-CAPACITY FUNCTIONS

$$D_E(x_D) = \frac{3}{x_D^4} \int_0^{x_D} \frac{x^3 dx}{e^x - 1}$$

$$D(x_D) = \frac{3}{x_D^3} \int_0^{x_D} \frac{x^4 e^{-x} dx}{(1-e^{-x})^2}$$

x_D	$1/x_D$	$D_E(x_D)$	$D(x_D)$
0.050	20.000000	19.627500	0.999875
0.100	10.000000	9.629999	0.999500
0.150	6.666667	6.299165	0.998876
0.200	5.000000	4.634995	0.998003
0.250	4.000000	3.637491	0.996882
0.300	3.333333	2.973317	0.995514
0.350	2.857143	2.499617	0.993902
0.400	2.500000	2.144962	0.992045
0.450	2.222222	1.869668	0.989948
0.500	2.000000	1.649926	0.987611
0.550	1.818182	1.470583	0.985037
0.600	1.666667	1.321539	0.982229
0.650	1.538462	1.195799	0.979190
0.700	1.428571	1.088369	0.975922
0.750	1.333333	0.995585	0.972430

x_D	$1/x_D$	$D_E(x_D)$	$D(x_D)$
0.800	1.250000	0.914699	0.968717
0.850	1.176471	0.843610	0.964787
0.900	1.111111	0.780684	0.960643
0.950	1.052632	0.724630	0.956290
1.000	1.000000	0.674416	0.951732
1.050	0.952381	0.629206	0.946974
1.100	0.909091	0.588316	0.942020
1.150	0.869565	0.551182	0.936875
1.200	0.833333	0.517331	0.931545
1.250	0.800000	0.486370	0.926033
1.300	0.769231	0.457963	0.920346
1.350	0.740741	0.431824	0.914489
1.400	0.714286	0.407709	0.908467
1.450	0.689655	0.385408	0.902286
1.500	0.666667	0.364738	0.895950
1.550	0.645161	0.345539	0.889467
1.600	0.625000	0.327672	0.882842
1.650	0.606061	0.311014	0.876079
1.700	0.588235	0.295459	0.869186
1.750	0.571429	0.280909	0.862168
1.800	0.555556	0.267280	0.855031
1.850	0.540541	0.254495	0.847780

x_D	$1/x_D$	$D_E(x_D)$	$D(x_D)$
1.900	0.526316	0.242487	0.840423
1.950	0.512821	0.231195	0.832963
2.000	0.500000	0.220564	0.825408
2.050	0.487805	0.210544	0.817763
2.100	0.476190	0.201091	0.810034
2.150	0.465116	0.192163	0.802227
2.200	0.454545	0.183724	0.794347
2.250	0.444444	0.175741	0.786400
2.300	0.434783	0.168181	0.778391
2.350	0.425532	0.161017	0.770327
2.400	0.416667	0.154224	0.762212
2.450	0.408163	0.147777	0.754053
2.500	0.400000	0.141654	0.745853
2.550	0.392157	0.135837	0.737619
2.600	0.384615	0.130305	0.729355
2.650	0.377358	0.125042	0.721067
2.700	0.370370	0.120032	0.712759
2.750	0.363636	0.115260	0.704436
2.800	0.357143	0.110713	0.696103
2.850	0.350877	0.106377	0.687764
2.900	0.344828	0.102241	0.679424
2.950	0.338983	0.098295	0.671088

x_D	$1/x_D$	$D_E(x_D)$	$D(x_D)$
3.000	0.333333	0.094527	0.662758
3.050	0.327869	0.090928	0.654440
3.100	0.322581	0.087489	0.646137
3.150	0.317460	0.084201	0.637854
3.200	0.312500	0.081058	0.629593
3.250	0.307692	0.078051	0.621359
3.300	0.303030	0.075173	0.613154
3.350	0.298507	0.072418	0.604983
3.400	0.294118	0.069780	0.596848
3.450	0.289855	0.067253	0.588753
3.500	0.285714	0.064832	0.580700
3.550	0.281690	0.062511	0.572692
3.600	0.277778	0.060286	0.564732
3.650	0.273973	0.058152	0.556823
3.700	0.270270	0.056105	0.548966
3.750	0.266667	0.054141	0.541165
3.800	0.263158	0.052255	0.533421
3.850	0.259740	0.050446	0.525736
3.900	0.256410	0.048707	0.518113
3.950	0.253165	0.047038	0.510554
4.000	0.250000	0.045434	0.503059
4.050	0.246914	0.043893	0.495631

x_D	$1/x_D$	$D_E(x_D)$	$D(x_D)$
4.100	0.243902	0.042412	0.488272
4.150	0.240964	0.040988	0.480982
4.200	0.238095	0.039618	0.473763
4.250	0.235294	0.038301	0.466616
4.300	0.232558	0.037034	0.459543
4.350	0.229885	0.035815	0.452544
4.400	0.227273	0.034642	0.445620
4.450	0.224719	0.033513	0.438772
4.500	0.222222	0.032425	0.432002
4.550	0.219780	0.031379	0.425308
4.600	0.217391	0.030371	0.418693
4.650	0.215054	0.029399	0.412157
4.700	0.212766	0.028464	0.405700
4.750	0.210526	0.027562	0.399322
4.800	0.208333	0.026694	0.393024
4.850	0.206186	0.025856	0.386807
4.900	0.204082	0.025049	0.380669
4.950	0.202020	0.024270	0.374612
5.000	0.200000	0.023519	0.368635
5.050	0.198020	0.022795	0.362738
5.100	0.196078	0.022097	0.356922
5.150	0.194175	0.021423	0.351186

x_D	$1/x_D$	$D_E(x_D)$	$D(x_D)$
5.200	0.192308	0.020772	0.345529
5.250	0.190476	0.020145	0.339953
5.300	0.188679	0.019539	0.334456
5.350	0.186916	0.018954	0.329038
5.400	0.185185	0.018389	0.323698
5.450	0.183486	0.017843	0.318438
5.500	0.181818	0.017317	0.313255
5.550	0.180180	0.016808	0.308149
5.600	0.178571	0.016316	0.303121
5.650	0.176991	0.015841	0.298169
5.700	0.175439	0.015382	0.293293
5.750	0.173913	0.014938	0.288493
5.800	0.172414	0.014509	0.283767
5.850	0.170940	0.014094	0.279115
5.900	0.169492	0.013693	0.274536
5.950	0.168067	0.013305	0.270031
6.000	0.166667	0.012930	0.265597
6.050	0.165289	0.012567	0.261234
6.100	0.163934	0.012216	0.256943
6.150	0.162602	0.011877	0.252721
6.200	0.161290	0.011548	0.248568
6.250	0.160000	0.011230	0.244483

x_D	$1/x_D$	$D_E(x_D)$	$D(x_D)$
6.300	0.158730	0.010922	0.240466
6.350	0.157480	0.010624	0.236515
6.400	0.156250	0.010335	0.232631
6.450	0.155039	0.010056	0.228811
6.500	0.153846	0.009785	0.225056
6.550	0.152672	0.009523	0.221364
6.600	0.151515	0.009269	0.217735
6.650	0.150376	0.009023	0.214168
6.700	0.149254	0.008785	0.210662
6.750	0.148148	0.008554	0.207215
6.800	0.147059	0.008330	0.203828
6.850	0.145985	0.008113	0.200500
6.900	0.144928	0.007903	0.197229
6.950	0.143885	0.007699	0.194015
7.000	0.142857	0.007501	0.190856
7.050	0.141844	0.007309	0.187753
7.100	0.140845	0.007123	0.184704
7.150	0.139860	0.006943	0.181709
7.200	0.138889	0.006768	0.178766
7.250	0.137931	0.006598	0.175875
7.300	0.136986	0.006433	0.173035
7.350	0.136054	0.006273	0.170245
7.400	0.135135	0.006118	0.167505

x_D	$1/x_D$	$D_E(x_D)$	$D(x_D)$
7.450	0.134228	0.005967	0.164813
7.500	0.133333	0.005821	0.162169
7.550	0.132450	0.005679	0.159572
7.600	0.131579	0.005541	0.157021
7.650	0.130719	0.005407	0.154515
7.700	0.129870	0.005277	0.152055
7.750	0.129032	0.005150	0.149638
7.800	0.128205	0.005027	0.147264
7.850	0.127389	0.004908	0.144933
7.900	0.126582	0.004792	0.142644
7.950	0.125786	0.004680	0.140395
8.000	0.125000	0.004570	0.138187
8.050	0.124224	0.004464	0.136019
8.100	0.123457	0.004360	0.133889
8.150	0.122699	0.004259	0.131798
8.200	0.121951	0.004162	0.129744
8.250	0.121212	0.004067	0.127727
8.300	0.120482	0.003974	0.125746
8.350	0.119760	0.003884	0.123801
8.400	0.119048	0.003796	0.121890
8.450	0.118343	0.003711	0.120014
8.500	0.117647	0.003628	0.118172
8.550	0.116959	0.003548	0.116279

x_D	$1/x_D$	$D_E(x_D)$	$D(x_D)$
8.600	0.116279	0.003469	0.114499
8.650	0.115607	0.003393	0.112750
8.700	0.114943	0.003318	0.111126
8.750	0.114286	0.003246	0.109443
8.800	0.113636	0.003175	0.107790
8.850	0.112994	0.003107	0.106166
8.900	0.112360	0.003040	0.104572
8.950	0.111732	0.002975	0.103005
9.000	0.111111	0.002911	0.101467
9.050	0.110497	0.002849	0.099956
9.100	0.109890	0.002789	0.098472
9.150	0.109290	0.002730	0.097015
9.200	0.108696	0.002673	0.095583
9.250	0.108108	0.002617	0.094177
9.300	0.107527	0.002563	0.092795
9.350	0.106952	0.002510	0.091438
9.400	0.106383	0.002458	0.090105
9.450	0.105820	0.002408	0.088796
9.500	0.105263	0.002359	0.087509
9.550	0.104712	0.002311	0.086245
9.600	0.104167	0.002264	0.085004
9.650	0.103627	0.002219	0.083784
9.700	0.103093	0.002174	0.082585

x_D	$1/x_D$	$D_E(x_D)$	$D(x_D)$
9.750	0.102564	0.002131	0.081408
9.800	0.102041	0.002089	0.080251
9.850	0.101523	0.002048	0.079114
9.900	0.101010	0.002007	0.077997
9.950	0.100503	0.001968	0.076900
10.000	0.100000	0.001930	0.075821
10.050	0.099502	0.001892	0.074761
10.100	0.099010	0.001856	0.073719
10.150	0.098522	0.001820	0.072696
10.200	0.098039	0.001785	0.071690
10.250	0.097561	0.001751	0.070701
10.300	0.097087	0.001718	0.069729
10.350	0.096618	0.001685	0.068774
10.400	0.096154	0.001653	0.067835
10.450	0.095694	0.001622	0.066912
10.500	0.095238	0.001592	0.066005
10.550	0.094787	0.001563	0.065113
10.600	0.094340	0.001534	0.064236
10.650	0.093897	0.001505	0.063374
10.700	0.093458	0.001478	0.062526
10.750	0.093023	0.001451	0.061693
10.800	0.092593	0.001424	0.060874
10.850	0.092166	0.001399	0.060068

x_D	$1/x_D$	$D_E(x_D)$	$D(x_D)$
10.900	0.091743	0.001373	0.059276
10.950	0.091324	0.001349	0.058497
11.000	0.090909	0.001325	0.057731
11.050	0.090498	0.001301	0.056977
11.100	0.090090	0.001278	0.056237
11.150	0.089686	0.001255	0.055508
11.200	0.089286	0.001233	0.054791
11.250	0.088889	0.001212	0.054086
11.300	0.088496	0.001191	0.053393
11.350	0.088106	0.001170	0.052710
11.400	0.087719	0.001150	0.052039
11.450	0.087336	0.001130	0.051379
11.500	0.086957	0.001110	0.050730
11.550	0.086580	0.001091	0.050091
11.600	0.086207	0.001073	0.049462
11.650	0.085837	0.001055	0.048843
11.700	0.085470	0.001037	0.048235
11.750	0.085106	0.001019	0.047636
11.800	0.084746	0.001002	0.047046
11.850	0.084388	0.000986	0.046466
11.900	0.084034	0.000969	0.045895
11.950	0.083682	0.000953	0.045333
12.000	0.083333	0.000938	0.044780

x_D	$1/x_D$	$D_E(x_D)$	$D(x_D)$
12.050	0.082988	0.000922	0.044236
12.100	0.082645	0.000907	0.043700
12.150	0.082305	0.000892	0.043172
12.200	0.081967	0.000878	0.042653
12.250	0.081633	0.000864	0.042142
12.300	0.081301	0.000850	0.041639
12.350	0.080972	0.000836	0.041143
12.400	0.080645	0.000823	0.040655
12.450	0.080321	0.000810	0.040175
12.500	0.080000	0.000797	0.039702
12.550	0.079681	0.000784	0.039236
12.600	0.079365	0.000772	0.038777
12.650	0.079051	0.000760	0.038325
12.700	0.078740	0.000748	0.037880
12.750	0.078431	0.000736	0.037442
12.800	0.078125	0.000725	0.037010
12.850	0.077821	0.000714	0.036585
12.900	0.077519	0.000703	0.036166
12.950	0.077220	0.000692	0.035754
13.000	0.076923	0.000681	0.035347
13.050	0.076628	0.000671	0.034947
13.100	0.076336	0.000661	0.034552

x_D	$1/x_D$	$D_E(x_D)$	$D(x_D)$
13.150	0.076046	0.000651	0.034164
13.200	0.075758	0.000641	0.033781
13.250	0.075472	0.000632	0.033403
13.300	0.075188	0.000622	0.033031
13.350	0.074906	0.000613	0.032665
13.400	0.074627	0.000604	0.032304
13.450	0.074349	0.000595	0.031948
13.500	0.074074	0.000586	0.031597
13.550	0.073801	0.000578	0.031251
13.600	0.073529	0.000569	0.030910
13.650	0.073260	0.000561	0.030574
13.700	0.072993	0.000553	0.030243
13.750	0.072727	0.000545	0.029916
13.800	0.072464	0.000537	0.029595
13.850	0.072202	0.000529	0.029277
13.900	0.071942	0.000522	0.028964
13.950	0.071685	0.000514	0.028656
14.000	0.071429	0.000507	0.028352
14.050	0.071174	0.000500	0.028052
14.100	0.070922	0.000493	0.027756
14.150	0.070671	0.000486	0.027465
14.200	0.070423	0.000479	0.027177
14.250	0.070175	0.000472	0.026893

x_D	$1/x_D$	$D_E(x_D)$	$D(x_D)$
14.300	0.069930	0.000466	0.026613
14.350	0.069686	0.000459	0.026338
14.400	0.069444	0.000453	0.026065
14.450	0.069204	0.000447	0.025797
14.500	0.068966	0.000441	0.025532
14.550	0.068729	0.000435	0.025271
14.600	0.068493	0.000429	0.025013
14.650	0.068259	0.000423	0.024759
14.700	0.068027	0.000417	0.024508
14.750	0.067797	0.000411	0.024261
14.800	0.067568	0.000406	0.024016
14.850	0.067340	0.000401	0.023775
14.900	0.067114	0.000395	0.023538
14.950	0.066890	0.000390	0.023303
15.000	0.066667	0.000385	0.023071
15.050	0.066445	0.000380	0.022843
15.100	0.066225	0.000375	0.022617
15.150	0.066007	0.000370	0.022395
15.200	0.065789	0.000365	0.022175
15.250	0.065574	0.000360	0.021958
15.300	0.065359	0.000355	0.021744
15.350	0.065147	0.000351	0.021533
15.400	0.064935	0.000346	0.021324

x_D	$1/x_D$	$D_E(x_D)$	$D(x_D)$
15.450	0.064725	0.000342	0.021118
15.500	0.064516	0.000337	0.020915
15.550	0.064309	0.000333	0.020714
15.600	0.064103	0.000329	0.020516
15.650	0.063898	0.000325	0.020321
15.700	0.063694	0.000321	0.020128
15.750	0.063492	0.000317	0.019937
15.800	0.063291	0.000313	0.019748
15.850	0.063091	0.000309	0.019563
15.900	0.062893	0.000305	0.019379
15.950	0.062696	0.000301	0.019197
16.000	0.062500	0.000297	0.019018
16.500	0.060606	0.000263	0.017343
17.000	0.058824	0.000244	0.015859
17.500	0.057143	0.000218	0.014539
18.000	0.055556	0.000196	0.013361
18.500	0.054054	0.000177	0.012307
19.000	0.052632	0.000160	0.011361
19.500	0.051282	0.000135	0.010510
20.000	0.050000	0.000122	0.009741
20.500	0.048780	0.000110	0.009046
21.000	0.047619	0.000100	0.008415
21.500	0.046512	0.000091	0.007841

x_D	$1/x_D$	$D_E(x_D)$	$D(x_D)$
22.000	0.045455	0.000083	0.007319
22.500	0.044444	0.000076	0.006842
23.000	0.043478	0.000070	0.006788
23.500	0.042553	0.000064	0.006416
24.000	0.041667	0.000059	0.006073
24.500	0.040816	0.000054	0.005757
25.000	0.040000	0.000050	0.005465

BIBLIOGRAPHY

Chapter 1

1. Gibbs, J. W., *Elementary Principles in Statistical Mechanics*, Yale University Press, New Haven, Conn., 1902.
2. Fowler, R. H., *Statistical Mechanics*, Cambridge University Press, Cambridge, 1st ed. 1929, 2nd ed. 1936.
3. Fowler, R. H. and Guggenheim, E. A., *Statistical Thermodynamics*, Cambridge University Press, Cambridge, 1939.
4. Tolman, R. C., *The Principles of Statistical Mechanics*, Oxford University Press, London, 1938.
5. Schrodinger, E., *Statistical Thermodynamics*, Cambridge University Press, Cambridge, 1964.
6. Sommerfeld, A., *Thermodynamics and Statistical Mechanics*, Academic Press, New York, 1964.
7. Mayer, J. E. and Mayer, M. G., *Statistical Mechanics*, John Wiley and Sons, New York, 1940.
8. Band, W., *Quantum Statistics*, D. Van Nostrand Co., New York, 1955.
9. Landau, L. D. and Lifshitz, E. M., *Statistical Physics*, translated by J. B. Sykes and M. J. Kearsley, Addison-Wesley Publishing Company, Reading, Mass., 1969.
10. Eyring, H., Henderson, D., Stover, B. J. and Eyring, E. M., *Statistical Mechanics and Dynamics*, John Wiley and Sons, New York, 1964.
11. Slater, J. C., *Introduction to Chemical Physics*, McGraw-Hill Book Company, New York, 1939.
12. Fermi, E., *Molecules, Crystals and Quantum Statistics*, W. A. Benjamin, New York, 1966.

Chapter 2

In addition to references for Chapter 1, see the following:

13. de Launay, "The Theory of Specific Heats and Lattice Vibrations" in *Solid State Physics*, Vol. 2, F. Seitz and D. Turnbull, Eds., Academic Press, New York, 1956, pp. 219–303.

14. Born, M. and Huang, K. *Dynamical Theory of Crystal Lattices*, Oxford, Clarendon Press, 1954.

Chapters 3 and 4

15. Raimes, S., *The Wave Mechanics of Electrons in Solids*, North-Holland Publishing Company, Amsterdam, 1963.

16. Smith, R. A., *Wave Mechanics of Crystalline Solids*, Chapman and Hall, London, 1961.

17. Wilson, A. H., *The Theory of Metals*, 2nd ed., Cambridge University Press, Cambridge, 1965.

18. Spenke, E., *Electronic Semiconductors*, McGraw-Hill Book Company, New York, 1958.

19. Ehrenberg, W., *Electric Conduction in Semiconductors and Metals*, Oxford, Clarendon Press, 1958.

20. Blatt, F. J., *Physics of Electronic Conduction in Solids*, McGraw-Hill Book Company, New York, 1968.

21. Smith, A. C., Janak, J. F., and Adler, R. B., *Electronic Conduction in Solids*, McGraw-Hill Book Company, New York, 1967.

Chapter 5

22. Krivoglaz, M. A. and Smirnov, A. A., *The Theory of Order-Disorder in Alloys* translated by Scripta Technica B. Chalmers, Ed., American Elsevier Publishing Company, New York, 1965.

23. Muto, T. and Takagi, Y., "The Theory of Order-Disorder in Alloys," *Solid State Physics*, Vol. 1, F. Seitz and D. Turnbull, Eds., Academic Press, New York, 1955, pp. 193–282.

24. Guttman, L., "Order-Disorder Phenomena in Metals," *Solid State Physics*, Vol. 3, F. Seitz and D. Turnbull, Eds., Academic Press, New York, 1956, pp. 146–223.

Chapters 6 to 9

25. A. C. Damask and G. J. Dienes, Point Defects in Metals, Gordon and Breach, New York, 1963.

26. van Bueren, H. G., *Imperfections in Crystals*, North-Holland Publishing Company, Amsterdam—Interscience, New York, 1960.

27. Girifalco, L. A., *Atomic Migration in Crystals*, Blaisdell Publishing Company, New York, 1964.

28. Shewman, P. G., *Diffusion in Solids*, McGraw-Hill, New York, 1963.

29. Girifalco, L. A. and Welch, D. O., *Point Defects and Diffusion in Strained Metals*, Gordon and Breach, New York, 1967.

30. Girifalco, L. A., "Diffusion in Solids at High Pressures" in *Metallurgical Society Conferences*, Vol. 22, Proceedings of IMD-TMS-AIME Conference in Dallas, Feb. 25–26, 1963, K. A. Gschneider, Jr., M. T. Hepworth, and N. A. D. Parlee, Eds., Gordon and Breach, New York, 1964.

31. Eshelby, J. D., "The Continuum Theory of Lattice Defects" in *Solid State Physics*, F. Seitz and D. Turnbull, Eds., Academic Press, New York, 1956, pp. 79–144.

32. Kroger, F. A. and Vink, H. J., "Relations Between the Concentrations of Imperfections in Crystalline Solids," *Solid State Physics*, Vol. 3, F. Seitz and D. Turnbull, Eds., Academic Press, New York, 1956, pp. 310–435.

33. Lazarus, D., "Diffusion in Metals" in *Solid State Physics*, Vol. 10, F. Seitz and D. Turnbull, Eds., Academic Press, New York, 1960, pp. 71–126.

34. Peterson, N. L., "Diffusion in Metals" in *Solid State Physics*, Vol. 22, F. Seitz, D. Turnbull, and H. Ehrenreich, Eds., Academic Press, New York, 1968, pp. 409–512.

35. Girifalco, L. A., "Vacancy Concentration and Diffusion in Order-Disorder Alloys," *J. Phys. Chem. Solids*, **24,** 323 (1964).

36. Schoijet, M. and Girifalco, L. A., "Theory of Diffusion in Ordered Alloys of the β-Brass Type," *J. Phys. Chem. Solids*, **29,** 481 (1968).

37. Schoijet, M. and Girifalco, L. A., "Diffusion in the Face-Centered Cubic Ordered Alloys," *J. Phys. Chem. Solids*, **29,** 497 (1968).

38. Schoijet, M. and Girifalco, L. A., "Diffusion in Order-Disorder Alloys of the Face Centered Cubic AB_3 Alloy," *J. Phys. Chem. Solids*, **29,** 911 (1968).

39. Liu, G. C. T., Girifalco, L. A., and Maddin, R., "Quenched in Electrical Resistivity of Dilute Binary Alloys," *Phys. Status Solidi*, **31,** 303 (1969).

INDEX